THE

S T O R Y

OF

MATHEMATICS

Other Avon Books by
Lloyd Motz and Jefferson Hane Weaver

THE STORY OF PHYSICS

THE
S T O R Y
OF
MATHEMATICS

LLOYD MOTZ AND JEFFERSON HANE WEAVER

AVON BOOKS ◆ NEW YORK

AVON BOOKS
A division of
The Hearst Corporation
1350 Avenue of the Americas
New York, New York 10019

The Plenum Press edition contains the following Library of Congress Cataloging in Publication Data:

Motz, Lloyd, 1910-
  The story of mathematics / Lloyd Motz and Jefferson Hane Weaver.
      p.   cm.
Includes bibliographical references and index.
1. Mathematics—History.   I. Weaver, Jefferson Hane.   II. Title.
QA21.M846   1994   93-26527
510'.9—dc20   CIP

First Avon Books Trade Printing: October 1995

To Minne and Shelley

Preface

To the reader, the title of this book, *The Story of Mathematics*, may seem unsuitable, for in essence, the book, to some extent, is a history. Why, then, not call it a "History of Mathematics"? The reason, of course, is that this book is not a history; a history is a detailed study of a subject in all its meanderings and twists and turns. For this reason a history of mathematics would require many volumes, because mathematics is like a huge maze with a main track from which many intricate channels branch off and lead to new, complex, self-sustaining fields of study. *The Story of Mathematics* differs from this in that it follows the main track of the maze leaving it only occasionally to explore some of the ideas that led to departures from the main track. We have done this only when and where the concepts that produced the new path in the maze contributed significantly to the development of mathematics.

In this "story of mathematics," we trace the growth of mathematics from the early Greeks, Pythagoras, Euclid, Archimedes, Apollonius, etc., to Newton and the great post-Newtonian mathematicians who laid the groundwork for what we now call modern mathematics. This story also sheds light on why mathematics did not develop steadily from stage to stage but did so in spurts. As a good example, we note in the text that Archimedes did not develop the calculus which Leibnitz and Newton did, some 2000 years later, even though Archimedes was on the verge of doing so. The reason was that nothing in Archimedes's time demanded the calculus, whereas Newtonian science did. This is but one example of how societal, technological, and scientific needs shaped the development of mathematics, which, in turn, revealed unsuspected scientific truths.

Lloyd Motz
Jefferson Hane Weaver

New York and Davie, Florida

Contents

CHAPTER 1   Mathematics and Mind: The Early Greeks and Alexandrians   *1*

CHAPTER 2   The Origins of Arithmetic   *31*

CHAPTER 3   The Origin of Abstract Mathematics: The Beginnings of Algebra   *57*

CHAPTER 4   From Geometry to Trigonometry   *79*

CHAPTER 5   Analytic Geometry: The Geometrization of Arithmetic and Algebra   *103*

CHAPTER 6   The Birth of the Calculus   *125*

CHAPTER 7   Post-Newtonian Mathematics   *151*

CHAPTER 8   The Golden Age of Mathematics   *181*

CHAPTER 9   The End of the Golden Age of Mathematics   *207*

CHAPTER 10   The Beginning of Modern Mathematics   *247*

CHAPTER 11   Modern Mathematics and the New Physics   273

CHAPTER 12   The Evolution of Modern Mathematics   295

CHAPTER 13   Whither Mathematics?   317

                    Bibliography   337

                    Index   339

CHAPTER 1

Mathematics and Mind
The Early Greeks and Alexandrians

When I trace at my pleasure the windings to and fro of the heavenly
bodies, I no longer touch the earth with my feet: I stand in the
presence of Zeus himself and take my fill of ambrosia,
food of the gods.

—PTOLEMY

For thousands of years philosophers and scientists have tried to discover and understand the basis of human knowledge and, ultimately, the nature of thought itself. Insofar as knowledge and thought in the form of structured thinking and reasoning are products of the brain, we must conclude that, to some extent, all living creatures think. Here, however, we must differentiate between human thought and the thinking of all lower forms of life because only humans, as far as we know, generate abstract thought—thought that is entirely a creation of the mind. Of all such intellectual creations, mathematics is both the most remarkable and most difficult to grasp. It is remarkable because its various branches can be developed from basic elements that need not be related to anything in our experience; it is difficult to grasp because we must confine our thoughts within the boundaries that are defined by rigid rules or laws. The mathematician may not allow his thoughts to ramble unrestrained but must direct them along a line that leads unambiguously and ineluctably from a group or set of ideas or concepts to another group or set, which, in general, may appear to be unrelated to the first. This is one of the beautiful and exciting features which mathematics offers to its practitioners. The direct line that leads from one set of mathematical ideas to another is a line of pure reasoning. The excitement lies in anticipating the final result obtained by this line of reasoning which also carries an element of surprise that contributes to the excitement. It may well be that the final result is different from the anticipated result.

This excitement is one of the reasons that people with what some might call a "mathematical mind" are drawn to mathematics. We do not insist, however, that the study of mathematics began this way, but something of this sort must have directed people in antiquity toward the pursuit of mathematics. It is the only purely mental activity that can generate new concepts; these concepts are pure thought having no relationship to any aspect of the world or to any physical experience. Because mathematical reasoning can thus lead to every possible kind of rational construct and because we believe that the universe is a rational entity, we may conclude that among these constructs will be all those that correspond to every possible feature of the universe. The problem is to find which of these constructs are relevant to nature and which are not.

This discussion is somewhat fanciful, yet a very prominent and influential school of theoretical physicists, cosmologists, and mathematicians has emerged today which promulgates this philosophy—that every mathematical synthesis has its counterparts in nature. With the discovery of the two great physical theories—the quantum theory and the theory of relativity—that guide us in probing the universe, theoretical physicists (particularly astrophysicists and cosmologists such as Eddington and Jeans) spoke of the "creator of the universe" as being a mathematician. That mathematics is a powerful, indeed indispensable, intellectual tool for the scientist does not mean that mathematics is science; it is the language, not the essence of science.

Returning now to the origins of mathematics, we emphasize its eclectic features because it impinges on all aspects of our activities, whether mental or physical, recreational or work related. Indeed, mathematics may have originated simultaneously from many different kinds of activities. We consider these in turn starting with two of the many recreational activities we pursue in almost every stage of our lives, namely games and puzzles. We play physical and mental games, both of which involve some mathematical aspects, which may be more or less pronounced, depending on the extent to which they are structured and subject to rules. Childhood games are very unstructured and so we play them with hardly any mathematical thoughts. But as we reach adulthood, our physical and mental games become more structured and we have to use some kind of mathematical reasoning to succeed.

It may not have appeared to us at the time or even appear to us now that our play had much to do with mathematics; some consideration indicates that as our games became more competitive, we compared our skills numerically so that numbers and, hence, arithmetic became part of our play. This was not as apparent in physical games as in mental games such as bridge, chess, and Go (the famous Japanese game), but physical games did introduce geometry to us because all physical games are spatial and temporal. The champions in such games are the players whose various senses and muscles become expert geometers, working harmoniously to perform impressive feats at which we marvel. The mathematical analysis of the trajectory of a tennis ball, a baseball, a football, or a hockey puck, taking into account all the external forces acting on it, is extremely complex. But the champion solves all the mathematical problems the physical game presents instantaneously, thus doing difficult mathematics unknowingly.

Mental games are, of course, different in that they require deliberate mental concentration and analysis instead of the automatic muscle response that physical games require. Among mental games we include puzzles of all kinds. Indeed, it may be that puzzles, which appeal to all ages, generated mathematical thinking, but not necessarily mathematics itself. By mathematical thinking we mean seeking and finding relationships among entities that constitute a puzzle to make the puzzle meaningful or to use these relationships to solve it. In any case, the mathematical mind never lets go of a problem or a puzzle but struggles with the problem until the mind discovers the path to a solution. Indeed, all such a mind requires is not a complete delineation of the path but the merest hint of such a path which permits the first step to be taken along it. Games have done more for mathematics than merely stimulating a casual interest in it. A whole branch of mathematics called "the theory of games" has evolved from games and is based on the same mathematical principles that are very important in other less purely mathematical areas such as economics, statistics, probability, gambling, and weather prediction as well as physics, statistical mechanics, and quantum mechanics.

The part of mathematics stemming from purely mental activities such as games and puzzles is clear. But the great initial drives behind the rapid development of mathematics from a purely intellectual toy or game to the most powerful intellectual tool in technology and science were prompted by:

1. The practical need for an accurate accounting procedure to track and keep records of the exchange of goods and services among people.
2. The demands of science and technology for the kind of analytical tool that mathematics finally became.

3. The intellectual pressures within mathematics itself, which led mathematicians to study mathematics relentlessly and to invent increasingly more complex branches of it.

Unlike science, which requires its practitioners to operate within its laws—the given constraints of the discipline—mathematics is open ended, with no boundaries. It thus grows and evolves at the whim of the mathematician who is constrained only by the rules of logic and the rules of the mathematical game he is playing. Today the pure mathematician pays no more attention to the needs of science and technology than the chess player does; his drive is the pleasure of proving a theorem that has been a mere speculation for years or developing a new branch of some phase of mathematics (such as fractals) that had not been considered previously.

The demands of industry for a useful mathematics stemmed from various technologies, but geometry certainly owes much of its early development to the building trades; they required a precise knowledge of spatial relationships to construct houses, churches, viaducts, bridges, and roads. There is no evidence that Euclid wrote his 13 volumes in geometry to help builders or that Archimedes deduced the value of pi (π) geometrically to further technology, but early technicians were quick to incorporate whatever mathematicians offered. That one or another esoteric branch of mathematics would be useful to technology or science was not always obvious when that branch first appeared. But no branch of mathematics has failed to become applicable to some phase of technology or science.

In the early stages of civilization the most important life-supporting activities were associated with animal husbandry, agriculture, commerce, shipping, and exploration. However primitive these activities may have been, their pursuit required some kind of mathematics in addition to simple arithmetic. Thus, knowledge of some elements of geometry was essential in agriculture and even in animal husbandry; animal shelters of determined sizes had to be built and bins for animal fodder constructed. Land farming stimulated the development of accurate surveying techniques and, hence, geometry itself. In early Egyptian history, the flooding of the Nile signaled the beginning of the planting season. The deposits of very fertile soil left behind on the banks of the Nile when its waters receded showed that the land was in its optimum condition for seeding. The Egyptians had discovered that the Nile flooding occurred when the star Sirius was rising in the early evening and so they kept careful records of the position of Sirius in the sky with respect to the position of the sun. These records required some knowledge of angles and probably marked the beginning of trigonometry.

Because commerce depended heavily on shipping, ocean navigating became a very important skill which required precise knowledge of the positions of the constellations from season to season and the ability to locate one's ship with respect to the north celestial pole. What we now call our latitude exactly equals the angle of the north celestial pole above our horizon (the altitude of the north celestial pole). Whether or not trigonometry—the geometry of angles—began with celestial navigation is not certain, but we do know that the early navigators knew a great deal about the measurement of angles. The need for precise navigation instruments led to the invention and design of various optical devices which in turn stimulated the application of mathematics, essentially trigonometry, to the study of light passing through glass. These studies culminated many centuries later with the inventions of the camera, the microscope, the telescope, and, most remarkable of all, the spectroscope, which has contributed more to our knowledge of the structure of atoms, of molecules, of matter in general, of the earth and planets, of the sun, of stars in general, and of the universe itself than all the other instruments combined. No instrument has given us so much information for so little investment.

When these instruments were first invented the models were primitive, but they evolved rapidly into very sophisticated devices with the application of mathematics to aid in the improvement of their designs. The development of mathematics, of course, was never limited by its great contributions to the growth of physics. Indeed, it was in turn greatly stimulated by physics which pointed out new directions along which mathematics could grow. This growth, stemming in part from the pressures of science and technology but primarily from its own internal pressures, has produced a new mathematics that is quickly departing from its scientific and technological aspects.

Returning to the origins of mathematics—we do not know whether it began in geometry or in some other phase, but we start our story with geometry because plane geometry is the first branch of mathematics that was almost fully developed by the ancient Greeks. It was presented by Euclid in his *Elements* which was used as a mathematical text for more than 1000 years. Euclid certainly based his writings on the empirical geometry that flourished in ancient Egypt some hundreds of years earlier and was influenced by Thales, Eudoxus, and Pythagoras, all of whom contributed significantly to early Greek geometry. Yet, in thinking about geometry, we generally consider Euclid first because he established geometry as a rigorous theoretical discipline based upon precise definitions, axioms, and theorems. Today the geometry taught in schools—which is still called Euclidean geometry—is taught along the general lines laid down by Euclid in his masterpiece.

Little is known about Euclid of Alexandria (330 BC?–275 BC?). Both the location of his birth and his nationality are uncertain, but he is thought to have been of Greek or Alexandrian origin. Some of the confusion about his nationality also stems from the fact that many scholars have tended to use his name interchangeably with that of Euclid of Megara, a Greek philosopher, who preceded our Euclid by nearly a century. In any event, Euclid probably studied mathematics and philosophy in Athens and supported himself by tutoring others in mathematics.

It was only after Euclid settled in Alexandria, then a Greek colony, and began teaching mathematics at the new university there that he began to develop his reputation as a scholar. By the age of 40 he had finished his *Elements*, perhaps the most famous and successful textbook of all time, having appeared in more than 1000 editions by the twentieth century. Although Euclid wrote books in a number of other areas such as astronomy, music, optics, and methods of solutions, it was his *Elements* which secured his reputation for posterity.

Euclid's *Elements* is a collection of a number of books composed at a time when all written works were recorded on long strips of parchment that were then rolled up around wooden pins. In the case of longer works such as *Elements*, the strips were usually cut into smaller parts called *biblia* or bibles (books). These books were then collected together under a single title.

That it is a masterpiece of mathematical reasoning is unquestioned, but *Elements* is as much a synthesis of the works of earlier mathematicians such as Pythagoras as it is a compendium of Euclid's own original work. Indeed, Euclid contributed many propositions for *Elements* but its fame as a text rests as much on what David Smith, a famous mathematician, called "its simple but logical sequence of theorems and problems."

Euclid's work was fairly unique among the writers of the ancient world because it was structured in a systematic and consistent way. The author himself apparently wanted his treatise to be both rigorous and detailed, a far cry from the rambling, often contradictory, philosophical works written by Plato and Aristotle. Accordingly, Euclid marshaled definitions, axioms, postulates, and propositions into a work of unprecedented intellectual tidiness. Although *Elements* did not use very much in the way of algebra, its organization imbued its readers with the sense that the study of geometry had been completed and that it was time to move on to more advanced areas of mathematics. It also revealed the logical underpinnings of the subject itself and signaled to future mathematicians that they would have to check both the logical rigor and the intellectual content of their work.

Euclid is said to have been modest in dress and in habit. But he was also impatient with those who sought shortcuts when learning mathematics, reportedly declaring that "there is no royal road to geometry." As a scholar, he was consumed with the abstract beauty of geometry, disdaining its practical applications. According to legend, one of Euclid's students asked what he would gain by studying geometry. Euclid responded that knowledge should be acquired for the sake of knowledge but ordered that his slave give the boy some copper coins "so that he might profit from his learning."

The importance of Euclid's work for the development of mathematics is that it transformed geometry from a purely utilitarian mental device to an abstract mental pursuit, governed by its own principles. Geometry differs from all the other branches of mathematics in that it must be incorporated into the laws that govern the universe. In a sense, geometry is a natural law just like the laws of motion of a body. For many centuries people believed that the geometry of the universe was Euclidean and hence Euclidean geometry was studied assiduously, not only as a beautiful logical structure, but as the spatial framework of the laws of nature. Geometry was of particular importance for astronomy because one could hardly picture the space of the stars without some geometrical frame of reference. But introducing such a frame involved such geometrical concepts as points, distances, directions, lines, surfaces (areas), and volumes.

These concepts must have been the starting point of Euclid's thinking because as he developed his geometry around these concepts he saw that they could not all be introduced on the same footing. Some had to be introduced as fundamental or elementary entities; the others would then be secondary entities or concepts, derived from the fundamental entities. Considering the six concepts listed above, we see that the "point" cannot be defined and therefore must be accepted as the most elementary of concepts in terms of which all the others can be defined. The point, the only undefinable element in Euclidean geometry, is thus an abstract concept which has no physical meaning because it, unlike a particle of matter, does not occupy space. Nevertheless, we picture all of space as consisting of points and describe geometrical figures in terms of points. This picture suggests that space is continuous since one point cannot be isolated. This conception is the basis of Euclidean and modern geometries. Euclid and the other great Greek mathematicians accepted the concept of geometric continuity without question. As far as physical space is concerned, however, the contemporary physicist reserves his opinion because whether or not it is continuous can be decided only empirically.

To develop his geometry, Euclid introduced two other abstract concepts: the definition and the axiom. We are all acquainted with the "definition" in its application to language because dictionaries are essentially collections of definitions. The mathematical "definition" implies more than language because, as we shall see, it also implies an operation or a construction of some sort. We note that language grows with the introduction of new words, which must be defined in terms of known words. But mathematicians introduce as few definitions as possible; new mathematical concepts then emerge as the products of pure reasoning, carrying their own meanings with them.

The axiom is one of a set of statements that are accepted as "self-evident truths" and which thus form the foundation of a logical structure such as geometry and other branches of mathematics. Euclid's great contribution to geometry and to mathematics in general was his precise differentiation of propositions into definables, definitions, axioms, and theorems. Before considering how these concepts are brought together to produce a complete system of knowledge such as Euclidean geometry, we analyze in some detail the relationship between undefinables and definitions as exemplified by or introduced into geometry. Here the only undefinable is the point. We can then define a line as a special continuous array or set of points. This means that an infinite number of points lies between any two points—however close—on a given line. Here we must be more precise and specify the line as a "straight" line. We then say that any two points define or determine one and only one "straight" line. But we now face the undefined concept of "straight." Is this concept to be accepted as an undefinable or can it be precisely defined? To define the concept of "straight" we must introduce the concept of distance or length. We then define a straight line passing through or connecting any two points as the shortest distance between those two points. But the distance concept removes us from the abstract domain and launches us into the real world so that a geometry built on this definition of a straight line requires a measurement operation.

To pursue this idea further we first note that the distance concept cannot be defined in terms of points (which are the only undefinables we have to work with), because to define the distance between two points as the number of points on the straight line connecting the two points is nonsense—that number is infinite and, in fact, not countable. Moreover, this task presupposes that we have some way of knowing which one of the many infinite number of lines that can be drawn between the two points is the shortest and therefore straight. In other words, we must have a criterion for "shortest" which necessarily involves a physical operation—measurement. Geometry thus forces us to become empirical. We shall see later how mathematicians in the nineteenth century set up a purely mathematical criterion for the shortest distance between two points which leads to or permits the non-Euclidean geometries that have had an enormous impact on physics, particularly cosmology.

The introduction of axioms and the use of them as a basis for the construction of a vast intellectual edifice such as geometry are among the greatest achievements of the human mind because they permit an infinitude of deductions or theorems. Euclid showed how to develop a geometry from a system of undefinables (points), definitions (straight lines), and axioms, but he achieved something more, because his synthesis of a geometry in this way was carried over by others to other intellectual, particularly mathematical, structures. In developing his geometrical synthesis Euclid had to consider the constraints that must be imposed on any system of axioms. Clearly, the number of axioms must be finite but no more nor fewer than those required to develop a complete geometry. In other words, the system of axioms must be complete in the sense that any theorem that may be proposed in the geometry can be deduced or derived from the definitions and axioms. If these constraints on the number of axioms are not fulfilled, the geometry itself reveals the flaw. If the number of axioms is too small, certain theorems cannot be derived; if the number of axioms is too large, one or more of the axioms are theorems in the sense that they can be deduced from the others. Another obvious constraint is that no two axioms contradict each other, which is the basis of all logic; indeed, all the rules of logic are incorporated in the axioms as are the rules of arithmetic and algebra.

Accepting these constraints, we now consider the general principles or the criteria that led Euclid to choose one statement rather than another as an axiom. In elementary geometry textbooks, this criteria is contained in the definition of an axiom as a "self-evident truth" and thus beyond question.

Euclid did accept his axioms as timeless truths. Elementary school students also accept this criterion and therefore that Euclidean geometry is the geometry that correctly describes the space of our universe. Insofar as Euclid's axioms deal with purely abstract concepts, with no specific applications to reality, we may indeed accept, without question, the idea of a self-evident truth, but if we deal with the geometry of the real universe, concepts such as "self-evident" and "truth" are not well defined.

To delineate the difficulty associated with this criterion for an axiom, we consider in some detail the most famous of Euclid's axioms—axiom number 5—which is probably the most famous of all statements in mathematics. This axiom states that given a straight line and a point outside this line, one and only one straight line (in the plane determined by the given line and point) can be drawn through the point parallel to the given line (not cutting the given line). Before criticizing this axiom as a basis for a real geometry of space, we note that it is essential for the proof of a basic theorem in Euclidean geometry: that the sum of the three angles of a triangle equals 180°. As long as we can deal with an abstract space, we can accept this axiom without question, but if we try to prove that the sum of three angles of a real triangle drawn in space equals 180° (which the theorem states) by measuring the three angles, we run into serious practical problems; points, straight lines, and planes are purely abstract concepts and therefore we cannot construct such a triangle. Let us suppose that we can draw a perfect triangle and picture it as three intersecting straight lines connecting three points. We now measure the three angles as accurately as we possibly can, only to find that the sum is not (and never can be) exactly 180° as demanded by the theorem. The reason for this discrepancy is that we can never draw perfectly straight lines nor can we measure any angle exactly. But the theorem demands that the sum be exactly 180°; it does not allow the minutest departure from 180°. Because we cannot obtain exactly 180° by measurement, we are thrown into a quandary; we do not know whether the difference between the measured sum and 180° is due entirely to measurement errors or arises partly from the non-Euclidean character of the geometry of real space. We return to this point later when we discuss the geometries of Gauss, Lobachevsky, and Riemann, but we note here that Euclidean geometry must always remain as an ideal geometry that can never be demonstrated to be applicable to any space.

Most people consider Euclid to be *the* outstanding mathematician of ancient Greece owing to his important contributions to geometry, yet he also advanced the theory of numbers by his investigations of perfect numbers and prime numbers. These studies led to some of the earliest discoveries in a realm of mathematics known as the theory of numbers that has attracted the greatest mathematicians in every mathematical epoch. This attraction stems from the difficulty of the problems in number theory, some of which are still unsolved and have challenged mathematicians for centuries. Unlike problems in almost all other branches of mathematics, these problems cannot be solved by mathematical analysis. Each problem in the theory of numbers must be solved in its own particular way.

A perfect number is defined as a number, the sum of its aliquot divisors (factors) equals the number itself; by aliquot we mean all the factors of the number including unity, but not the number itself. The number 6 is the smallest perfect number; the sum of its aliquot divisors 1, 2, and 3 equals 6. The next perfect numbers in increasing magnitudes are 28, 496, 8128, and 3,355,036; no odd perfect numbers are known. Although the mathematics of perfect numbers does not constitute a separate branch of the theory of numbers, mathematicians have been attracted to them because they stand out owing to the unique property of their factors and to their scarcity. They challenge mathematicians to predict their frequency in the number system by finding how many such numbers lie between any two integers. This problem has never been solved although Euclid knew the formula for even perfect numbers which, expressed algebraically, is $(2^{n-1})(2^n - 1)$, where the exponent n is any positive integer such that $(2^n - 1)$ is a prime number (any integer other than 0 or ± 1 that is not divisible without remainder by any other integers except ± 1 and $\pm$ the integer itself). This statement is true for $n = 2, 3, 5, 7, 13$, etc., but we have no general rule for finding all such integers. That those values of n we have listed are themselves primes should not lead us to the false conclusion that $2^n - 1$ is a prime if n itself is a prime [primes are known for which $(2^{\text{prime}} - 1)$ is not a prime].

Perfect numbers were studied by the Greeks long before Euclid and by the early Hebrews; they were of particular interest to ancient theologians. No less a religious philosopher than St. Augustine wondered about the "perfect" number 6 in arguing that "God created all things in 6 days because this number is perfect." In a similar vein, the mathematician Nichomachus stated that the perfect numbers 28 and 496 are "beautiful and excellent" as compared to the "prolific ordinary (nonperfect) but ugly numbers" because perfect numbers "are rare and easily counted." In any case, mathematicians have devoted a great deal of time and effort to the study of perfect numbers

since the time of Euclid. As we proceed in our story we highlight the progress made in this study of perfect numbers from one mathematical epoch to
the next.

Euclid contributed to another branch of the theory of numbers which is
far more important, for physics as well as for mathematics, than perfect
numbers—namely, the theory of prime numbers. As mentioned above, a
prime number is defined as one that has only 1 and itself as factors; it cannot
be written as a product of numbers smaller than itself. The first few primes
are 1, 2, 3, 5, 7, 11, 13, 17, and 19. The number of primes in any numerical
interval of ten integers decreases as the integers themselves become larger.

Prime numbers constitute one of the most challenging branches of the
theory of numbers because they present mathematicians with problems that
have not been solved since prime numbers were first studied. Yet Euclid
made one of the most important discoveries in the theory of numbers when
he proved that there is an infinite number of prime numbers or, alternatively
stated, that there is no single largest prime number. We know that the larger a
composite (nonprime) number, the larger is the number of its factors. The
number 60, for example, has the factors 1, 2, 3, 4, 5, 6, 10, 12, 15, 20, and
30. One may then argue that the chance that a number has a factor and is,
therefore, not a prime number increases as the number increases. One may
then conclude, erroneously, that beyond a certain very large value all
numbers are composite. Euclid disproved this hypothesis in a very simple
way and showed that the number of primes is infinite.

Euclid's proof, a type of reasoning that can be applied to the solution of
different kinds of problems whether they are mathematical or not, is
generally called "*reductio ad absurdum*"; one shows that the assumption
that the theorem to be proved is false leads to an absurdity. Euclid assumed
that a largest prime p does exist and then considered the product $1 \times 2 \times 3$
$\times 5 \times \cdots \times p$ of all the primes up to and including p. This product, of
course, is not a prime, but the product plus 1—the number $(1 \times 2 \times 3 \times 4 \times$
$\cdots \times p) + 1$ must either be a prime or divisible by a prime larger than p.
But this product is not divisible by any of the primes up to and including p
since the division by any one of these primes leaves the remainder 1. Thus
either this quantity is itself a prime or divisible by a prime larger than p. In
either case, a prime larger than p exists, which proves the theorem, since the
contrary assumption leads to a contradiction.

Euclid was but the first among equals in the hierarchy of early Greek
mathematicians, famous more for his organization of geometry into a self-

consistent, logical discipline than his own original contributions to mathematics. Indeed, historians have argued that included in some of his 13 volumes was the work of others, such as Eudoxus of Cnidus, who flourished in the fourth century BC and developed the first method of measuring the circumference of a circle. Euclid himself was predated by Thales and Pythagoras. According to the Greek historian Herodotus, Thales predicted (in 590 BC) the year of a solar eclipse and was the founder of Greek geometry, which he used in his astronomical studies. Unfortunately, very few of Thales's original manuscripts have survived to the present.

The story of Pythagoras (569 BC?–500 BC?) is quite different. Even though none of his writings has survived, his philosophy and mathematics are known from the philosophical school, the Pythagoreans, he founded, which was based on the principle that number is everything. Not only do numbers express the relationships among natural phenomena, but, in a sense, they cause these phenomena to manifest themselves. In propounding this principle, Pythagoras was probably greatly influenced by the Egyptian mysticism he acquired while in Egypt during his early years. His philosophy that the world is ruled by harmony led Pythagoras to conclude that this harmony can best be represented by numerical relationships; all proportions, order, and harmony in the universe could be represented by numbers. Pythagoras was the first to represent musical harmonies in terms of the numerical ratios of the pitches (frequencies) of the musical notes in these harmonies. He pointed out that in a stringed instrument in which the strings are all under the same tension the most harmonious sounds are produced if the lengths of the strings are related to each other by simple ratios such as 1:2, 2:3, etc.

Much of what we know about Pythagoras's life is shrouded in myth but it is possible to chronicle his years with large brush strokes. As a young man, Pythagoras studied philosophy with Pherecydes and Anaximander. He then traveled throughout Asia Minor, spending long periods of time in Egypt and Babylonia. After moving to Sicily in 529 BC, he founded a school at Croton in the south of Italy. Pythagoras was a very charismatic and popular lecturer; he often attracted large crowds of men and women, especially citizens of the upper classes. From these audiences he selected his disciples (Pythagoreans) who were to be privy to Pythagoras's most important discoveries. Even though the Pythagoreans were fed a mixture of theory, speculation, and sheer nonsense by their teacher, they adored Pythagoras with a fervor that bordered on fanaticism. Indeed, one of his most devout followers was a beautiful young woman named Theano, the daughter of his host Milo. Theano was so taken with Pythagoras that she later became his wife.

Pythagoras made important contributions to mathematics but his investigations of physical phenomena were of little value. He offered a bizarre numerology to the world which he believed was the foundation of reality. He also instructed his followers in philosophy and politics, swearing them to secrecy. As time went on, the Pythagoreans began to resemble a cult, adopting their own ethical and moral code. Their belief in the transmigration of souls caused the Pythagoreans to embrace vegetarianism; they also were not allowed to eat beans. According to Pythagoras, the purification of the soul could be achieved only through the observance of mathematically inspired rituals and a strict dietary and physical regimen.

The Pythagoreans gradually closed themselves off from the outside world, giving rise to unfounded rumors that they engaged in ritualistic orgies and black magic. Pythagoras's own insistence that his disciples seek pleasure did little to discourage these rumors. Hence, the Pythagoreans served as a lightning rod for the suspicions of the surrounding community and they became the focus of Pythagoras's political opponents. Ultimately, Pythagoras's enemies instigated a mob attack on the Pythagoreans and murdered Pythagoras and many of his followers. Their triumph was short-lived, however, as many of Pythagoras's surviving disciples established a new society at Tarentum (now Taranto, Italy) where they flourished for many years.

The myth of Pythagoras was also enhanced by his refusal to allow his teachings to be recorded in writing by his followers. He believed that his discoveries should be made known only to his followers and thus hidden from any outsiders. Moreover, Pythagoreans of subsequent generations were instructed to attribute any of their own discoveries to Pythagoras; the reliance on a purely oral record helped to augment the apparent significance of Pythagoras's own intellectual legacy. So great was the insistence on secrecy that it was not unknown for Pythagoreans who publicly boasted of their own discoveries, such as Hippasus, to be put to death. As the sect became more dispersed, however, these secrecy controls lessened and authors began to document the mathematical, philosophical, and ethical teachings of Pythagoras. Even with the dissemination of these works, however, it was almost impossible, in the opinion of the mathematician Carl B. Boyer, "to separate history and legend concerning the man, for he meant so many things to the populace—the philosopher, the astronomer, the mathematician, the abhorrer of beans, the saint, the prophet, the performer of miracles, the magician, the charlatan."

Even though it is difficult to separate the teachings of Pythagoras from those of his followers, it does appear that Pythagoras himself was the first to use deductive reasoning in grappling with geometrical problems. Indeed, his arrangement of the central propositions of geometry in a logical order undoubtedly influenced Euclid's *Elements*. According to W. W. Rouse Ball's *A Short History of Mathematics*, Pythagoras was also the first to raise arithmetic above the needs of merchants: "It was their boast that they sought knowledge and not wealth, or in the language of one of their maxims, 'a figure and a step forwards, not a figure to gain three oboli [Greek coins].' " Pursuant to this idealization of mathematics, Pythagoras, in the words of the mathematician Eric Temple Bell's *Men of Mathematics*, "imported proof into mathematics" because "geometry [before Pythagoras] had been largely a collection of rules of thumb empirically arrived at without any clear indication of the mutual connections of the rules, and without the slightest suspicion that all were deducible from a comparatively small number of postulates." That the proof itself is the *raison d'être* of mathematics today is taken for granted by both mathematicians and their students; such an assumption would have been utterly foreign to the mathematicians who preceded Pythagoras.

Pythagoras is best known as a geometer for his famous theorem that the square of the hypotenuse of any right triangle equals the sum of the squares of the other two sides; its algebraic equation is $h^2 = a^2 + b^2$, where h is the length of the hypotenuse and a and b are the lengths of the other two sides. One need not be a mathematician to see that this equation must be correct; simple reasoning demonstrates its validity. In any algebraic equation every term that represents a measurable physical entity must represent the same such entity. Thus if one term in the equation is a distance, every term must be a distance and if one term is an area (square of a distance) every term must be an area. If the terms in an equation represent numbers with no physical significance, we may have first powers, squares, or cubes of these terms in the equation without violating any basic principle. Returning to the theorem of Pythagoras for a right triangle, we see that h^2, the square of a distance, is an area and therefore can depend only on some algebraic combination of a^2 and b^2, which are also areas. The simplest combination is $a^2 + b^2$ and so, as a first approximation, we write $h^2 = a^2 + b^2$. This is certainly correct in stating, as it does, that the hypotenuse of a right triangle increases if either a or b increases. We note, by the way, that the equation $h = a + b$ is incorrect; the hypotenuse of a triangle must be less than $a + b$, the sum of the other two sides, because by definition, the hypotenuse is the shortest distance between the two end points of a and b.

One might, of course, try the equation $h^2 = pa^2 + qb^2$, where p and q are just numbers *different from* 1, but it is easy to show that both p and q *must each equal* 1. The quantities p and q must be the same number because h^2 must depend on a^2 and b^2 in exactly the same way; a and b can be interchanged without altering their relationship to h^2. The equation must then be of the form $h^2 = pa^2 + qb^2$. Now this relation must hold no matter how small the side h is so we may place $b = 0$, obtaining $h^2 = pa^2$. But for $b = 0$ the hypotenuse equals the side a and this can be true only if $p = 1$. We thus obtain the theorem of Pythagoras $h^2 = a^2 + b^2$.

The Pythagorean triangle has the remarkable property that h, a, and b can be an integer triplet, the simplest example of which is 3, 4, 5, where 5 is the hypotenuse of a right triangle whose other two sides are a and b. We see that $5^2 = 25 = 3^2 + 4^2 = 9 + 16 = 25$. If one triplet of integers satisfies the Pythagorean equation, we can obtain an infinite number of such triplets by multiplying each number of the first triplet by any integer we please. Thus the triplets 6, 8, 10 and 9, 12, 15 also satisfy the Pythagorean theorem. We now show that if we are given any pair of integers m, n, we can find a Pythagorean integer triplet h, a, b by simple algebra. We place h equal to $m^2 + n^2$ and a equal to $m^2 - n^2$ (where $m > n$) so that b must then equal $2mn$ (twice the product of m and n). From the Pythagorean theorem we obtain

$$
\begin{aligned}
b^2 = h^2 - a^2 &= (m^2 + n^2)^2 - (m^2 - n^2)^2 \\
&= m^4 + 2m^2n^2 + n^4 - m^4 + 2m^2n^2 - n^4 \\
&= 4m^2n^2 \text{ or } b = 2mn.
\end{aligned}
$$

As an example of how this calculation works, we choose $m = 2$ and $n = 1$ so that $h = m^2 + n^2 = 4 + 1 = 5$, $a = m^2 - n^2 = 4 - 1 = 3$, $b = 2mn = 4$; thus we obtain the 3, 4, 5 right triangle.

Pythagoras and his students developed an esoteric cult around integers, endowing them with a mystical quality which has no meaning for us at all today. Some people still practice this sort of numerology to "explain" physical phenomena they do not understand in terms of arbitrary numerical relationships. Knowing that physical phenomena can be described or expressed in terms of measurable numerical entities, these practitioners of numerology believe that if they can concoct any combination of integers

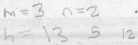

that is close to the measured value of the physical entity, their numerology is justified. Discounting the mysticism of the Pythagoreans we must credit them with the founding of the Theory of Numbers, one of the most beautiful branches of modern mathematics.

The Greek mathematician Eudoxus (408 BC?–355 BC?) is often overlooked in treatises on the mathematics of the early Greeks, perhaps because he devoted much of his time to the study of the apparent motions of the planets, which he explained by his ingenious theory of homocentric spheres—a system of concentric spheres, with the earth at their common center. This theory shows clearly that Eudoxus stands in the foremost ranks of Greek mathematicians because his success in fitting the observed motions of the planets to the assigned rotations of his spheres (a different sphere was assigned to each planet) showed his great mathematical skill. His analysis of these motions led him to propose a solar cycle of four years: three years, each consisting of 365 days and one year consisting of 366 days (leap year), which, some 300 years later, was introduced by Julius Caesar as the Julian calendar.

Born in Asia Minor (now Turkey) Eudoxus grew up in extreme poverty. He studied under the Pythagoreans at Tarentum and then traveled with Plato to Egypt. He founded a school of philosophy at Cyzicus but later moved to Athens. His dire financial circumstances prevented him from living near Plato's Academy. Hence, Eudoxus was forced to walk back and forth each day from Piraeus, a port in Athens, where he could obtain food and lodging for a pittance. Although Plato himself was not really a mathematician, he recognized Eudoxus's immense talent for mathematics and took him under his wing. Unfortunately, Eudoxus's mathematical abilities soon far outstripped those of his famous teacher, thus causing a falling out between the two. Because of Plato's hostility toward him, as well as the xenophobia of the Athenian citizenry, Eudoxus left Athens for Cnidus shortly before his death.

Like Archimedes, Eudoxus was a versatile man, achieving prominence in areas ranging from astronomy and medicine to geometry and law. In his *History of Mathematics*, David Smith writes that Eudoxus was the first to give a description of the constellations and that he brought the theory of the motions of the planets from Egypt to Greece. Eudoxus also devoted much time to making models of the solar system and calculating the diameter of the sun, which he estimated to be nine times that of the earth. As a physician, Eudoxus brought the same problem-solving mentality to medicine that he used in the physical sciences. He was thus reluctant to accept long-held superstitions and speculations at face value and instead checked them against observations and experiments. He may have also invented a then-new type of sundial. Like many other early scientists, what little we do know about Eudoxus has been obtained from the writings of others as all of his original works have been lost.

The importance of Eudoxus's work for mathematics lies in his descriptions of the apparent motions of the planets, sun, and moon. These descriptions required a precise understanding of the concept of angle, which, even today, is poorly or incorrectly defined. Because angle is the basis of trigonometry, the work of Eudoxus may be described as the birth of trigonometry. The angle concept as introduced in the study of triangles is a static idea as compared to the way Eudoxus and other early Greek astronomers conceived it because its full physical or dynamical significance is revealed only in representing the changes in the direction from the observer to a moving object. We arrive at this dynamical aspect of an angle by considering a point moving along the circumference of a circle and following the radius of the circle attached to this point. As the point moves, the direction of this radius changes; the change in this direction is called the angle between the initial and final positions of the radius. To illustrate this differently, consider the spokes of a rotating wheel. As the wheel turns, the direction of each spoke changes by exactly the same amount in a given time. The amount the wheel turns in this time is defined as the angle formed by the intersection of the direction of any spoke in its initial position (initial moment) and its direction at the final moment. Angle is thus a measure of turning or rotation; the wheel makes one complete rotation when every spoke returns to its original direction, and we then define that movement as 360 units of turning; each unit is called one degree (1°). Angle is to rotation as distance is to translation, but angle is a pure number whereas distance is a spatial dimension.

The degree, an arbitrary unit of angular measure, was probably introduced by the early Greek astronomers as a measure of the change, in one day, of the sun's apparent position with respect to the fixed stars. Because the sun appears to complete its journey among the stars in 365¼ days, the daily change in its apparent celestial position is slightly less than 1°, but the degree, as a unit of angular measure, was universally adopted. Another unit of angular measure, the *radian* is used exclusively in mathematics because it emerges quite naturally from the geometry of the circle.

As far as the use of mathematics to elucidate astronomical phenomena is concerned, the work of Aristarchus of Samos (310 BC?–230 BC?) deserves special attention. He was the first to propose a heliocentric (sun centered) model, based on the apparent dimensions of the sun and moon and their relative distances from the earth. He carefully measured the angular separation between the sun and moon when the side of the moon facing the earth was half illuminated (half moon). Noting that this angle is very close to 90°, he argued correctly that the sun's distance from the earth greatly exceeds the moon's distance and therefore that the size of the sun greatly exceeds that of the moon and, hence, that of the earth. That the actual sizes he calculated from his observations are quite erroneous does not detract from the importance of this work for science and mathematics. It not only stimulated Copernicus to propound his heliocentric solar system but it also demonstrated the great usefulness of mathematics in deducing new phenomena from empirical data and constructing mathematical models of physical systems. Unfortunately, only remnants of Aristarchus's writings remained after his death and his ideas survived only because Archimedes, his younger contemporary, incorporated them into his own treatises. But Archimedes used false arguments against the acceptance of Aristarchus's heliocentric solar system, which consequently remained dormant for more than 1500 years owing to Archimedes's great influence and authority.

Whereas Aristarchus pursued and mastered mathematics as a tool in his astronomical investigations, Apollonius of Perge (260 BC–200 BC), one of the greatest of the early Greek mathematicians, studied mathematics for its own sake and was the first to develop the theory of conic sections. This subject had been treated by Euclid in a book that was, by this time, gone. The great Alexandrians, Hipparchus and Ptolemy, credit Apollonius with developing the mathematics they needed for their epicycle models of planetary motions. Though incorrect, these models demonstrate a masterful application of geometry and trigonometry.

Apollonius studied in Alexandria for many years and gave numerous lectures. Unlike the gracious Archimedes, Apollonius is reported to have been self-centered and insecure, always ready to depreciate the discoveries and academic accomplishments of his colleagues. Apollonius later taught at Pergamum in Pamphylia at a newly established university modeled after the one in Alexandria. At Pergamum, Apollonius met Eudemus and Attalus; they were among the few persons to receive a book from the author of his treatise on conics along with an explanatory letter. Apollonius eventually returned to Alexandria where he remained until his death.

The appreciation of Apollonius's treatise on conics is not in its reading;

its proofs are long and difficult. But W. W. Rouse Ball asserts that "Apollonius so thoroughly investigated the properties of these curves that he left but little for his successors to add." His method seemed to indicate some familiarity with geometrical coordinate systems and, hence, analytical geometry. Apollonius was apparently motivated to write the treatise, which was laden with 400 propositions, by a geometrician who had stayed with him in Alexandria. Apollonius also managed to complete a number of other works on geometrical planes, regular solids, and unclassed incommensurables, such as the circumference of a circle and its radius. He also wrote a book outlining a method of arithmetic calculation that is not unlike that offered by Archimedes in his *Sand Reckoner*.

Of all of the early Greek mathematicians, Archimedes (287 BC–212 BC) was probably the most brilliant and inventive, contributing to the physical sciences as well as to geometry, arithmetic, and trigonometry. Though most famous for his discovery of the basic hydrostatic principle (known as the Archimedes principle), that the loss of weight of a body immersed in a fluid equals the weight of the fluid displaced by the body, Archimedes's greatest contributions to humanity's intellectual development were in pure mathematics. His inventions in mechanics and hydraulics, such as the hydraulic or Archimedes screw, and machines based on the lever principles, were only incidental to Archimedes's devotion to mathematics—particularly to the geometry of the circle. This devotion led him to a very clever method of approximating the number π (pi), which enters into the formula for the circumference of a circle, $2\pi r$, and its area, πr^2, where r is the radius of the circle. This remarkable number, one of a set called transcendental numbers, a subclass of a larger group called irrational numbers, enters into all branches of mathematics, theoretical physics, chemistry, and astronomy. In a sense, the universe could not operate properly without the number pi, whose exact value in terms of our number system can never be known; but mathematicians have discovered ways of approximating its value to any accuracy that may be desired.

Archimedes was born in Syracuse, a relative of the ruling family and the son of an astronomer. He probably studied in Alexandria with students of Euclid, but Archimedes did not travel very much, spending most of his life at his home in Syracuse. Like Euclid, Archimedes was much more interested in discovering mathematical principles than devising practical applications. Archimedes was also a prolific inventor whose devices helped the citizens of Syracuse resist a brutal Roman seige for two years. He designed huge catapults that tossed stones over the city walls at the Roman legions and large mirrors that set fire to the Roman ships anchored offshore. Indeed, the city might not have ever fallen to Rome if its citizens and soldiers had not become drunk while celebrating a religious holiday. As the Roman troops looted the city, Plutarch writes, a Roman soldier approached Archimedes—who was busy working out a mathematical problem in the sand with a stick—and ordered the elderly mathematician to accompany him back to the Roman camp. Archimedes, lost deep in thought, waved him away with his hand. The enraged soldier killed Archimedes with his sword, despite the orders of the Roman general Marcellus, that the life of the famous mathematician be spared. According to Plutarch, Marcellus himself was greatly moved by the news of Archimedes's death; he not only bestowed great favors upon Archimedes's family but also declared the one who killed Archimedes to be a murderer. According to Eric Temple Bell, the death of Archimedes marked "the first impact of a crassly practical civilization upon the greater thing which it destroyed—Rome, having half-demolished Carthage, swollen with victory and imperially purple with valor, falling upon Greece to shatter its fine fragility."

Even though he was more concerned with mathematical principles than with technological achievements, Archimedes composed several works, such as *On the Equilibrium of Planes*, which explored the relationship between mathematics and mechanics using a process of reasoning that moved from propositions to deductions much like the process employed by Euclid in his *Elements*. Archimedes established himself as the father of mathematical physics with his *On Floating Bodies*, a book which included the hydrostatic principle—the principle of buoyancy—the discovery of which supposedly caused Archimedes to leap from his bath and run home naked shouting "Eureka!" Archimedes also used his pulleys and levers to help Syracuse's King Hiero launch a massive warship that would have otherwise remained stuck in its dock. Indeed, Archimedes is reported to have boasted that he could move the earth if he had a sufficiently long lever and a place to stand. And the citizens of Syracuse were not the only ones who witnessed his genius. While living in Egypt, Archimedes invented the Archimedian screw to raise water from the Nile River to irrigate surrounding lands.

Archimedes was one of the first mathematicians to show the utility of very large numbers. He calculated in his *Sand Reckoner* that it would take 10^{63} grains of sand to fill completely the universe. Archimedes arrived at this number by estimating the number of grains of sand that would fit in a poppy seed, the number of poppy seeds that could be laid across the width of a finger, the number of finger-lengths in a stadium, and the number of stadia in a universe of the size calculated by the Greek astronomer Aristarchus. Archimedes's computations not only showed the ease with which the mathematician can handle unimaginably vast numbers but also underscored the distinction between very large numbers and infinity itself.

Archimedes never allowed a meal or a social obligation to stand in the way of his finding a solution to whatever problem he happened to be working on. When visiting the homes of acquaintances and family members, he often spent his time scratching out the solution to a particular problem on the earthen floor, forgetting all about the people around him. He seems to have been something of a loner even though he did consider Conon and Eratosthenes—both of whom were mathematicians—to be good friends. But Archimedes was distinguished from most of his contemporaries and predecessors by the fury with which he attacked mathematical problems. Unlike Plato, who believed that mathematics should be studied according to certain very limited rules, Archimedes declared that mathematicians should employ any means and weapons at their disposal to conquer mathematics.

When asked to rank the three greatest mathematicians in history, almost all scholars place Archimedes at the head of the list, followed by Newton and Gauss. To this day, Archimedes remains unsurpassed in his ability to apply mathematics to mechanics. Even more impressive is that Archimedes essentially invented integral calculus and anticipated many of the important ideas of Newton and his fellow scientists, such as differential calculus, at a time when science was more mysticism and theology than a self-consistent rigorous discipline. Pliny the Elder called him a mathematical god and Voltaire asserted that Archimedes had a greater imagination than Homer. Even the Romans paid homage to Archimedes, erecting a lavish tomb in his honor, upon which was engraved, according to W. W. Rouse Ball, "the figure of a sphere inscribed in a cylinder, in commemoration of the proof he had given that the volume of a sphere equals two-thirds that of the circumscribing right cylinder and its surface area equals four times the area of a great circle."

Archimedes knew that the radius of a circle is not contained in its circumference a rational number of times (an integer or an integer plus a fraction) nor can this number be expressed in terms of roots of integers such as $\sqrt{2}$, $\sqrt[3]{5}$, which, themselves are irrational numbers, but far less interesting than π, which Archimedes encountered as soon as he began to study the geometry of a circle. We have already noted that Euclid and his contemporary mathematicians had to introduce the concept of the angle in their development of geometry, which is a measure of the change of direction or of turning. The unit angle, one degree (1°), which we discussed previously, is the 360th part of a complete turn (360°); it is an arbitrary unit which does not relate the radius of a circle directly to the angle through which the radius turns as its free end moves from point to point on the circle's circumference. Early geometers therefore found it convenient to introduce a unit of angle that is directly related to the radius and to pi.

To define this unit we consider an arc of a circle whose length measured along the circumference of the circle exactly equals the radius of the circle. The angle formed at the center of the circle by the two radii from the center to the two ends of the arc is the unit we want; it is called one radian because it is the angle subtended at the center of any circle by an arc whose length is exactly one radius. This unit is very useful because it relates the length of any arc of a circle directly to the radius of the circle and the angle (measured in radians) covered by the arc: length of arc = angle in radians times radius. The arc length must be expressed in the same units of length as that of the radius. If the radius of a circle is of unit length (e. g., one inch), then the angle covered by the arc, in radians, exactly equals the length of the arc in the same units (e. g., inches). The problem that Archimedes solved was to find the exact relationship between the radian and the degree, or, put differently, the number of times the circumference of a circle is larger than its radius. Because the circle's circumference covers 360° (one complete turn) the number of radians contained in 360° equals the number of times the radius is contained in the circumference. If we call this number 2π, then one radian equals 360° divided by 2π.

Archimedes knew these facts but this information alone does not tell, nor did it tell the numerical value of pi. One can try, as the early Greeks did, to determine pi empirically—by actually measuring the circumference of a circle and comparing that measurement with the measured length of the radius. Although this shows that pi is slightly larger than 3 so that one radian is slightly less than 60°, this empirical procedure can give us only an approximate value for pi, which Archimedes found unsatisfactory. He therefore used the method of "exhaustions" to obtain a formula for π which gives its numerical value to any desired accuracy.

This method, the precursor of the integral calculus, was developed by the early Greeks, in particular Eudoxus, who used it to obtain the areas of various plane figures. The reason for the curious name "exhaustions" applied to this method is that the given area to be calculated is covered by a mesh of small known areas (squares, triangles, etc.) which does not quite cover the entire area being considered. The mesh is therefore made finer and finer so that successive meshes "exhaust" more and more of the area to be covered until a mesh containing an infinite number of infinitesimal known areas covers ("exhausts") the entire given area.

Archimedes applied this method to obtain a formula for the area of a circle and thus to obtain the numerical value of π, but instead of covering the circle with a mesh, he inscribed in the circle a series of regular polygons which approximated the circle more and more closely. A regular polygon is a plane figure in which all sides are equal. An equilateral triangle and a square are such figures. If we inscribe such a figure in a circle so that each vertex of the figure touches the circumference, the center of the figure coincides with the center of the circle and the line from the center of the circle (also the figure) to any vertex equals the radius of the circle. The area of the regular polygon is always smaller than the area of the circle, but the polygon area approaches the area of the circle as the number of sides n of the regular polygon increases. Finally, the two areas coincide as n becomes infinite. One can now use this method, as Archimedes did, to obtain the value of π by equating the area of a circle given as πr^2 in terms of π and the radius r of the circle to the polygon area given in terms of the square of the radius of the circle and the number n. The equality holds only if n is made infinite; the r^2 on both sides of the equation then cancels out and an arithmetic formula (actually an infinite numerical sum) for π remains. Many such infinite numerical series for π have been discovered by mathematicians since Archimedes—evidence of the growth of mathematics during those years.

Although Euclid, Archimedes, Eudoxus, Apollonius, and other early Greek mathematicians were born in Greece, they were all educated in Alexandria and spent most of their productive years at the great Museum of Alexandria, founded and financially supported by the Ptolemaic dynasty. Historians therefore divide this early period of mathematics into the Greek and Alexandrian periods. This is not a meaningful division, however, since the Alexandrian school is justifiably famous for three of its outstanding mathematician–astronomers: Eratosthenes, Hipparchus, and Ptolemy, who used the early Greek mathematics to establish astronomy as a science. Up to that time, astronomy had been primarily observational, but done casually, with no attempt to make accurate measurements or to correlate these measurements mathematically and thus to construct some kind of mathematical model to account for the observed motions of the planets. Aristarchus had done this very same work some centuries earlier, in presenting his heliocentric model of the solar system, but this model had been ignored or dismissed as purely speculative when Hipparchus began his remarkable career as an astronomer. Eratosthenes (276 BC?–194 BC?), however, predated Hipparchus as an accurate astronomical observer and mathematician. Born in Cyrene and educated in Athens and later in Alexandria, he became the librarian of that great museum, where he did most of his important astronomical and mathematical work. Eratosthenes was a polymath who excelled in many fields ranging from history and mathematics to poetry and astronomy. Only after he had reached middle age, though, was Eratosthenes asked by the Egyptian monarch Ptolemy III to tutor the king's young son in Alexandria. He was a close friend of Archimedes and, like that Syracuse native, constructed useful scientific instruments. Eratosthenes urged that what is now known as the Julian calendar be adopted so that an extra day would be added to every fourth year. His admirers considered him to be "the second Plato"; an admittedly exaggerated view of his importance to learning even though he was among the greatest scholars of Alexandria. But Eratosthenes was cursed with ophthalmia, and slowly went blind. When he could no longer read, Eratosthenes committed suicide.

He is most famous as a mathematician for his discovery of what is known as the sieve of Eratosthenes—a systematic way of identifying prime numbers—which may be explained as follows. From a list of all the integers, starting from 2, cross out every second integer excluding 2. Then, starting from 3, cross out every third integer excluding 3. Return to the first uncrossed number after 3 (the number 5) and cross out every fifth number excluding 5 and so on. The primes are all the uncrossed numbers: 1, 2, 3, 5, 7, 11, 13, etc. Each of these numbers has no factors other than 1 and itself; all numbers with more than two factors have been crossed out.

As an astronomer and geographer Eratosthenes is best remembered for his calculation of the circumference of the earth. His method for doing this calculation, which indicates that he was a very skillful mathematician with a mastery of geometry and trigonometry, is based on the simple relationship between the arc of a circle and the angle this arc covers at the center of the circle. If the arc length is 1 and the angle it subtends at the center is 1°, the length of the circumference is 360° × 1 (a distance 1 times the number of degrees subtended by a complete circumference). Eratosthenes measured the arc length on the surface of the earth between Alexandria and Syene and found the angle that the vertical (plumb line) at Syene makes with the vertical at Alexandria. Because this angle is contained in 360° as many times as the Syene–Alexandria arc length is contained in the earth's circumference, he equated these two quotients to obtain the earth's circumference, which is simply the arc length multiplied by the first quotient. His result, 24,662 miles, is remarkably close to the true distance. Eratosthenes's procedure is called measuring the length of a degree of latitude. Eratosthenes's great contribution to the growth of mathematics was to show its great practical importance to geography and navigation, which helped to ensure the continued support of mathematical research by heads of states in future eras.

Hipparchus of Nicaea in Bithynia (180 BC?–125 BC?) was yet another mathematician who was educated in Alexandria. But Hipparchus's writings have not survived and almost all of what we know about his work in mathematics and astronomy is found in Ptolemy's *Almagest*, a treatise that brought together the astronomical observations and speculations of earlier astronomers. Were it not for Ptolemy, we might not know that Hipparchus calculated such things as the length of the year, the inclination of the ecliptic to the equator, and the eccentricity of the solar orbit. He also devised a system of longitude and latitude coordinates for identifying specific locations on the surface of the earth. His astronomical investigations also led him to the study of trigonometry.

But Hipparchus was less concerned with the practical application of mathematics to earthly affairs than with its application to astronomy. In terms of his remarkable astronomical observations he must be ranked as the greatest of the Greek astronomers of the pre-Christian era. His most brilliant contribution to astronomy was probably his discovery of the precession of the equinoxes which showed that the direction of the earth's axis of rotation is not fixed but changes continuously, because the earth behaves like a huge spinning top whose axis of rotation itself is rotating once every 26,000 years. This discovery required a deep knowledge and mastery of both geometry and trigonometry. He was the first to apply trigonometry to astronomical calculations. He had to deduce a trigonometric formula for the length of a chord drawn from one point on the circumference of a circle to another point on the circumference in terms of the cosine of the angle subtended by the chord at the center of the circle. To prepare a table of chords, as Hipparchus did, he must have known the formula $2\sqrt{\sin(\theta/2)}$

for the length of a chord that subtends the angle θ at.the center of the circle of radius r.

Hipparchus's contribution to the development of mathematics lies not so much in any new mathematical discoveries he made, as in his introduction of mathematics to astronomy. He laid down a procedure for drawing important astronomical conclusions from observations which was utilized by all serious astronomers who followed him. He applied his mathematical skills to the development of the theory of epicycles to account for the apparent motions of the planets; his mathematical epicycle model was so cleverly constructed that few doubted its validity. Hence, the geocentric theory of the solar system was universally accepted and dominated astronomical theory until Copernicus upset it in 1543.

The Christian era in mathematics and astronomy was introduced by the great Alexandrian astronomer Ptolemy (100 AD?–170 AD?), who lived some 200 years after Hipparchus. One may then wonder at the dearth of astronomical and mathematical activity during the two centuries separating Hipparchus and Ptolemy. This lapse can perhaps be explained by the hegemony of the Roman culture which during that time replaced Greek theory with Roman pragmatism. The Greeks were great theoreticians who delighted in pure thought whereas the Romans were concerned with more earthly matters such as military conquest, commerce, construction, and the maintenance of a vast empire. The Roman era was not entirely devoid of mathematical activity, but the application of mathematics to architectural and engineering projects, rather than the development of mathematics itself, occupied the Romans. Such Roman scholars as Pliny the Elder and Seneca wrote extensively on astronomical matters but they contributed little to mathematics. Yet, we cannot completely dismiss the Roman influence on mathematics, because their great success in engineering projects shows that they were good—though not creative—mathematicians. This emphasis stimulated the engineers who followed the Romans to develop the practical aspects of mathematics to their fullest scope.

Ptolemy's *Almagest*, which played such a crucial role in ensuring the immortality of Hipparchus and a host of other early Greek astronomers, contained little in the way of original science. But it did preserve much of what we know about the astronomical discoveries of the early Greeks. Although Ptolemy lived in Alexandria, he was probably a Roman citizen. He also taught in Athens, becoming thoroughly imbued with the magnificent Greek legacy in astronomy and mathematics. It may have been in Greece that Ptolemy developed the analytical skills that were needed to bring together the disparate works of the Greek astronomers into a single volume that served as the standard textbook on astronomy for more than 1000 years. Because the *Almagest* was not an original work, however, some historians have tended to dismiss Ptolemy as little more than a dilettante. But it was in the *Almagest* that Ptolemy developed his theory of epicycles to explain the apparent movements of the sun, moon, and planets in a geocentric solar system. He also introduced a method for classifying stars that continued to be used until the nineteenth century.

Ptolemy did not confine himself solely to astronomy and mathematics. He wrote an influential treatise entitled *Geography*, which dominated the thinking of geographers and cartographers for many years. According to Carl B. Boyer, "the *Geography* of Ptolemy introduced the systems of latitudes and longitudes as used today, described methods of cartographic projection, and catalogued some 8000 cities, rivers, and other important features of the earth." But because Ptolemy lacked the means to determine longitudes, his *Geography* was riddled with errors. He also greatly underestimated the size of the earth, a mistake that later figured greatly in the strategies of sailors such as Columbus who tried to reach India by sailing westward. Had Columbus known that a westward voyage would entail crossing two oceans and sailing around an entire continent, he probably would have followed the eastern routes used by the Portuguese around the southern coast of Africa.

Returning now to Ptolemy, who developed Hipparchus's theory of epicycles to its fullest extent, we see that mathematics can be used to prop up an incorrect theory as well as to support and illuminate a correct theory, thus showing that mathematics in itself must not be confused with scientific theory. A closer analysis of Ptolemy's geocentric model of the solar system reveals why it was accepted for some 1500 years, almost without question. Ptolemy's unique epicycles predict the future positions of planets with surprising accuracy. This accuracy can be explained by a mathematical discovery that was made in the eighteenth century by the French mathematician Fourier, who showed that complex motion of any kind can be expressed mathematically by a sum of simple circular motions. This is what Ptolemy did with the apparent motions of the planets. He replaced the observed complex motion of a planet by a sum of epicycles (one circular motion on top of another, with the radius of each new circle smaller than that of the previous one).

From the era of the earliest Greeks (Thales, Pythagoras, etc.) to the

Christian era, geometry and trigonometry were extensively applied, but no real growth in Greek mathematics occurred. The reason for this stagnation may well be that science hardly developed at all during this period. The Greeks had discovered many facts about the world around them during the millennium from Thales to the post-Ptolemaic era, but they failed to discover a single physical law. Indeed, this state of affairs remained unchanged until Newton propounded his laws of motion in the last quarter of the seventeenth century. The second law, stated as a simple algebraic equation $F = ma$ (force equals mass times acceleration), was a watershed event in the history of mathematics as it was in science. The development of mathematics from that discovery to the beginning of the twentieth century was far greater during that period than in the thousands of years before its discovery.

We have discussed the geometry and trigonometry of the Greeks and Alexandrians without mentioning their arithmetic or algebra in any detail. Insofar as arithmetic as a computational process is concerned, it was used extensively in all the ancient civilizations but its development as a rigorous mathematical discipline paralleled the development of the theory of numbers. Like all other branches of mathematics, the theory of numbers—which deals with the properties and interrelationships of numbers—is based on a set of definitions and axioms from which the theorems of the theory can be deduced. These definitions and axioms govern the mathematical operations we may apply to arithmetic. We shall treat this topic in more detail in a later chapter but we note here that arithmetic in its present form stemmed from the Hindus who introduced the modern number system in the second century AD.

That the Greeks were interested in number theory is indicated by Euclid's proof that the number of primes is infinite and by Eratosthenes's procedure for listing all the primes. The numeral systems used by both the Greeks and Romans were clumsy; they impeded rapid numerical calculations and made the introduction of fractions practically impossible. The ancient Greeks used two parallel numeral systems: the first based on the initial letters of the names of numbers such as "delta" for 10 and "chi" for 1000; the second based on assigning the first nine letters of the alphabet to the numbers 1 to 9 and the second 9 letters to the decades 10 to 90, and so on. The Roman numerals are easy enough to remember but forbidding as far as arithmetic operations go. How, for example, can anyone do any arithmetic with DCCCLXXXVIII, the Roman equivalent of 888?

As we have already noted, we can trace the origins of modern arithmetic back to the second century AD Hindus, but the current numeral system, with its integers, goes back to the Hindus of the third century BC, who used it extensively in their computations. However, the credit for using the position of an integer in a many-digit number as the measure of its contribution to the numerical value of the number goes to the Arabs of the seventh and eighth centuries AD. Considering the drawbacks of the numerical system used by the Greeks, we must marvel at their mathematical and astronomical achievements, which required extensive arithmetical computations.

Though the ancient Greeks were not acquainted with algebra as we know it today, they certainly used algebraic notation and relationships to express some of their mathematical concepts. In writing the Pythagorean theorem in the form $h^2 = a^2 + b^2$, the Greeks were using algebra, but they did not pursue algebra as a logical system governed by a set of rules as they did geometry. The idea of setting up equations to be solved for the unknowns in these equations did not occur to them. However, the first treatise on algebra was written by Diophantus, a Greek mathematician of Alexandria. Diophantine equations or analysis became an important branch of the theory of numbers.

In considering the totality of early Greek mathematics, one wonders why it was not developed more fully as a tool to construct mathematical models of natural phenomena and thus probe nature deeply. The Greek astronomers did this to a limited extent and Archimedes applied his knowledge of mathematics to analyze the lever and to the study of hydraulics and optics, but no new mathematics grew out of these applications of Greek mathematics. Mathematics thus remained static for centuries even though Greek mathematics was accepted as a standard part of the curriculum in universities and even in lower grades. We pass over this period with only brief references and proceed to the modern era which began with the Renaissance.

CHAPTER 2

The Origins of Arithmetic

In mathematics I can report no deficiency, except it be that men do not sufficiently understand the excellent use of the Pure Mathematics.

—FRANCIS BACON

The story of mathematics is much more difficult to write in a sequential way than, for example, the story of physics because mathematics, unlike physics, did not evolve sequentially, with each phase dependent on previous mathematical discoveries. Indeed, many different phases of mathematics evolved and grew independently of each other; yet this does not mean that the various branches of mathematics are unrelated to each other. They are all part of a vast intellectual fabric woven together by strands of logic that permit one to pass from one branch to any other through mathematical reasoning. This is one of the reasons that mathematics attracts us as the unchallenged domain of pure reason. In our mathematical linkage of two apparently disparate fields, new mathematical properties of the two fields are revealed that are not evident without the linkage. Thus, certain numbers, called "Fermat numbers," because they were first investigated by the great seventeenth-century French mathematician, Pierre de Fermat, founder of the modern "theory of numbers," can be shown to be related to geometric properties of regular polygons inscribed in a circle. We recall from our discussion of the number π in the preceding chapter, that Archimedes used the areas of such polygons to deduce the numerical value of π to any desired accuracy. This is an example of the linkage of one of the most esoteric, abstract branches of mathematics to geometry and to a number that appears in every branch of physics.

Historically, we should have started our story of mathematics with arithmetic and algebra—both predate geometry—but we have chosen to begin with plane geometry, the development of which is associated with the names of the great early Greek mathematicians, because it stands by itself as a self-consistent closed system. It can be developed to any desired degree, as it was by the Greeks, without involving or calling upon any other branch of mathematics. It rests on a few undefinables (points, lines, etc.), some definitions, and a set of axioms. From these modest materials, an infinitude of theorems can be deduced so that plane geometry is in a sense open-ended. But it is a closed discipline in that we are not permitted to go outside these axioms in proving any theorems.

Though plane geometry could have been developed without explicit reference to arithmetic or algebra, these were the mathematical tools of the geometers that enabled them to develop their geometry rapidly. Both arithmetic and algebra are open-ended in the sense that they are governed only by a few rules and definitions and not by a body of axioms. These rules are broad enough to allow both arithmetic and algebra to expand far beyond the realms they originally spanned. Thus, arithmetic in its youth dealt only with integers and signified quantity only and soon expanded into the realms of fractions and irrational and complex numbers, and became associated with order, organization, and information. Hence, computers today are concerned far less with pure arithmetic than they are with information. The rules of arithmetic can be applied not only to individual numbers but to collections of numbers or things and to infinitely large numbers. The first of these extensions of arithmetic is called set theory and the second is called the theory of transfinite numbers.

The rules of algebra are essentially the same as those of arithmetic. But algebra expanded far beyond its original domain, which had been similar to that of arithmetic. Its numbers were replaced by letters with the understanding that these letters were numbers in disguise. Though algebraic expressions were written in letters, they had meaning only when the letters were replaced by the appropriate numbers. Algebra has evolved to such an extent today that the basic algebraic entities are far removed from numbers although they may refer to numbers. Indeed, the algebraic entity may be a quantity called a matrix or a mathematical operator which is used to transform one algebraic entity into another. Moreover, some of the rules themselves that govern numerical operations may be altered to introduce special branches of algebra. As an example, the multiplication rule may be altered so that the product of two algebraic entities depends on the order in which the multiplication is performed. But algebra has evolved even beyond that stage and has been applied to logic and language in a form called Boolean algebra, after the nineteenth century mathematician George Boole. Here in our discussion of algebra, we limit ourselves to elementary algebra and leave the story of modern algebra for a later chapter.

Arithmetic was the first and most basic branch of mathematics to emerge from the daily intercourse among people in the earliest societies. No great names are associated with its creation nor was arithmetic recognized as an intellectual enterprise when it began as a purely utilitarian mental device some millennia before the Christian era. In evolving from its limited initial role as a method of counting and keeping records of transactions to the "mathematics" we now call "arithmetic," it was necessary for a number system to be introduced, as we briefly discussed in the preceding chapter, where we considered the relationship of Greek geometry to arithmetic.

Even the most primitive societies required some kind of a number system. The historical evidence suggests that the more primitive the society, the closer its number system tended to approximate sets of straight vertical lines. In short, the system itself consisted of a single digit—the vertical line—and quantities. Egyptian hieroglyphics dating from 3400 BC contain groups of vertical straight lines to designate quantities of things; these are also present in the earliest writings of the Mesopotamians. The most important advancement in numerals was made in the third century BC by the Hindus who used a straight line for one but adopted different symbols for quantities greater than one. This numbering system was introduced to the Arabs around the eighth century AD and first appeared in European arithmetic in 976 AD, when the nine digits 1, 2, 3, 4, 5, 6, 7, 8, 9 were first used. Although this was a tremendous advance in arithmetic, it was incomplete because it lacked the digit 0 for zero, without which certain important arithmetic operations such as subtraction are not always possible.

The symbol 0 for zero did not appear until a thousand years after the Babylonians first introduced written symbols for numbers, but even then it was not used in arithmetic operations but merely as a marker. The Maya first used 0 in the first century AD; 500 years later it was adopted by the Hindus and then it finally appeared in European treatises on arithmetic. An enormous advancement occurred in the number system in the seventh century AD when the position of a digit in a many-digit number such as 5378 was first used as a mark of the quantitative value of the digit. The last digit to the right represents units, the next to the last represents tens, the third from the last represents hundreds, and so on. This development immediately simplified all arithmetic operations and placed arithmetic on a sound foundation.

With the acceptance of the symbol 0 as a number with many of the properties of the other nine digits, the decimal-based number system was born and arithmetic began its journey from a lowly accounting tool to its present status as the "Queen of Mathematics," a designation offered by Karl Friedrich Gauss, perhaps the greatest of all mathematicians. The digit 0 was first introduced to differentiate between two numbers such as 63 and 603 but it acquired importance of its own as a digit when negative numbers were introduced; it is neither positive nor negative and so it separates the positive from the negative numbers. As arithmetic evolved, the importance of 0 as a digit grew; it became clear that arithmetic itself would be incomplete without the number 0. It differs from all other digits in four important arithmetic properties:

- Adding or subtracting 0 from any number leaves the number unchanged.
- Raising any number to the power 0 always gives unity (1).
- Multiplying any number by 0 always gives 0.
- Dividing any number by 0 is always infinite and therefore forbidden as an arithmetic operation since infinity is not a number.

The digit 0 acquired importance beyond its purely digital property with the introduction of graphical methods of representing arithmetic operations, because it is used to designate the starting point (the origin) of a graph or a system of coordinates in analytic geometry. It thus assumed the role of representing the beginning of a series of physical phenomena and positions in space. By including 0, one obtains a ten-digit (decimal) number system which then allows one to represent any number as a sum of powers of 10, each power multiplied by one of the 10 digits. The digit 0 is also necessary to develop a complete algebra for it leads to algebraic equations and to the concept of the roots of an equation. Finally, we note that without 0, calculus could not have been developed because calculus is based on a mathematical operation which involves allowing certain entities in a mathematical formula to go to 0 (to become 0).

Arithmetic evolved as a decimal system with the introduction of 0 as the tenth digit along with certain basic rules. A collection of digits becomes an arithmetic only when rules are introduced which define the four basic operations: addition, subtraction, multiplication, and division. These operations were used in the practical application of arithmetic to daily activities long before they were defined in a mathematically rigorous way. The digits represented to the ancient merchants of Greece, Egypt, and Mesopotamia quantities of real things which might increase or decrease from day to day. To the mathematician, however, the digits were and are abstract symbols to which one can apply rules that in turn enable one to combine these symbols to obtain other symbols.

The first of these rules is called addition and is represented by the cross "+" (the plus sign) which is placed between any two or more symbols, as in the examples $1 + 1 + 1$ and $2 + 3 + 4$. The "=" sign (the equality sign) is used to express the result of this operation which is called a sum. All addition can be reduced to the simplest example $1 + 1 = 2$. Thus $3 + 2 = (1 + 1 + 1) + (1 + 1) = 5$, because each digit can be expressed as a sum of ones. The sum $0 +$ any digit $=$ any digit; 0 added to any number does not alter the number. We note that in these examples the digits do not refer to any objects or collections of objects. They are just meaningless figures which obey the rule that if any one of these numbers is followed by the two symbols $+ 1$, they are to be replaced by the number that is one unit to the right of the original number on a number line. Thus $6 + 1$ or $1 + 6$ is to be replaced by 7, and so on. This abstract way of introducing addition has nothing to do with size or magnitude. It is merely a rule for going from one symbol to the next one on its right if the digits are all arranged from left to right (0, 1, 2, 3, 4, 5, 6, 7, 8, 9 . . .) so that $0 + 1 = 1$, $1 + 1 = 2$, etc. For practical applications each of these symbols represents a quantity, which increases from left to right so that 7 means one more of anything than does 6. This is implicit in the addition rule itself since $6 + 1 = 7$ or $(1 + 1 + 1 + 1 + 1 + 1) + 1 = 7$. Subtraction, denoted by the "−" sign (the minus sign), is the inverse of addition so that $7 - 6 = (1 + 1 + 1 + 1 + 1 + 1 + 1) - (1 + 1 + 1 + 1 + 1 + 1) = 1$. The addition and subtraction operations, when first introduced, were applied only to positive integers. Fractions came into use, as numbers in their own right, when it became clear that fractions were required in daily commercial activities, as well as in construction, surveying, and navigating. The incorporation of fractions into the number system was an enormous advancement in arithmetic because fractions made it possible to define fully the division operation. But it was not clear how fractions should be labeled. One could not, as with the integers, start with the "first fraction" and label it with a symbol like 1, because no first fraction exists in the sense that a first integer exists; 1 is the smallest integer but no smallest fraction exists. Moreover, using a whole new set of symbols to label fractions would not suffice because such symbols would not convey to the reader the essential difference between a fraction and an integer. The modern symbols for fractions can be traced back to the ancient Egyptians who used the same symbols as for the integers but with special markings

FRACTIONS

(like dots) on top of these symbols to distinguish them from integers. Thus 1̇1̇1̇ might represent one-third. But the transition from this rudimentary and quite inadequate system of symbols to the modern symbols was made in the seventh century AD by the Hindus, who were the first to use combinations of the ten integer symbols 0, 1, 2, . . ., 9 to represent fractions. This was probably the single most important advancement in the evolution of arithmetic because it placed fractions on exactly the same numerical footing as integers, governed by the same arithmetic operations (addition, multiplication, etc). But the use of combinations of digits to represent fractions enlarged the domain of arithmetic enormously so that it could be applied to the demands of precision in technology, which depends so heavily on accurate measurements. Fractions permit the engineer, the surveyor, the builder, and the scientist to express observations and make measurements to as high a degree of accuracy as may be desired; this cannot be done with integers.

To see how the digits 1, 2, 3, . . ., 9 were introduced and are still used to express fractions, we consider a unit length on a table (e.g., 1 centimeter) and follow a tiny bug crawling from one end of this length to the other. At each point on its trip, the bug completes a fraction of its total journey of 1 centimeter. How should we express that fraction? We first picture the bug at its halfway point. Because the centimeter is then divided into two equal parts, the symbol that represents this fraction should contain the digits that express this quantity; the only two digits that convey this information are the numbers 1 and 2. We therefore write this fraction as $\frac{1}{2}$; it represents a unit entity 1 (the centimeter) divided into two equal parts. Note that this representation of a fraction carries with it the concept of division.

We can now carry this symbolism over to represent any fraction we please with the understanding that the number above the line in the fraction (called the numerator of the fraction) is the quantity that is to be divided and the number below the fraction line (called the denominator) is the number of parts into which the numerator is to be divided. As originally understood, the fraction $\frac{25}{75}$, for example, means that 25 identical objects are to be divided into 75 equal parts. But the present concept of the fraction, as used in modern arithmetic and algebra as well as in its application to measurements, has diverged considerably from its primitive connotation. The evolution of the concept of the fraction from its basic definition to its present role in mathematics is an excellent example of the growth of mathematics, a growth that is driven by its own internal dynamics.

Fractions, of course, are governed by the same operational rules as are integers. We can add them and subtract them in exactly the same way as we do integers, but when fractions were first introduced into the number system, certain difficulties arose concerning how fractions with different denominators were to be added and subtracted. We shall discuss these difficulties in full after we introduce the multiplication operation. If we have two fractions, such as $\frac{1}{3}$ and $\frac{2}{3}$ or $\frac{2}{7}$ and $\frac{3}{7}$, their sums are written as $\frac{1}{3} + \frac{2}{3}$ and $\frac{2}{7} + \frac{3}{7}$ just as for the sum of integers. The rule here is that we simply add the numerators, without changing the denominators, obtaining $\frac{1}{3} + \frac{2}{3} = \frac{3}{3} = 1$ and $\frac{2}{7} + \frac{3}{7} = \frac{5}{7}$.

The difficulty arose when the sum of two or more fractions with different denominators, such as $\frac{2}{3}$ and $\frac{3}{7}$, was required. We shall see later how this problem was solved with the aid of multiplication. The subtraction of two fractions was defined in the same way as that of integers. Thus the subtraction of $\frac{2}{7}$ from $\frac{3}{7}$, written as $\frac{3}{7} - \frac{2}{7}$, was defined as the operation $(3 - 2)/7 = \frac{1}{7}$, which is just the inverse of addition of fractions.

We now define the operations of multiplication and division as applied to integers and then to fractions. But we can attach no dates to the introduction of these operations. Indeed, it may be meaningless to speak of such dates because the operations of addition and subtraction imply multiplication and division; these latter operations had no date of origin distinct from that of addition and subtraction. We return now to the elementary operation of addition which stemmed from combining entities into groups. This idea was extended or generalized to arrange entities into groups containing equal numbers of entities and counting the number of such groups, e.g., (111) + (111) + (111) + (111) which is a sum of four groups, each containing three entities. We therefore write this quantity as 4×3 and call this operation multiplication. When we add the four groups of three, the sum is 12 so that we say the product of 4 and 3 equals 12. The symbol $\times$ was introduced to represent a product. As this concept evolved, the dot and the parenthesis became as common as the $\times$ for expressing the multiplication operation. The product was thus introduced to compress addition.

Mathematicians have always been concerned about rigor in mathematics. Euclid was perhaps the first mathematician to attempt to place one branch of mathematics—plane geometry—on a firm foundation by marshaling an array of definitions, axioms, and theorems. Other Greek mathematicians spelled out rigorous rules for arithmetic even though people did arithmetic without worrying about whether they were following these rules, which are few in number, simple, and so "self-evident" that many nonmathematicians wonder why they were introduced at all. It seemed that these rules were just "common sense" applied to arithmetic. Simple as these rules (axioms) might appear, however, they are essential to all mathematics. Only two rules apply to addition—the commutative and the associative law. The commutative law simply states that the order in which we add two numbers does not affect the outcome of the operation so that $2 + 3 = 3 + 2 = 5$. This rule applies to a sum of any number of numbers. The associative law of addition states that we may add numbers by grouping them together in any way we desire so that $3 + 4 + 7 + 8 = (3 + 4) + (7 + 8) = 3 + (4 + 7) + 8 = 3 + (4 + 7 + 8)$. The same two rules apply to multiplication, but a third rule, the distributive rule, must also be added. It states that $3 \times (4 + 5) = (3 \times 4) + (3 \times 5)$.

The commutative law is of particular interest in multiplication because one may invent an arithmetic in which this law does not apply. The entities that may be used in this particular type of arithmetic are not ordinary numbers. We shall return to this idea in a later chapter and see that the noncommutative law is very important in the mathematics of atomic physics.

We saw that multiplication was introduced into arithmetic shortly after addition was developed. Similarly, division, which is the inverse of multiplication (but a more subtle idea since it does not suggest itself as readily as multiplication), was formulated only after subtraction itself had been invented. Multiplication is an extension of addition and division is an extension of subtraction. Division is the subtraction of groups of entities of the same numerical size from a given number of entities. If we consider the division of 12 by 4, for example, we see that the division of twelve entities //////////// by the group of 4 entities //// initially involves the subtraction of //// from //////////// which leaves eight entities ////////. If we carry out the subtraction of the same group //// from the eight entities, we will be left with four entities ////. If we repeat this operation once again, we will be left with nothing so we know that the group //// is found three times in the group ////////////. This operation may be expressed more conveniently as $12 \div 4 = 3$, where the symbol "$\div$" stands for division, but this symbol is rarely used today. It may instead be replaced by $\frac{12}{4}$.

This introduction of division as an arithmetic operation helped arithmetic to evolve from a purely pragmatic calculating device to the theory of numbers, a rigorous mathematical discipline. An important consequence of the division of one number by another is the concept of factors. A factor of a number is any other number that divides the given number exactly so that there is no remainder. If we know one factor of a number, we immediately know another factor because the number, when multiplied by the known factor, equals the given number. If 15 is a known factor of 45, for example, we know that 3 is also a factor since $15 \times 3 = 45$. Finding the factors of a number became one of the outstanding exercises in the theory of numbers, which deals with problems involving prime numbers and perfect numbers— both of which require knowledge of the factors of a number.

Two other arithmetic operations which greatly simplified arithmetic itself and its symbolism began to appear in the arithmetic formulas of Arab mathematicians about 900 AD; these applications involve raising a number to a power and taking the root of a number. Raising a number to a power stemmed from the considerations of multiple products such as $5 \times 5 \times 5 \times 5$. This cumbersome way of writing such a product was replaced by the elegant expression 5^4, which is described as "raising 5 to the fourth power." The number 4 is called the exponent of the power. An important application of this operation is raising 10 to any power, which is simply the digit 1 followed by a number of zeros equal to the exponent. Thus $10^1 = 10$, $10^2 = 100$, $10^3 = 1000$, and so on. We also note that any number raised to the 0 power is 1; thus $3^0 = 25^0 = 100^0 = 1$.

The inverse operation of raising a number to a power is that of taking or finding the root of a number such as the square root, which was encountered in many practical applications of arithmetic. The square root of a number is familiar to all of us because it is defined as the number, the second power of which (the square) equals the given number. Thus the square root of 25, written as $\sqrt{25}$, is 5, since $5^2 = 25$. But very few integers for a given range of numbers have integer square roots. Of all the integers between 1 and 100, only 1, 4, 9, 16, 25, 36, 49, 64, 81, and 100 have integer square roots. In general, the chance that any integer, chosen at random, has an integer square root is very small. For that reason, arithmetic logarithms for obtaining the square root of any number were introduced. The most famous of these algorithms was offered by Euclid. Taking the root of a number is not restricted to the square root; we can take the cube root, the fourth root, and so on. These various roots of a number are written as $\sqrt[3]{}$, $\sqrt[4]{}$, etc. Algorithms have been developed for finding the cube root of a number, but general algorithms for higher roots are not known. (The development of logarithms in the seventeenth century, however, eliminated the need for any kind of algorithm to obtain any root of any number to any desired numerical accuracy.)

With the definitions of multiplication, division, powers, and roots of integers complete, we now go back to fractions for which we defined the operations of addition and subtraction in a restricted form, limited to fractions with the same denominators. Because this was a restriction that mathematicians could not tolerate, simple rules were developed for allowing the addition and subtraction of any two (or more) fractions. Mathematicians were thus forced to develop techniques for the multiplication and division of fractions. These operations probably arose from an extension of the multiplication and division of a fraction by an integer. As a simple example, we consider the multiplication of the fraction $\frac{2}{3}$ by 2, which we may write as $2 \times \frac{2}{3}$. This means that we take $\frac{2}{3}$ twice or $\frac{2}{3} + \frac{2}{3}$, which is equal to $\frac{4}{3}$. This product is obtained by multiplying 2 by the numerator 2 of the fraction. This is an example of the general rule that the product of any fraction multiplied by any number is a fraction with exactly the same denominator but with a numerator that is the product of the number and the numerator of the original factor. The denominator plays no rule in this operation; it remains the same.

This multiplication operation was extended to the multiplication of two proper fractions such as $\frac{2}{3}$ and $\frac{4}{5}$ by generalizing the multiplication of two fractions such as $\frac{1}{2}$ and $\frac{1}{2}$, each with 1 in the numerator. We recall that $\frac{1}{2}$ means dividing 1 into 2 equal parts. If we divide each of these parts into 2 equal parts, we obtain 4 equal parts of 1, which we represent as $\frac{1}{4}$, so that $\frac{1}{2} \times \frac{1}{2} = \frac{1}{4}$. This procedure led to the general rule that in multiplying two fractions with 1 in each numerator, the numerator is still 1, but the denominator is the product of the two denominators of the two fractions. Thus $\frac{1}{3} \times \frac{1}{4} = \frac{1}{12}$. The next step in this story of the product of fractions was the combination of the two rules described above for the multiplication of two fractions, each with a different numerator and a different denominator, such as $\frac{2}{3}$ and $\frac{4}{5}$. The rule is that both the two numerators and the two denominators are to be multiplied to give the product of the two fractions. Thus $\frac{2}{3} \times \frac{4}{5} = \frac{8}{15}$. This rule applies not only to the product of two fractions but also to the product of any number of fractions; it led immediately to the operation of raising a fraction to a power, which is the same in every respect as raising a number to a power. This procedure, introduced as shorthand for writing the multiple product of a given fraction, such as $\frac{2}{5} \times \frac{2}{5} \times \frac{2}{5}$ as $(\frac{2}{5})^3$, equals $\frac{8}{125}$. Both the numerator and the denominator of the fraction are raised to the power to which the entire fraction is raised. This placed the multiplication of fractions and raising them to a given power on the same footing as those operations applied to whole numbers.

But one more step was required before mathematicians could assign fractions to their rightful role in the number system—the introduction of the rule for the division of fractions. We recall that division is the inverse of multiplication when applied to integers. The same must be true when these operations are applied to fractions. Division is the reciprocal of multiplication for fractions as it is for integers. This statement is true for the division of any two fractions such as $\frac{2}{3}$ and $\frac{4}{5}$ which, by the inversion rule, equals $\frac{2}{3} \times \frac{5}{4}$ = $\frac{10}{12}$. Despite the apparent simplicity of this step, it represented a very important advance in the evolution of mathematics because it illustrated the power and importance of generalization. Although we cannot be sure that generalization, either in mathematics or in science, is always correct, generalizations have been the most productive of all techniques in both mathematics and science.

Because taking the root of a number is the inverse of raising a number to a power, the rule for taking a root that applies to integers also applies to fractions. The square root of the fraction $\frac{2}{3}$, written as $\sqrt{\frac{2}{3}}$, means $\sqrt{2}/\sqrt{3}$; this statement applies to all roots, which is the same as the power law for fractions as described above. When mathematicians make such generalizations, they go to great pains to prove them rigorously but this degree of analytical rigor came long after the rules were introduced and accepted as pragmatic necessities. Much of the growth of arithmetic that we have described stemmed from the need for practical mathematical techniques to handle daily needs. Yet these techniques acquired a general validity which permitted the mathematician to apply them to all phases of mathematics.

With our discussion of the multiplication of fractions in hand, we can complete our examination of the addition of fractions which we had previously defined only for fractions with the same denominators, such as $\frac{2}{5}$ and $\frac{3}{5}$. We are then dealing with the addition of similar entities as would be the case if we added three apples to two apples to get five apples. But adding two fractions such as $\frac{2}{3}$ and $\frac{4}{5}$ is an entirely different matter because thirds and fifths are different entities as are apples and oranges, which cannot be added. But adding thirds and fifths differs in one very important respect from adding apples and oranges: we cannot change apples and oranges to the same kind of entity but we can change thirds and fifths to the same kind of mathematical entity. Mathematicians discovered this when arithmetic was still in its early stages and it was probably known to Pythagoras. We now characterize this process as expressing each fraction with the same denominator or finding a common denominator. This can be done with any number of fractions, all with different denominators. This clever idea is carried out by multiplying each fraction to be added by another fraction, but this other multiplication fraction must not change the value of the fraction being multiplied. This is achieved if the numerical value of the multiplying fraction is 1 since multiplying by 1 does not change the numerical value of anything. Using the simple example of adding the two fractions $\frac{2}{3}$ and $\frac{4}{5}$, we multiply $\frac{2}{3}$ by $\frac{5}{5}$, which is 1, to obtain $\frac{10}{15}$, and multiply $\frac{4}{5}$ by $\frac{3}{3}$, which is also 1, to obtain the fraction $\frac{12}{15}$. We thus find $\frac{2}{3} + \frac{4}{5} = \frac{10}{15} + \frac{12}{15} = \frac{22}{15}$. This example shows us how concepts and operations in mathematics form chains or linkages which permit mathematicians to extend and enlarge their mathematical domains almost endlessly. In all such generalizations, however, all mathematical rules must be obeyed and rigor must be maintained. Increasing both the numerator and denominator of a fraction by multiplying them both by the same factor is the inverse of the operation called reducing a fraction to its lowest terms. This involves canceling out all common factors in the numerator and denominator. Thus the fraction $\frac{15}{25}$ is reduced to its lowest terms by canceling out the factor 5 which is common to the numerator and denominator; the reduced fraction is then $\frac{3}{5}$.

As the arithmetic of fractions evolved and grew, mathematicians introduced a clever scheme that enables anyone to determine how much larger one fraction is than another merely by inspection. This scheme is what we now call the decimal representation of a fraction, which stems from the ten-digit base of the number system. Expressing a fraction as a decimal simply means rewriting the fraction with 10, 100, 1000, etc., in its denominator and then dropping the denominator and placing a dot and zeros at the appropriate position in the numerator. The dot and the zeros following it then represent the denominator as illustrated by the following examples: $\frac{1}{10} = .1$, $\frac{1}{100} = .01$, $\frac{1}{1000} = .001$, etc. Although it is clear how one writes a decimal in these simple examples, it is not evident how one can represent fractions such as $\frac{2}{7}$ or $\frac{1}{4}$. Considering $\frac{1}{4}$, for example, we see that it can be written as $(\frac{100}{4})/100$, a fraction whose numerator is $\frac{100}{4}$ and whose denominator is 100. Since $\frac{100}{4}$ is 25, the fraction $\frac{1}{4}$ has the decimal form .25.

From such simple examples, mathematicians introduced the following general rule for expressing any fraction as a decimal: place a dot after the last digit of the numerator, adding as many zeros as may be desired after the dot, and then divide the denominator into the numerator, keeping the dot in place. The dot is called the decimal point, and, as already stated, it represents the denominator of the fraction; the number of digits, including zeros, following the decimal point indicates whether the decimal represents tenths, hundredths, thousandths, etc. Thus, .1, .10, .100, .1000 mean one-tenth, ten one-hundredths, 100 one-thousandths and 1000 ten-thousandths, which are all equal and represent the same fraction $\frac{1}{10}$. Placing a zero in front of the decimal point does not alter the value of the fractions either; thus 0.1 is the same as .1.

Decimal representations of fractions led mathematicians to a remarkable discovery about fractions that is not obvious from the standard form of the fraction itself. This difference is illustrated in the decimal representations of the two fractions $\frac{1}{4}$ and $\frac{1}{7}$, which are 0.25 and 0.142857. . ., respectively; the fraction $\frac{1}{4}$ ends as a decimal with its two nonrepeating digits whereas the decimal representation of $\frac{1}{7}$ never ends, with the digits 142857 repeated over and over again in that order. Mathematicians have therefore divided fractions into two groups—those with nonrepeating decimal representations and those with repeating decimal representations. Note that the decimal 0.25 may be said to be a repeating decimal if we write it as 0.250000. . ., adding as many zeros after the 5 as we please, but mathematicians exclude such decimal representations of fractions from their definition of repeating fractions.

The question as to why some decimals terminate without a remainder, some repeat the first digit power forever, others repeat the same digit endlessly after one or more different starting digits, and still others repeat cycles of digits endlessly (like the cycle 142857 in the fraction $\frac{1}{7}$) was studied in great detail by mathematicians through the ages and became part of the theory of prime numbers in the vast field of the theory of numbers. Another important question which arose in this area of arithmetic was the following: What kind of number does a nonrepeating decimal represent if none of its endless digits repeat in any cycle? Mathematicians called this group of numbers irrational numbers; we shall discuss these numbers later and see the important role they play not only in arithmetic but in every phase of mathematics.

The introduction of the decimal representation of fractions was probably one of the greatest advances in the history of arithmetic, and perhaps in mathematics itself, not only from a mathematical point of view but also from a pragmatic and scientific point of view because it simplified numerical calculations enormously and permitted engineers, economists, surveyors, architects, and scientists to express their measurements to any desired degree of accuracy they pleased. We are aware of the advantage of working with decimals rather than with the standard expressions for fractions when

we see how quickly decimals permit us to determine the numerical difference between two fractions that are close in value to each other. To see how
much $\frac{1}{11}$ is larger than $\frac{1}{13}$ we write them in their decimal forms as
0.090909. . . and 0.076923076923. . . . Subtracting this decimal from the
preceding one reveals a difference of 0.02408. . . so that $\frac{1}{11}$ is larger than
$\frac{1}{13}$ by the amount 0.02408. . . .

The introduction of decimals greatly simplified the arithmetic of
fractions, which became exactly the same as that of integers, as illustrated
by a few simple examples. Adding two fractions such as $\frac{1}{4}$ and $\frac{2}{5}$ becomes
the sum 0.25 + 0.40 = 0.65, and multiplying them becomes the product
0.25 × 0.40 = 0.100. Division is also simplified because we can divide $\frac{2}{5}$
by $\frac{1}{4}$ by inverting $\frac{1}{4}$ and then multiplying the two fractions to get the result $\frac{8}{5}$.
In these operations with decimals, one has to keep track of the decimal
points, making sure to keep them lined up in addition and subtraction, and
shifted by the correct number of digits to the left after multiplication. But
these rules become familiar over time. Decimals also simplified arithmetic
operations in commerce, finance, merchandising, medicine, and in scientific measurements. Thus our entire pricing system would be chaotic to most
people if prices were expressed in fractions rather than in decimals. The
price of an item expressed as "0.35" (35 cents) is easier to grasp than if
expressed as "7/20ths of a dollar." Unlike fractions, decimals permit the
shopper to compare prices very easily. Moreover, it would be almost
impossible for most shoppers to obtain the cost of many items if the prices of
things were not expressed as decimals.

In financial transactions, the concept of rates (the flow of money per
unit time), expressed as percentages or as fractions with 100 in the denominator is readily understood by almost everyone. A bank interest rate of 5
percent, expressed in decimal form as 0.05 and $\frac{5}{100}$ or $\frac{1}{20}$ as a fraction, means
that the principal on deposit increases by 0.05 times the principal in the time
specified by the rate. Rates, as percentages, are also used in mortgages,
loans, and taxes, to name but a few more areas in which decimals have
simplified routine financial activities.

In chemistry, pharmacology, and medicine, decimals in the form of
percentages are used extensively; thus concentrations of particular constituents of a solution are written as percentages. This usage has been carried
over to describing contents of foods, vitamins, beverages, wines and liquors.
Demographers, statisticians, and meteorologists all use percentages extensively as do those who record sports of all kinds. In the physical sciences,

such as physics, decimals permit the scientist to express measurements with the precision that ordinary fractions do not allow. Thus, to say that a certain physical quantity has the measured value 1.056 ± 0.005 means that its true value lies somewhere between 1.051 and 1.061; to write this expression using fractions is very cumbersome.

Fractions were a natural outgrowth of the extension of practical arithmetic to deal with parts of objects in the daily activities of people. Thus fractions were necessary to describe equal parts of a distance, a time, an apple, or even a population. To the nonmathematician, a fraction was, and still is, a real entity associated with the physical or mental division of an object or thing. The mathematician, however, eschews this point of view and treats fractions as abstract symbols governed by very definite rules which relate them to integers. These rules, the addition, subtraction, multiplication, and division of fractions, allow one to pass from fractions to integers and from integers to fractions. Thus the symbols $1, \frac{1}{2}, +$ and $=$ are related to each other by the sum rule $\frac{1}{2} + \frac{1}{2} = 1$, and the symbols $1, 2, \frac{1}{2}, \div,$ and $=$ are related to each other by the division rule $1 \div 2 = \frac{1}{2}$. Considering fractions from this point of view, we see that they enlarge arithmetic enormously. But this kind of abstraction, which mathematicians cherish, relieves us of the need to associate a fraction with a precise division of an entity into equal parts, which can never be verified. Thus, to say that we have two halves of an apple is meaningless if by "halves" we mean exactly equal parts because we can never verify this equality. This is true of all measurable entities, and particularly so of scientific measurements, which are never precise. Recognizing this truth, the scientist therefore expresses his measurements as lying within a certain small range of values which is an estimate of his error. This treatment of fractions is an excellent example of how the mathematician sweeps away any concreteness (measurement or quantity) associated with a mathematical entity and, instead, considers it as a pure, abstract entity.

The evolution of the concept of fractions from their concrete meanings to their purely mathematical states occurred over many centuries, during which time they became increasingly important in mathematics. This became increasingly evident when the ordinal aspect of a number—its ordering property as distinct from its cardinal or quantitative property—was introduced into mathematics because this ordinal aspect is the bridge that connects arithmetic to geometry. To understand this geometry–arithmetic relationship we consider points on a line which we discussed in the preceding chapter where we defined a point as the intersection of two lines, or conversely, a line as a continuous collection of points. The mathematician accepts these definitions but only as pure mathematical abstractions. A point has no concrete meaning because it has no physical dimensions. Yet we can speak of a point in some concrete sense if we have some way of specifying its location or position. The ordinal property of numbers permits us to perform this task. We consider a line and specify any point on it with the digit 0, which we may call the origin, anticipating our later introduction and discussion of coordinate systems. This "point" 0 is still an abstract concept, not a concrete entity, but it was the first step that mathematicians took to relate arithmetic and, ultimately, algebra to geometry. Starting from 0, mathematicians then labeled equally spaced points to the right and left of 0, with the integers 1, 2, 3, . . ., called positive to the right of 0 and the integers −1, −2, −3, . . ., called negative to the left of 0. The distance between any two neighboring points, such as 1 and 2, was called the unit distance. Arithmetic thus became an adjunct of the practical art of measurement. Note that distance itself cannot be defined in terms of any more elementary concepts; it acquires its meaning only through a measurement operation. Since measuring a distance requires two points, which themselves are abstract concepts, so, too, is distance; hence, to speak of a precise distance between two abstract points is meaningless. The apparent meaninglessness of such concepts does not bother mathematicians who apply concrete numbers, and therefore all the laws of arithmetic, to abstract distances.

The introduction of the negative integers −1, −2, −3, . . . was an enormous advance, not only in arithmetic, but in all of mathematics because the concept of subtraction could then be replaced by the addition of negative numbers. The concept of "negativeness" here is itself abstract because the points to the left of 0 on the line labeled with negative integers are in no way different from the points to the right of 0; labeling one set negative and the other set positive is purely arbitrary. But we see here that the concepts of positive and negative numbers play an important geometric role for they define direction. If we are at 0, we know that we must move 2 units of distance to the right if we want to reach the point 2, but 2 units of distance to the left of 0 to reach the point −2.

To the scientist, particularly the physicist, negative numbers are very important because they permit him to distinguish between two different species of the same physical property such as electric charge, for example, which is characterized by very definite physical features. Two species of electric charge have been known since the times of the ancient Greeks; in terms of their electric properties alone we cannot distinguish one kind of charge from the other. Looking at a single charge cannot tell us to which of the two species it belongs, but if we bring any two charges close to each other, they either repel each other or attract each other. Electric charges that repel each other are identical and belong to the same species; if they attract each other, each charge belongs to a different species, one of which physicists call negative and the other of which they call positive. This negative–positive designation of electric charge is purely arbitrary, but is extremely convenient for the description of collections or aggregates of positive and negative charges such as atoms and molecules. As seen from the outside, an atom appears to be electrically neutral, with zero electric charge. Yet it consists of equal numbers of negative and positive charges so that these charges appear to cancel each other as equal positive and negative numbers should when added to each other. The opposite electric charges in the atom do not actually cancel each other; they are so close together, however, that the sum of the electric effects is essentially zero. This is a beautiful example of the way a purely abstract arithmetic concept (negative numbers) plays an important role in enabling us to construct models of very important real physical objects.

The next big step, taken by mathematicians, that brought arithmetic closer to geometry was the recognition that integers alone cannot label all the points on a line; this conclusion led to the introduction of fractions which, at first sight, seemed to do the trick. Fractions seemed to solve the problem of locating or labeling points on a line, because an infinite number of fractions can be found between any two fractions no matter how closely they approximate each other. Thus an infinitude of fractions, each smaller than $\frac{1}{2}$ and larger than $\frac{1}{3}$, lies between $\frac{1}{3}$ and $\frac{1}{2}$. Similarly, an infinite number of fractions, each larger than $\frac{1}{1001}$ and smaller than $\frac{1}{1000}$ lies between $\frac{1}{1001}$ and $\frac{1}{1000}$. Because points on a line can be described in the same way—given any two points, an infinite number of points lies between these two given points—it seemed reasonable to assume that all the points on a line can be covered by or correspond to the fractions. This assumption is wrong but this was not discovered until centuries after fractions were first introduced. Mathematicians, however, did recognize that fractions are densely distributed on a line. They described this property of fractions by the statement that "fractions lie everywhere dense on the line."

One might think, owing to this property of fractions, that if one pricked a line with numbered points using an infinitely thin needle, the needle would always hit a fraction. But this assumption is incorrect because the number of points that are overlooked by fraction labeling is infinitely larger than the number of those points that are labeled. In time it became clear that this infinite set of points, which covers the line completely, must be identified with a set of numbers called irrational numbers such as $\sqrt{2}$ and the number π. The roots of numbers such as $\sqrt{2}$, $\sqrt{3}$, $\sqrt[3]{5}$ belong to a family or set of numbers which mathematicians call algebraic numbers and numbers like π are called transcendental numbers; irrational numbers are so named because they cannot be written as fractions. We shall return to these numbers in a later chapter when we discuss a branch of arithmetic that is known as transfinite number arithmetic. Although a number like $\sqrt{2}$ cannot be written as a fraction (as one can prove), one can easily show that it labels a very definite point on a line whose points are related to an origin 0 by a definite unit of length. Like $\sqrt{2}$ and all other algebraic numbers, all transcendental numbers are associated with points on the line. This discovery stemmed from still another discovery that every irrational number can be approximated by an appropriate sum of fractions that is, in general, a nonrepeating decimal.

The limitation of the study of the natural numbers—all negative and positive numbers, rational and irrational—to points on a line cannot reveal the beautiful relationship between arithmetic and geometry. If we remove this limitation by considering all the points on two lines that intersect or are parallel, we move into geometry. Because two such lines define a plane, we can now locate all the points in such an area by assigning sets of numbers to them that are related in a very definite way to the numbers associated with or assigned to the points of the two lines. We can now construct all kinds of geometric figures on this plane by connecting points on the plane with lines of various shapes. We can draw an infinite number of lines on the plane connecting any two points. The shortest of these lines is called a straight line. The operation of assigning doublets of numbers to points in a plane and thus locating these points is one of the most imaginative and fruitful ideas in mathematics; it ultimately led to the concept of the coordinate system which the French philosopher and scientist René Descartes made the basis of his analytic geometry.

The arithmetic of points on two intersecting lines leads us to the arithmetic of points on the plane on which the two lines lie if we note that from any point on this plane we can draw lines to various points on the two intersecting lines. Because there are two infinite sets of such lines (one for each of the two intersecting lines), these sets of lines cannot be used to locate the point, but two lines, one chosen from each set, can locate the point for us. This is most obvious if we choose two perpendicular lines (they intersect forming four 90° angles) and locate points on the plane with respect to these lines. From any point on the plane we draw two straight lines which are perpendicular to each other and intersect at two points on the two chosen lines. We then locate the point on the plane by assigning to it the two numbers of the intersection points on the two chosen lines.

The extension of the number system to the plane was not only a very important advance in mathematics but also a breakthrough in the application of mathematics to practical work in science, economics, and social research. We refer here to the introduction of the graph which is used extensively to present and store knowledge and information about every type of human activity in which numerical relationships between two different sets of quantities are involved.

The arithmetic of the plane led to the concept of dimensionality and mathematical manifolds. A point has zero dimensions, a line is a one-dimensional manifold, a plane is a two-dimensional manifold and all of space is a three-dimensional manifold. As mathematics became more complex and sophisticated, mathematicians extended the manifold concept to any number of dimensions they pleased and developed the mathematics necessary to describe such n-dimensional manifolds (n larger than 3) even though these manifolds or spaces are purely fanciful and, in general, have no application to our universe. In a later chapter, however, we shall see that the four-dimensional geometry that mathematicians introduced in the nineteenth century is very useful in the mathematical formulation of the theory of relativity.

We now discuss a few more mathematical consequences of the arithmetic of the plane. One of the most important is the introduction of the angle concept which we examined in the preceding chapter where we noted that it is a measure of the change in direction or of turning from one direction to another. This idea has meaning, of course, only if we establish a reference direction on our plane, which we can do by drawing a straight line through any point 0 we choose on the plane. If we draw another straight line through the point, we then define the angle between these two lines as the amount we must turn when, while standing at the point 0 of intersection of the two lines, we change our line of sight from the direction of one line to that of the other.

This angle can be expressed in various ways; these varied expressions are the basis for trigonometry. The introduction of angles and, therefore, directions on the plane, ultimately led to the vector concept and the mathematics of vectors, which is used extensively in all branches of physics.

The extension of arithmetic to the plane may lead the reader to wonder, incorrectly, whether the uncountable infinity of points on the plane is greater than the uncountable infinity of points on a line which is itself larger than the countable infinity of the fractions and integers. Every point on the plane can be designated by two numbers, each identified with a point on one of the two lines that meet at the origin 0 on the plane. Thus the number of points on the plane is the same as the number of points on two lines, which is the same as the number of points on a single line, which mathematicians call the continuum.

Some other consequences of the arithmetic of a plane are the concepts of dimensionality and area. The dimensionality of a manifold is defined as the number of numbers required to designate or locate a point on the manifold. Every point on a plane can be located with just two numbers. Therefore a plane is described as a two-dimensional manifold; this characterization is true of any surface. Considering points on the surface of the earth, a sphere, we see that they can be located by pairs of numbers—the latitude and longitude. We thus see that points on a plane can be associated with multiplication. If we multiply the two numbers that locate any point on a plane, their product is the area of the rectangle whose lower left-hand corner is the origin 0 and whose upper right-hand corner is the point in question. Because the area is the product of two lengths on the plane, the area is expressed as the square of a length—square feet, for example. Thus the arithmetic of rectangles on the surface of a plane is known as multiplication.

The concept of dimensionality can be extended to three dimensions for the points in space and thus to the volumes of spatial figures. Such volumes are expressed as the cubes of distances—cubic meters, for example. Mathematicians have gone beyond the geometry of three-dimensional real space and developed the geometry of space of any number of dimensions (n-dimensional manifolds), which is of great importance in theoretical mathematics and physics, for example, in the mathematics of Einstein's theory of relativity. The geometries of n-dimensional manifolds were developed not because they might be of practical use or applicable to scientific theories but because mathematicians found them intellectually challenging. Mathematicians did not limit themselves to the geometries of flat, Euclidean spaces but also considered non-Euclidean, curved spaces that, at the time, appeared to be applicable only to two-dimensional surfaces, such as spherical and elliptical surfaces. The idea that such n-dimensional geometries might be used to describe the space of the real universe was dismissed out of hand because many mathematicians asked how three-dimensional real space could be curved if no higher dimensional space exists into which real space can bend. A line on a two-dimensional surface can be twisted and curved on the surface in any way desired and a two-dimensional surface can be bent into various shapes in our real three-dimensional space. But no real four-dimensional spatial manifold exists into which our real space can be bent or distorted. This difficulty did not deter mathematicians, however, who developed the most general theories of many-dimensional non-Euclidean geometries which were later applied to the real world by Einstein and his followers.

In their development of many-dimensional geometries, mathematicians introduced geometries which have nothing to do with real distances, areas, or volumes. In these geometries, which have an infinite number of dimensions, a purely fictitious kind of distance is introduced as well as fictitious angles and fictitious directions. Nevertheless, in spite of the fanciful nature of the spaces described by this geometry, these spaces are the domains of one of the most remarkable developments in the history of physics: the quantum mechanics.

We complete our discussion of the arithmetic of the plane by considering the role of the plane in the arithmetic of complex numbers. The concept of the complex number was introduced into arithmetic after imaginary numbers were discovered. Here we have another example of a mathematical concept that appeared to have no practical use when it was first introduced but which, in time, led to the development of one of the most remarkable and beautiful branches of mathematics—the theory of functions of complex variables. Without this apparently "useless" mathematics, the development of certain mathematical features of the theory of relativity and the basic equations of quantum mechanics could not have been discovered. The mathematics required to elucidate a particular phase of physics always preceded the conceptual discoveries of the physicist. It is as though the development or emergence of a certain branch of mathematics indicates the direction along which physics will ultimately evolve. No constraints except the logic and inner consistency of the mathematics itself limit the mind of the mathematician. His intellectual creations do not have to conform to any preconceived notions of the structure of the universe. One must therefore wonder about the conformity of all features of the universe with one or another branch of mathematics. Indeed, the wonder is if the universe conforms with mathematics at all.

We return now to the discussion of complex numbers and show how our knowledge of the properties of these numbers was greatly expanded by representing the numbers by points on the plane. We saw that every real number, positive or negative, rational and irrational, can be represented by a point on a line once we have chosen a reference point or origin 0 on the line. But no such point corresponds to the square root of any negative number such as $\sqrt{-3}$. Thus, the square roots of all negative numbers form a class or set of "imaginary" numbers that stands by itself. In discussing this set of numbers we find it convenient to separate the minus sign from the number itself. To that end mathematicians write $\sqrt{-5} = \sqrt{(-1)5} = (\sqrt{-1})\sqrt{5}$. This expression separates the imaginary number $\sqrt{-5}$ into a pure imaginary factor $\sqrt{-1}$ and a real factor $\sqrt{5}$. Mathematicians use the letter i to represent the imaginary factor $\sqrt{-1}$; i is called the imaginary unit which changes any real number into a pure imaginary number by multiplying that number by i. The next step in the evolution of the theory of complex numbers was the definition of a complex number as the sum of a real number and a pure imaginary number such as $4 + i8$. All the rules of arithmetic (addition, multiplication, etc.) were carried over to complex numbers; this process began the vast branch of mathematics called the theory of functions of a complex variable.

An important advance in the mathematics of complex numbers occurred in the mid-nineteenth century when the great German mathematician Karl Friedrich Gauss used the geometry of the plane to represent complex numbers geometrically. Because a complex number consists of two parts, each of which is a number, and each point on a plane is identified by two numbers, an exact correspondence can be established between the points on a plane and the complex numbers. Gauss established this correspondence by using two straight lines at right angles to each other on the plane. He represented all the real numbers, positive and negative, by points on one of these lines (e.g., the horizontal axis) and all the imaginary numbers on the vertical line perpendicular to the horizontal line. The point of intersection of the two lines is labeled 0. All positive real numbers are represented by points to the right of 0 and all negative numbers are represented by points to the left of 0. All positive imaginary numbers are represented by points above 0 on the vertical line and all negative imaginary numbers are represented by points below 0. The horizontal line is called the real axis and the vertical axis is called the imaginary axis. The plane itself is called the complex plane; all arithmetic properties of complex numbers can be deduced from the geometry of the complex plane. The same unit of distance is used to represent the spacings between the pure imaginary numbers on the imaginary axis as that used to represent the spacings between real numbers on the real axis. A complex number such as $4 + i7$ is a point on the plane we reach by moving four distance units to the right of 0 along the real axis and seven distance units above 0 along the imaginary axis.

We complete this chapter by defining the logarithm of a number; this concept was introduced into arithmetic by the Scottish nobleman and mathematician John Napier (1550–1617) in the seventeenth century. The discovery of the concept of logarithms came at a very propitious time for astronomy because it coincided with Johannes Kepler's calculations of the positions of the planet Mars which led him to the discovery of the laws of planetary motions. These laws, in turn, led Newton to the discovery of the law of gravity.

Napier himself was born at Merchiston Castle, now located in the city of Edinburgh, Scotland; he would also draw his last breath within that very same structure. As the Laird of Merchiston, Napier would achieve his own prominence in a family whose members had long obtained positions of importance in government and industry through diligence and hard work. As a boy, Napier was a witness to the ongoing struggle between Protestants and Catholics in Scotland. His family, perhaps seeking to insulate him from the violence of the country's religious bouts, sent him to the University of St. Andrews in Scotland when he was 13 years old. Napier left school before receiving his degree, electing to travel abroad before returning to Scotland in his twenty-first year. His heart hardened against Rome by the death and destruction in his own land, Napier composed a popular theological work, *A Plaine Discouery of the Whole Reuelation of Saint Iohn*, which attacked the Catholic Church and asserted that the Pope was the antichrist. The book met with varied responses; many of the landed gentry and government officials lauded it, but the common people were, on the whole, much less receptive.

Despite the notoriety which surrounded Napier's religious writings, his fame was assured only after he composed his *Descriptio* which outlined his method for calculating logarithms. Although it was not as lively as Napier's treatise on Catholicism, the *Descriptio* was immediately recognized as an outstanding achievement following its publication in 1614. Henry Briggs, a professor of geometry at Oxford, was among the most enthusiastic of Napier's readers, and visited Napier at his castle to discuss ways in which the system of logarithms could be improved. Briggs did suggest that the base 10 be used; Napier readily agreed. Upon his return to Oxford, Briggs produced the first of several tables of logarithms. Within a few years, mathematicians throughout Europe were using these charts of logarithms, which greatly increased the ease and speed of making complex computations. Hence, Briggs's name would forever be linked with Napier's logarithms because Briggs was one of the first distinguished mathematicians to take up the gauntlet on behalf of Napier's system. Briggs also managed to convince Kepler of the utility of these logarithms in performing laborious calculations; Kepler's enthusiasm greatly facilitated the acceptance of logarithms in Europe's universities.

When Napier was not calculating logarithms, he managed the family estate and continued to keep abreast of new discoveries in mathematics and science. He also completed his *Rabdologia*, which was published in 1617, the year of his death, and introduced, according to W. W. Rouse Ball, "an improved form of arithmetic by the use of which the product of two numbers can be found in a mechanical way, or the quotient of one number by another." Napier also invented two other rules for obtaining square and cube roots. But the *Rabdologia* was to be little more than a footnote in the story of mathematics, completely overshadowed by Napier's *Descriptio* and his system of algorithms, which has continued to grow in importance in science and industry to the present day.

The logarithm concept is quite simple and extremely ingenious. Napier got the very clever idea of doing ordinary arithmetic not with the numbers themselves but with exponents of other numbers that can be used to represent the numbers with which the arithmetic is to be done. Any given number can be expressed as a power or a root of some other number. Thus 8 is the cube of 2 (or 2 is the cube root of 8) and 100 is the square of 10, and so on. Thus we can represent any number (except 0) real or complex as some power of 10. Napier proposed that all numbers in arithmetic operations higher than addition and subtraction be represented by their powers of 10, called logarithms, and that the operations be expressed in terms of arithmetic operations with these powers. The number 10 is called the base of the logarithms but any other number can be used as a base. The introduction of logarithms reduces the level of the arithmetic operation; thus multiplication is reduced to addition, division is reduced to subtraction, raising a number to a power is reduced to multiplication, and taking the root of a number is reduced to division.

The concept of logarithms can be illustrated with a few simple examples. We note that the numbers 1, 10, 100, 1000, etc., can be represented by 0, 1, 2, 3, etc., because the latter numbers are the respective logarithms of the former numbers. In short, $1 = 10^0$, $10 = 10^1$, $100 = 10^2$, $1000 = 10^3$, etc. Because the logarithms are smaller than the numbers they represent by powers of 10, doing simple arithmetic, such as addition and subtraction, with the logarithms, corresponds to higher arithmetic with the actual numbers. Because $100 \times 1000 = 100,000$ or $10^2 \times 10^3 = 10^5$, we see that the product of numbers is equivalent to adding their exponents (logarithms). Similarly, $100,000/1000 = 100$ or $5 - 3 = 2$ so that division is given by the subtraction of logarithms. We note further that raising any number to some power, such as $(20)^{45}$, and finding a root, such as $\sqrt[5]{20}$, can be easily done with logarithms: the logarithm of $(20)^{45}$ is $45 \log 20$ and logarithm of $\sqrt[5]{20}$ is $\frac{1}{5}\log 20$. Because the logarithms of numbers between 1 and 10 are given in tables and the log of 20 is $\log 10 + \log 2 = 1 + \log 2$, the log of 20 is known.

The logarithms of all numbers between 1 and 10 are decimals less than 1 (irrational numbers) and the logarithms of all positive proper fractions (numbers between 0 and 1) are the negative of the logarithms of the numbers between 1 and 10. The logarithms of the negative numbers are, in general, complex numbers.

Napier introduced the base 10 logarithms because our number system is based on the 10 integers 0 to 9, but any base can be used for logarithms. Of particular importance are the logarithms based on the transcendental number called e, which plays an important role in all branches of mathematics. The logarithms on the base e are called natural logarithms. The discovery of logarithms led to the development of a very simple mechanical device, the slide rule, for doing multiplication and division based on logarithms. All such devices virtually disappeared, however, with the invention of the electronic computer.

Logarithms emerged in mathematics at a most propitious time for astronomy. Kepler was then immersed in long arithmetic calculations, which he detested, to obtain the complete orbit of Mars around the sun. Napier's discovery simplified Kepler's arithmetic enormously. Yet Kepler still had to fill some 800 folio pages with dense arithmetic computations before he obtained his laws of planetary motions.

CHAPTER 3

The Origin of Abstract Mathematics
The Beginning of Algebra

Algebra begins with the unknown and ends with the unknowable.

—ANONYMOUS

Mathematics differs drastically today from its youthful appearance because of its current great abstraction. Early arithmetic and geometry dealt with real things and actual societal activities. The transition in our concept of numbers from their purely numerical features as representing quantities of things we deal with from day to day, to our present concept of numbers as abstract symbols which we utilize according to certain rules was a tremendous step in the development of mathematics. Numbers and arithmetic began acquiring their abstract qualities when negative numbers, which still puzzle many people, were introduced. Fractions, irrational numbers and complex numbers enhanced this abstractness which finds its fullest expression in the theory of numbers. Geometry also evolved from a discipline that dealt with real manifolds, shapes, and objects, to an abstract mathematics with the introduction of such concepts as points, lines, and planes and many dimensional non-Euclidean spaces. But certain branches of mathematics began as abstract disciplines with no initial history of reality. Algebra was the first of these branches and it was the body of knowledge from which all other abstract mathematics emerged. Indeed, we may say that all branches of modern mathematics are, in a sense, extensions of the basic elements of algebra that one learns in school.

Most people find algebra more difficult to master than arithmetic because algebra is abstract whereas arithmetic, whose elements are numbers, is considered to be concrete. After all, what can be more concrete than a number, which represents a definite quantity, and what can be more abstract than a letter x or y that can stand for anything one pleases? Such reasoning indicates a misunderstanding of what is meant by "real" and "abstract." Because the numbers of arithmetic are symbols with no more reality than the letters x and y that appear in algebra, arithmetic, in a sense, is no less abstract than algebra. The symbols 0, 1, 2, . . ., 9, which define the number system of arithmetic, acquired concreteness only when quantities were associated with them, but they retained an important element of abstractness since a quantity has no real meaning until we specify the physical or mathematical nature of the quantity. The sum $3 + 4 = 7$ is as abstract as the algebraic sum $x + y = z$, since the statement that three things plus four things equals seven things tells us very little in a concrete sense. A sum of numbers remains abstract unless each term in the sum refers to things of the same kind: three apples plus four pears does not equal seven of anything. We may make the sum more explicit by placing the symbol x next to each term in the sum and write $3x + 4x = 7x$, where x may stand for any kind of quantity—including a number—we please. Thus arithmetic leads to algebra.

Although arithmetic and geometry grew quickly in Greece and Alexandria, very little algebra was produced in ancient Greece. But the development of geometry required the application of the elements of algebra. The statement of the Pythagorean Theorem in its most general form cannot be understood with arithmetic alone as a geometric relationship among the three sides of a right triangle. The geometrical essence of the theorem can, of course, be stated as follows: The square on the hypotenuse equals the sum of the squares on the other two sides; but one is then naturally led to replace each word by a symbol and write $h^2 = a^2 + b^2$, where h stands for the length of the hypotenuse and a and b stand for the lengths of the other two sides. The theorem thus becomes an abstract algebraic equation until lengths are specified. The symbols h, a, and b need not stand for any geometric quantity at all but just numbers.

Algebra seems to have predated the Greek era, as references are found to it in ancient Egyptian literature, but algebra as a mathematical discipline was first described in the third century AD by Diophantus, a Greek mathematician who lived in Alexandria. Even though only six of the thirteen volumes of his *Arithmetica* have survived, he was regarded as the "Father of Algebra." However, the European mathematicians of the early Christian era first learned about algebra in the twelfth century through Latin translations of Arab mathematicians, chiefly Al-Khwarizmi. The word "algebra" comes from the Arabic "al-jabra," literally, the reduction. The first European (Renaissance) treatise on algebra was written by the Friar Luca Paciola in

1494. In that period, mathematicians were trying to obtain solutions in closed forms of cubic and quartic algebraic equations.

Primitive algebra arose when it became clear that all the rules of arithmetic with some minor revisions can be extended to symbols or letters. At first, algebra consisted only of the set of rules which define addition, subtraction, multiplication, division, raising a power, and taking a root of a quantity. These are exactly the same as those of arithmetic, except that they apply to letters which may represent numbers or any possible combination of numbers or any possible combination of letters. To eliminate any confusion that might arise in the interpretation of combinations of letters, mathematicians supplemented the arithmetic rules with clarifying definitions. The plus and minus signs were retained to express addition and subtraction as in $x + y$ and $x - y$, but the multiplication sign $\times$, though still permitted, was generally dropped to avoid confusion; the juxtaposition of two letters next to each other, such as xy, means that the numerical entities for which the two letters stand are to be multiplied in the usual way. The division sign was also eliminated and replaced by a slash (solidus) between the dividend and the divisor as in x/y, where x is to be divided by y, which is the extension to letters of the concept of a fraction.

Because algebra involves numbers as well as letters, the role of a number or numbers in conjunction with a letter or letters had to be elucidated before the rules of algebra could be applied unambiguously in algebraic calculations. Writing $x + 3$ simply means adding 3 to whatever numerical quantity x represents and the combination $3x$ means multiplying the numerical quantity x by 3. Here 3 is called the coefficient of x, but x can just as well be called the coefficient of 3. The combination $x/3$ means that x is to be divided by 3, or, written as $(\frac{1}{3})(x)$, that we are to take one-third of x.

Because the mathematicians who developed algebra were very careful to avoid any misinterpretation of any possible combinations of letters and numbers, they introduced parentheses to represent inclusiveness or to indicate that certain combinations of letters (or numbers) are to be considered as a single entity as in the combination $3(a + b + c)$. This expression means that either 3 is to multiply each term in the parentheses, after which the products are to be added, or the terms in the parentheses are to be added first and the sum is then to be multiplied by 3. We may apply this rule whether we have 3 in front of the parentheses or the letter y, which may stand for anything we please. Thus the rule of addition as applied to algebra is fully illustrated by the simple example $(a + b + c)(x + y + z) = ax + ay + az + bx + by + bz + cx + cy + cz$. This example tells us that we may equate the product of one sum of terms (the sum in the first set of parentheses) and another sum of terms (the sum in the second set of parentheses) to a sum consisting of terms, each of which is a product of one term in the first parentheses and one in the second parentheses, so that, in this example, nine such products appear. This is a simple example of the general rule that the number of distinct products that must appear on the right-hand side of an equation equals the product of the number of terms in the first parentheses and the number of terms in the second parentheses. If we multiply three parentheses, each consisting of a sum of individual terms, we obtain a sum of individual products of triplets, each triplet product consisting of three different factors—one factor from each parentheses; the total number of such triplets equals the product of the number of terms in each parentheses.

As algebra developed over time, special rules were introduced that define the role of the minus sign. In arithmetic the combination $4 - 3$ means that 3 is to be subtracted from 4, and this interpretation of the minus sign applies to algebraic quantities as well. It is convenient in algebra to separate the minus sign from the letter in front of which it stands by treating it as -1; thus the term $-y$ means that y is multiplied by -1. When the minus sign is used in conjunction with parentheses, as in the example $-(a - b + c)$, every term in the parentheses is to be multiplied by -1 so that this expression equals $-a + b - c$. This is a simple example of the general rule that a minus sign in front of parentheses changes the signs of all the terms in the parentheses when it is brought inside the parentheses.

The rules concerning raising algebraic expressions to powers are the same as in arithmetic. Thus x^n means a product of n factors each of which is x, as, for example, $x^3 = xxx$. The expression $(x + y)^n$ means a product of n factors $(x + y)$ as in $(x + y)^3 = (x + y)(x + y)(x + y)$. An exponent applied to a product of different factors such as $(xyz)^n$ is the same as the product of each factor raised to the power, so that $(xyz)^n$ is the same as $x^n y^n z^n$.

These various arithmetic rules were not carried over to algebra immediately but were introduced over a period of many years. Yet even with the acceptance of these rules, modern algebra did not begin until the late eighteenth century and the early nineteenth century when the French mathematicians Lagrange and Galois began to investigate the general solutions of algebraic equations. Because one of the most important roles of algebra is to find the solutions of algebraic equations, the concept of the "algebraic equation" is basic to algebra. Although algebraic equations, in their modern abstract forms, were introduced relatively recently, the primitive form of the algebraic equation was probably an outgrowth of the practical need to find answers to simple problems in arithmetic involving sets of quantities, some of which were known and some of which were unknown. In commercial transactions, a merchant finds the price of one item from his knowledge of the cost of n such items by dividing the cost by the number of items; he then writes down the general formula price = cost/ number or $p = \text{cost}/n$ which is an algebraic equation. Various equations of this type are used constantly by people in their daily activities. One can also calculate the time required for an automobile trip by dividing the distance to be traveled by the speed of the automobile.

This type of simple equation is remarkable because it describes all kinds of physical phenomena and even expresses important physical laws. Thus Newton's second law of motion, which is the basis of classical dynamics, is expressed in the equation $F = ma$, where F is the force required to impart the acceleration a to a body of mass m. Similarly, Einstein's famous equation that expresses the equivalence of energy E and mass m is $E = mc^2$, where c is the speed of light. As two other examples, we have the relationship among the pressure P, the temperature T and volume V of a gas: $P = (R)T/V$, known as the gas law (where R is a numerical constant); and the relationship (called Ohm's Law) among the current I in a wire, the voltage V across the ends of the wire and the resistance R (to the current) of the wire: $I = V/R$. This simple equation is the basis of all electrical circuit theory. A simple example illustrates its great usefulness to electricians. To determine the gauge or thickness of the copper wire that the electrician will use to bring a certain size current to a house whose line voltage is V, he divides the voltage V by the current I to obtain the resistance R of the wire, which in turn depends on the gauge of the wire. The electrician can thus determine the gauge and order the wire he needs for the house to be safely wired.

The basic algebraic form of these various equations is $y = a/x$, which we altered in form but not in content as we applied it to the various physical examples. These alterations and examples express the basic principle we must follow in manipulating an equation. We may do anything we please (algebraically or arithmetically) to one side of an equation provided we do exactly the same thing to the other side. If we start with Newton's law of motion $F = ma$, for example, we may divide both sides by m to obtain $F/m = ma/m = a$, or divide both sides by a to obtain $F/a = m$, or divide both sides by F to obtain $F/F = ma/F = 1$ since $F/F = 1$.

The importance of this simple principle, as illustrated by Newton's law, cannot be overestimated. New physical principles and phenomena can be discovered simply by manipulating this law mathematically without violating algebraic rules. We can apply a force F to a body of mass m at rest and thus displace the body through a distance d in the direction of the force. Physicists define this physical operation as doing work on the body; they then say that the body acquires energy—owing to the work done on the body—which is exhibited in its motion. If the body is free to move (no friction hindering its motion), the force imparts the acceleration a to it according to Newton's law $F = ma$. If we now multiply the left-hand side of this equation by the distance d to obtain Fd (the work done on the body), we must also multiply the right-hand side by d to obtain the expression mad. Because this manipulation does not alter the equation, we have, according to the rules of algebra, $Fd = mad$. Additional algebraic manipulations of the right-hand side show that it equals $(\frac{1}{2})mv^2$, where v is the speed the body acquires due to the work done on it. This quantity $(\frac{1}{2})mv^2$ is called the kinetic energy of the body; it shows us through simple algebra that work and energy are equivalent. This is part of one of the most important laws in nature—the conservation of energy.

We have treated this simple algebraic exercise in some detail to illustrate a very important truth as well as the scientific power of algebra. The truth is that if we start with a basic law of nature expressed in some algebraic form, any deduction we obtain mathematically (algebraically or arithmetically) from this law must also be a law. A law of motion is thus transformed into a law of energy, thereby illustrating the great power of algebra. As we continue with our story of mathematics, we shall point out many instances in science where relatively simple and elementary algebra was used to express important scientific principles. Later mathematicians couched these principles in more complex and sophisticated mathematics, which was fruitful for the practicing scientist but tended to obscure the basic physical principles involved and to repel those people who did not possess great mathematical skills.

As long as algebra was used only as a mathematical tool in daily activities, its growth was limited. Once it threw off these constraints and became the intellectual toy of the pure mathematician, however, it grew phenomenally, sending out shoots into every realm of mathematics from which new branches of algebra grew. This vast growth began with the study of what mathematicians call algebraic forms or polynomials, which are sums of quantities such as $2x$, $4x^2$, etc., each of which is an algebraic quantity raised to some power and multiplied by some number (coefficient). Any such term can be represented as $a_n x^n$, where n is any integer and a_n is the numerical coefficient multiplying x^n. In such a sum, x is called the variable, which may have any numerical value we wish to assign to it; the numerical value of the sum depends on the value of x so that the sum varies as x varies. This simple idea led to the concepts of the algebraic function and the algebraic equation to which most eighteenth-century mathematicians devoted themselves.

In its most general form, a functional relationship between two quantities is a scheme or a formula which enables one to establish a one-to-one correspondence between the individual members of two different sets of quantities. If we have two sets of numbers, one of which consists of the numbers 1, 2, 3, 4, etc. (which we call x) and another one of which consists of the numbers 2, 4, 6, 8, etc. (which we call y), then we see that we obtain the set y by multiplying each member of the set x by 2. Mathematicians then say that y is a function of x and write $y = f(x)$. The introduction into algebra of this concept of a function, which describes how one set of quantities depends on or varies with another set was one of the most important events in the history of mathematics because it spread into all branches of mathematics thus stimulating mathematical research and growth. It was particularly important in the development of the calculus. All branches of science, particularly physics, benefited greatly from the use of the function concept, which appears in every phase of physics literature. Thus one speaks of the pressure in a gas as being a function of its temperature, we describe the behavior of an electron in an atom in terms of its "wave function," and so on.

Returning now to the example of a functional relationship given above, we can describe it algebraically as $y = 2x$. But this way of describing this functional relationship generalizes it because it goes beyond the restriction or constraint imposed on the relationship by the two sets of numbers. The function is not limited to integer values of x; the functional relationships hold no matter what values of x we choose. In setting up a function, mathematicians are very careful to define the domain of the function (the range of the function) over which the function is defined. The concept of the function was later extended to entities different from numbers and so the domain of a functional relationship might extend over sets of geometric figures or even sets of propositions as in the algebra of logic. In extending the concept of functions from its simple algebraic domain to more general domains, mathematicians introduced additional constraints and definitions which we discuss later. Yet any functional relationship raises questions regarding whether the function is single valued and whether it has an inverse. The first question asks whether only one value of y is associated with a single value of x or whether two or more values of y are associated with a single value of x. If the y, x values are one-to-one, the function is said to be single valued. If the correspondence is not one-to-one, the function is multiple valued. The second question asks whether the relationship between y and x can be reversed. If y is a function of x, can we say that x is also a function of y? This cannot always be done, but in the simple example above we can also write x as a function of y: $x = y/2$.

The introduction of the function concept led to the concept of the algebraic equation, which is now one of the largest and most fully developed branches of algebra. As stated before, the theory of algebraic equations began its remarkable growth in the late eighteenth and early nineteenth centuries primarily through the research of Lagrange and Galois. Starting from our simple example of a functional relationship $y = 2x$, we see that it is the forerunner of the algebraic equation since it suggests questions concerning the values of x that give certain values of y. What value of x gives $y = 10$? This is a very simple algebraic problem, indeed, because all we need do to find the answer is divide 10 by 2. The general solution of the algebraic equation $y = 2x$, which stems from this y, x functional relationship, is then $x = y/2$. This becomes a simple problem in arithmetic as soon as we give y a numerical value. Pursuing this thought further we then subtract $2x$ from each side of the equation $y = 2x$ to obtain $y - 2x = 2x - 2x$; this leads to the equivalent equation $y - 2x = 0$, which we recognize as the elementary algebraic equation to which we were introduced in our first class in algebra, except that y was replaced by a number such as 10. We thus obtain the simplest type of algebraic equation $10 - 2x = 0$ or $2x - 10 = 0$, or $x - 5 = 0$. This is called a linear or first degree algebraic equation; it is linear because, as we shall see later, it is the equation of a straight line. It is a first degree equation because x appears in the equation only to the first power.

Treated as an algebraic equation $x - 5 = 0$ has all the features of all algebraic equations so we can discuss and illustrate these features by referring to this equation which contains only the unknown x. The principal goal in pursuing the theory of algebraic equations is developing methods for finding solutions of the algebraic equation. A solution of an algebraic equation is a numerical value of the unknown x which makes the left-hand side of the equation 0; this value of x is also called a root of the equation. The solution of our simple equation $x - 5 = 0$ is $x = 5$, so that 5 is a root (the only root) of this equation.

Starting from this simple branch the tree of algebraic equations grew rapidly into its present overwhelming structure; this evolution proceeded by an extension of the function concept, which we recall began with the general consideration of the dependence of one set of quantities on another. This was extended by using sums of powers of a numerical quantity x to represent one of these sets and then writing $f(x)$ as the other set, the simplest example of which, $y = f(x) = 2x$, we discussed above. Mathematicians enlarged this mathematical structure by adding additional terms to obtain higher-order functions, or polynomials.

To see how mathematicians proceeded, we start from the elementary function of x we discussed above but replace the numerical coefficient 2 by a_1, writing $y = f(x) = a_1x$. We now add to this expression a number a_0, with no x attached to it so that we obtain the function $f(x) = a_1x + a_0$. This is the most general way of writing what mathematicians call a linear or first degree function of x. By adding additional powers of x to the linear function one obtains functions of any desired degree. Leibnitz was the first to study the general function of x or polynomials of the nth degree (n stands for any number) which he wrote as $f(x) = a_nx^n + a_{n-1}x^{n-1} + \ldots + a_1x + a_0$. Here the subscripts attached to the coefficients a_n, a_{n-1}, $\ldots$ merely identify them as those that multiply the various powers of x; they have no arithmetic significance. Leibnitz was the first to propose the following problem: Given the $n + 1$ numerical values x_0, x_1, $\ldots$, x_n of x and the corresponding $n + 1$ numerical values y_0, y_1, $\ldots$, y_n of the function $f(x)$, what numerical values must the $n + 1$ coefficients a_n, a_{n-1}, $\ldots$, a_0 have for this correspondence between the two sets of numbers x and y to hold? Lagrange solved this problem some 50 years later by developing what is now known as the Lagrange interpolation formula, the discussion of which is beyond the scope of this book. Lagrange's discovery of his interpolation formula was the first step in the history of the theory of algebraic equations to which the young French mathematician Galois contributed greatly.

Before we leave the theory of algebraic equations, we must consider the algebraic polynomial in greater detail. Because we may assign any numerical value we please to x that appears to various powers in the polynomial, the numerical value of the polynomial for a given numerical value of x depends on the numerical values of the coefficients. For a set of coefficients having definite values, we change the value of the polynomial or function by changing the value of x. In the standard polynomial of the nth degree which leads to the standard algebraic equation of the nth degree, the coefficients $a_0, \ldots, a_n$ are all real, rational numbers (coefficients such as $\sqrt{2}$, $\sqrt{-1}$, and π do not appear, although one can certainly imagine equations in which such coefficients do appear).

In dealing with polynomials we may treat them as arising from a product of two or more polynomials or as obtained by dividing one polynomial by another. Although the multiplication of one polynomial by another is an algebraic process—the extension of the arithmetic process of multiplying the sum of one set of numbers by the sum of another set—the algebraic division of one polynomial by another is not as easily related to the arithmetic division of one sum of numbers by another. Mathematicians therefore developed a simple algorithm for dividing one algebraic polynomial by another which made such division a simple exercise. We shall see that such division simplifies finding the roots of an algebraic equation or factoring a polynomial, which means finding polynomials of smaller degree (factors), which, when multiplied together, give the original polynomial.

We pursue this concept of factoring a polynomial by considering simple polynomials first and proceeding then to more complex polynomials; this simply means going from polynomials of lower degree to those of higher degree. The degree of the polynomial, as already stated, equals the highest power of x that appears in the polynomial. If x does not appear at all, so that $f(x) = a$, where a is a constant, then the polynomial is of 0 degree, and the function is a constant; it does not depend on x at all. If $f(x) = ax + b$, where a and b are constants (fixed numbers), the polynomial is of the first degree since x appears only to the first power. The polynomial $f(x) = ax^2 + bx + c$ is a second-degree polynomial, and so on.

The polynomial $ax + b$ is its own factor but the polynomial $ax^2 + bx + c$ has two factors; the only way we can obtain a term ax^2 is by multiplying two factors, each of which contains x. Thus $ax^2 + bx + c$ can be written as $(fx - g)(hx - k)$, where f, g, h, and k are numbers that can be found by multiplying out the two factors and comparing the expression obtained with the original polynomial. We can go on in this way and consider polynomials of ever increasing degrees. The number of factors of a polynomial equals the degree of the polynomial. The recognition of this simple relationship was the first step in the construction of the theory of algebraic equations.

Before discussing algebraic equations and how they arose from the polynomial concept, we emphasize an important distinction between the function of a variable x as treated by the mathematician and the function concept as treated by the physicist. The mathematician is not concerned with the physical meaning of x but treats it merely as a pure number. In fact, all powers of x are pure numbers to the mathematician. The physicist, however, considers the functional relationship between two sets of quantities that are physically different. In the theory of gases, for example, the pressure in a gas is described as a function of the temperature. Because temperature and pressure are qualitatively different physical entities, however, one cannot simply express this functional relationship as pressure equals a number times the temperature. If one writes that pressure equals a times temperature, a cannot be a pure number; it must be a physical entity that changes temperature to pressure. Another illustration of this important restriction which we must place on physical functional relationships is that between the distance a person travels in a vehicle and the time spent in the vehicle. The longer the time that elapses, the greater the distance traveled will be. We may write this relationship as $s = at$, where s is the distance and t is the time. In such an equation, however, a cannot be a pure number since a distance cannot equal a pure number times time; a must therefore be a physical entity that can change time to distance. Speed or velocity v is such an entity and our equation is physically correct if we place $a = v$ and write $s = vt$. This is a first-degree functional relationship but we can extend this reasoning to a second-degree function by writing $s = vt + bt^2$. The variable b must then be an entity that changes the square of a time to a distance. Acceleration performs such a task so that b is an acceleration, which is defined as a distance divided by the square of time.

We have pursued the study of the polynomials to lay the basis of the general algebraic equation of any degree. Returning to our general polynomial, we recall that an important question that mathematicians considered was the numerical value that had to be assigned to x for the polynomial to have a given numerical value. The step later taken by mathematicians was to find a value or values of x which make the value of the polynomial 0. This quest was the beginning of the theory of algebraic equations.

Starting with the simple first-degree polynomial $ax + b$, where a and b are numbers (x is the unknown), we equate it to 0 to obtain the first degree algebraic equation in one unknown $ax + b = 0$. We see at once that for $x = -b/a$, the left-hand side is 0 so that $x = -b/a$ is the solution or the root of this simple equation. We now consider the second-degree polynomial $ax^2 + bx + c$, where a, b, and c are numbers. If we set this equation equal to 0, we obtain the second-degree equation in one unknown $ax^2 + bx + c = 0$. This is the famous "quadratic equation" which is probably the most important equation in algebra because it appears so often in various physical problems; its extensive study is the content of the intermediate algebra high school courses. The left-hand side of the second-degree equation can always be written as the product of two factors, each containing x to the first power, $(x - x_1)(x - x_2) = 0$, so that the numbers x_1 and x_2 are two different arithmetic combinations of the coefficients a, b, and c, which can be found by multiplying out the two factors $(x - x_1)$ and $(x - x_2)$. Both x_1 and x_2 are thus roots of the quadratic equation because the first parenthesis is 0 for $x = x_1$ and the second parenthesis is 0 for $x = x_2$. The quadratic or second-degree equation thus has two roots, which may be equal for special values of the coefficients. The two roots are either both real or both complex.

Linear and quadratic algebraic equations and methods for solving them were known to the Egyptians more than 4000 years ago and the general solution (the two roots) of the quadratic equation was probably known very early in the history of algebra. This solution (exhibiting both roots) is written

$$x = \frac{-b \pm \sqrt{b^2 - 4ac}}{2a}$$

where the $+$ sign gives one root and the $-$ sign gives the other root. The two roots are expressed entirely in terms of the coefficients a, b, and c; the numerical value of b^2 compared to the product $4ac$ indicates the nature of the two roots. If $b^2 = 4ac$, the square root vanishes and both roots equal $b^2/2a$. If $b^2 > 4ac$, the square root is a real number and the two roots are real. If $b^2 < 4ac$, the square root is imaginary and the two roots are complex numbers, with one of them the complex conjugate of the other.

Several centuries elapsed before mathematicians discovered formulas for the roots of the cubic and quadratic algebraic equations (third- and fourth-degree equations). In the sixteenth century, the Italian algebraists Niccolo Tartaglia and Geronimo Cardano found the general formula for the three roots of the cubic equation expressed in terms of the coefficients and, some years later, the Italian Ludovico Ferrari found a formula for a special case of the quartic equation. The general solution of the cubic equation (its three roots) is given by the Cardan (Cardano) formula which is very complicated. Fortunately, most of the problems that we meet in science can be expressed as quadratic equations. But even where higher-degree equations occur, we do not use the formulas to find the roots; the solutions, which are only approximate, are obtained arithmetically with computers.

Mathematicians tried in vain to find formulae for equations of degree higher than the fourth degree for many years—after deriving the formulae for the roots of the cubic and quartic algebraic equations—until they discovered, early in the nineteenth century, that such formulae do not exist. The proof was first given by the Italian mathematician Paolo Ruffini and later by the Norwegian mathematician Hendrick Abel. They proved that no formula involving the coefficients of the equation and containing a finite number of algebraic operations (addition, multiplication, etc.) exists which expresses the roots of the equation in terms of the coefficients if the degree of the equation is higher than the fourth.

Attempts to find formulae for the roots of higher-degree algebraic equations stimulated Lagrange and later Galois to develop general methods for determining which algebraic equations of any degree with specific coefficients have algebraic solutions. We can understand the difficulties these men encountered in finding the roots of the general algebraic equation of the nth degree—expressing the roots as algebraic formulas involving the coefficients—if we note that that is equivalent to expressing the polynomial, which is the basis of the equation (the left-hand side of the equation), as a product of n factors. Each of these factors is of the form $(x - x_1)$ where x_i is one of the n roots (i is just an identification subscript). Because this factor is 0 when $x = x_i$, this value of x makes the entire polynomial zero so that $x = x_i$ is a root. But factoring the polynomial in this way, where the roots x_i are algebraic combinations of all the n coefficients $a_0, a_1, \ldots, a_n$ without specifying the numerical values of these coefficients is impossible if the degree of the polynomial is greater than 4 in that it has five or more coefficients.

Where mathematicians confronted this problem, they first had to convince themselves by rigorous mathematical logic that every nth-degree polynomial in the variable x has exactly n roots (the polynomial has the value zero for n values of x, which may be the same). This statement is called the fundamental theorem of algebra and was first proved rigorously by the German mathematician Karl Friedrich Gauss in 1799. Now it may seem obvious to us that a polynomial $a_n x^n + a_{n-1} x^{n-1} + \ldots + a_0$, where the coefficients a_n, a_{n-1}, ..., a_0 are definite numbers, is zero for n values of x (n roots) but such conclusions are of little comfort to mathematicians. All mathematicians detest such statements and are happy only when they can prove rigorously what is obvious. We therefore indicate in a general way the nature of such a proof which is not as rigorous as mathematicians may like but is good enough for our purposes.

Considering the nth-degree polynomial above with definite numbers for the a's, we note that if we substitute in it a definite numerical value for x, the entire polynomial (the sum of all the terms in it, each with a definite number), takes on a definite numerical value which happens to be larger than zero (positive). We can then find another value of x, either smaller or larger than the first value, which makes the polynomial less than zero. Clearly some x, whose numerical value lies between the two values of x already chosen, exists, which makes the polynomial zero. We thus obtain one root of the polynomial or algebraic equation. If we call this root x_1, then $(x - x_1)$ must be a factor of the original polynomial which we call $f(x)$. By dividing $f(x)$ by $(x - x_1)$, we obtain another factor of $f(x)$ which is of degree $n - 1$. If we call this factor $g(x)$, we then have $f(x) = (x - x_1)g(x)$. But we can go on in this way until we have reduced the original polynomial $f(x)$ into a product of n factors $(x - x_1)(x - x_2) \ldots (x - x_n)$ and thus uncovered the n roots. This result proves the fundamental theorem of algebra.

We may reverse this reasoning and construct a polynomial in x of any degree by multiplying together as many factors as we please, each of which is linear (first power) in x such as $(x - x_i)$ where x_i is any number we choose and thus one of the roots of the polynomial. Thus the product $(x - x_1)$ $(x - x_2) \ldots (x - x_n)$ is a polynomial of degree n (the product contains exactly n factors, each linear in x) whose n roots are exactly $x_1, x_2, \ldots, x_n$; if x is chosen equal to any one of these numbers, the factor containing that number, and, therefore, the entire polynomial, vanishes. If all these n factors are multiplied together, we obtain the explicit form of the polynomial with each power of x multiplied by its proper coefficient. Each coefficient is thus some arithmetic combination of the n roots. By comparing this explicit form of the nth-degree polynomial with the general expression for the nth-degree polynomial $x^n + a_{n-1} x^{n-1} + \ldots + a_0$, mathematicians have deduced algebraic expressions for all the coefficients $a_{n-1}, \ldots, a_0$ in terms of the roots x_1, x_2, x_3, . . ., x_n. We thus see that a_0 is the product of all the roots taken with a plus sign if n is even and with a minus sign if n is odd. In the same way, one finds that a_{n-1} is the sum of all the n roots and the coefficient a_{n-2} is the sum of the products of the roots taken two at a time and so on.

These simple algebraic truths may be illustrated by constructing the fourth-degree polynomial whose roots are 1, 2, 3, and 4. This polynomial is the product $(x - 1)(x - 2)(x - 3)(x - 4)$, which is the sum of four different powers of x: $f(x) = x^4 - 10x^3 + 35x^2 - 50x + 24$, which can be obtained by multiplying out the four factors. We see that -10, the coefficient of x^3, is $-(1 + 2 + 3 + 4)$, the negative of the sum of the roots; that 35 is the sum of all possible double products of the roots such as 1×2, 2×3, etc.; that 50 is the negative of the sum of all triple products of the roots such as $1 \times 2 \times 3 \times 4$; and that 24 is the product of the roots.

We consider a few more simple examples of polynomials to illustrate the occurrence of complex roots, which can be present even though the coefficients are all real. These are best revealed by what mathematicians call the roots of one. Beginning with the linear equation $x - 1 = 0$, we have the root $x = 1$. Going on to the second-degree equation $x^2 - 1 = 0$, we exhibit the two roots by factoring the left-hand side to obtain $(x - 1)(x + 1) = 0$. Accordingly, the two roots are $x = 1$, $x_2 = -1$, which are both real numbers. If we consider the cubic equation, $x^3 - 1 = 0$, however, we see that $x = 1$ is, indeed, one root, but the other two roots cannot be real because no real number, other than 1, equals 1 when cubed. The other two roots must therefore be complex numbers and one of them must be the complex conjugate of the other. This is a very special example of a general rule about roots: complex roots of an algebraic equation must come in pairs and the members of such a pair are the complex conjugates of each other. Later we shall return to the roots of 1 and see that these roots are an important branch of trigonometry. As complex roots must come in pairs, every polynomial of odd degree must have at least one real root.

Because finding the roots of an algebraic equation of degree higher than four is generally quite difficult, mathematicians developed methods for obtaining approximate solutions of definite algebraic equations (equations with definite numerical coefficients). Outstanding examples of this technology include Newton's and Horner's methods, both of which depend on knowing how the numerical value of the polynomial changes with slight changes of the variable x. If, for a given algebraic polynomial (a polynomial in x with definite numerical coefficients), the numerical value of the polynomial for a given value of x (let us say $x = 2$) is found by substituting 2 for x in the polynomial, we can then find how the polynomial changes if we increase x from 2 to a slightly larger value. If the numerical value of the polynomial also increases, then we must take a smaller value than 2 for x to get closer to a root. In this way, starting with a given value of x and slightly increasing and then decreasing it by stages as demanded by the polynomial, we can close in on a root. We can then use this root to reduce the degree of the polynomial by 1 to obtain a new polynomial of lower degree which we treat in the same way. We can thus close in on one root after another until we obtain them all. This is only a rough description of the general way of approximating the roots. The various methods of obtaining the roots have been greatly refined over time so that roots of equations of high degrees can be obtained easily using electronic computers.

The theory of algebraic equations has been extended considerably since their early history to encompass entities which themselves are functions of other quantities. In the development of geometry and trigonometry, quantities must be considered which themselves are functions of other quantities. The direction of the motion of a particle on a plane, for example, may change from moment to moment. This change in direction can be described by the angle which a line from the particle to a given point on the plane makes with a line passing through this point. The motion of the particle can then be described by an algebraic equation which involves functions of this angle. These functions are called trigonometric functions so that we obtain algebraic equations of trigonometric functions. These equations are governed by the laws of algebra but the roots themselves depend on other quantities. As another example of the extension of algebra to entities that are not ordinary numbers, we cite the algebra of vectors (quantities that have direction), which were introduced into physics in the late nineteenth century to simplify the laws of motion of particles subjected to forces. These are all variations or extensions of Newton's basic laws of motion. The second law is the basis of Newtonian dynamics and is merely a statement of how the motion of a body of a given mass is affected by the force acting on the body. But before the law can be applied, it must be written in such a way that it takes into account the effect of the force on the direction of the motion of the body as well as on its magnitude. These effects collectively are called acceleration, which is a vector that is a directed quantity; the law of motion must be an algebraic equation of vectors. We recall that Newton's second law of motion relates the force F acting on a particle to its mass m and its acceleration a induced by the force ($F = ma$). Because force has direction as well as magnitude, it is a vector, hence, the left-hand side of the equation describing Newton's law of motion is a vector quantity. Therefore, the right-hand side of the equation must also be a vector. In any event, the acceleration a is a vector and its direction must be exactly the same as the direction of the force.

The algebraic equations in physics that deal with the dynamics of bodies must be vector equations because our space is three-dimensional and motions in space of two or more dimensions have directions; such motions can therefore be described fully only by vectors. We can choose three directions in space arbitrarily—no two of which are parallel to each other—and replace the single vector equation by three separate nonvector (ordinary algebraic) equations, one for each of these directions. This operation is generally performed when solving vector equations. Yet we must examine all algebraic operations carefully to see whether they still apply, without change, when we carry algebra over from one domain to another, as in developing vector algebra. That the ordinary rules of addition and multiplication do not apply to vectors is apparent. Two vectors must point in exactly the same direction (or exactly opposite directions) so that their magnitudes can be added or subtracted in the usual arithmetic way. Otherwise, a special rule must be introduced for these operations. As far as the multiplication of vectors is concerned, the usual rule does not apply because the product of two vectors may or may not be a vector, which means that special rules for the multiplication of vectors must be introduced or defined.

As mathematics grew, mathematicians became more and more concerned with maintaining its intellectual purity against the inroads and practical demands of the sciences, particularly physics. Mathematicians introduced entities that were as far removed from numbers as possible; this approach led them to develop special algebras that can be applied to such nonnumerical entities, which in turn gave birth to the "algebra of sets" and with it much of modern mathematics. Although modern mathematics stemmed from the function concept and the calculus as introduced by Leibnitz, the "algebra of sets" was the first application of algebra to nonnumerical entities. The theory of sets, in its modern form, was created in the nineteenth century by the German mathematician Georg Cantor. The set in its most abstract form is a collection or aggregate of entities, all of which have similar properties that mark them specifically as belonging to the particular set in question. All even numbers, for example, belong to a definite set, as do all odd numbers, all propositions of a given kind, and all points on a line. These are but a few examples of the infinitude of sets that one can imagine. Because sets encompass all kinds of aggregates, mathematicians reasoned that all mathematics can be deduced from the theory of sets and this point of view prompted mathematics teachers to base their pedagogical approach to mathematics on set theory. This approach was not very fruitful, because most students who were not interested in becoming mathematicians could not see the relationship between abstract set theory and the concrete operations of arithmetic. To the mathematician, however, the algebra of sets is the purest form of mathematics.

In the algebra of sets, the basic algebraic operations that are used in the algebra of numbers have to be redefined to conform to the abstract quality of a set. Insofar as a set is an aggregate of similar objects (objects possessing the same property such as evenness if the objects are numbers) we can easily conceive of combining two or more similar sets in certain ways to obtain a new set of the same kind. Such combinatorial operations are extensions of the ordinary arithmetic operations of adding and multiplying with certain important revisions. We mention these operations here without discussing them any further to indicate how the algebra of sets emerged as an extension of ordinary algebra.

The theory of sets generated new branches of mathematics, some of which have important applications to physics and the solutions of problems in physics. Some of the most important of these branches include the theory and algebra of groups which have played a central role in the development of the mathematics of physics. The name "group" implies an aggregate of objects and therefore a set. A "group" is, indeed, a "set" but one with special properties which are contained or defined in the multiplication operation. Consider a set of objects which are such that if any two of the objects are multiplied according to our rule of multiplication; the product so obtained is a member of the group. Suppose also that one member of the group (called the unit member) is such that it leaves every member unchanged when it multiplies that member. Finally, let us assume that for every member of the group another member exists (the inverse) which when multiplied by the member gives the unit member. These three restrictions on the properties of the group members give the group important characteristics which make groups of various kinds very important and useful in physics and mathematics. Thus, the theory of groups became a powerful analytical tool in the hands of physicists and mathematicians and the algebra of groups became an extensive branch of algebra in its own right. The brilliant French mathematician Evariste Galois applied the theory of groups to algebraic equations to divide such equations into those whose roots can be expressed in closed algebraic forms and those whose roots cannot be expressed in closed algebraic forms.

Algebra began from the concept of the algebraic polynomial in a single variable x and the function $f(x)$ of the variable x; these concepts led to the algebraic equation in a single variable and to the roots of that equation— the values of x that make that polynomial or function 0. But one might just as well consider values of x that give the function any value one may desire. The need to develop algebra along these lines led the French mathematicians Fermat and Descartes in the seventeenth century to geometrize algebra by using curves in a plane to represent the function or polynomial. This branch of mathematics is called analytical geometry and was the precursor of calculus. An important question that arose in connection with the values of a polynomial for a given value of x dealt with those values of x for which the polynomial was either a maximum or a minimum. Such questions are important in many applications of algebra to real problems; the calculus solves such maxima and minima problems.

The next step in the development of algebra was the introduction of polynomials in two variables x and y or in three variables x, y, and z. One can place such polynomial equals to 0. But speaking of the roots of such an equation—by which we mean finding the values of the pair of variables x, y (or the triplet x, y, z) which make the polynomial equal to 0—is meaningless because infinite numbers of such pairs or triplets make the polynomial equal to 0. The simple example $2x + 3y = 4$ illustrates this point. No matter what value we choose for x, we can always find a value for y that makes the left-hand side of the polynomial of the equation 0. Mathematicians say that the solution of such an equation (a single set of values of x and y) is not defined or does not exist. An equation in x and y, where x and y can vary independently of each other, is called a linear equation in two unknowns. In pure mathematics, x and y are pure numbers which have no concrete meanings. If such an equation deals with real entities, however, x and y have to be the same in kind. In short, if x is a distance, y must also be a distance, and so on. In any case, an equation with two or more unknowns cannot be solved.

Another equation must be imposed on or set down for the two unknowns before their values can be found. This step led to the algebra of simultaneous equations which play extremely important roles in almost all phases of mathematics and science. Returning to the equation $2x + 3y = 4$, we see that if we impose the additional equation on x and y that $4x - 3y = 5$, we can solve the two simultaneous equations for x and y. If we add the equations (sum the two left-hand sides and the two right-hand sides) we obtain $6x - 3y + 3y = 9$ or $6x = 9$ so that $x = \frac{3}{2}$ and $y = \frac{1}{3}$.

Linear simultaneous equations with two unknowns are the simplest type of simultaneous equations. As we increase the number of unknowns, we must add additional equations to our simultaneous set, one for each unknown, so that solving them for all of the unknowns becomes increasingly more difficult and tedious. Yet the general principles for obtaining all the solutions are the same whether we are dealing with two or ten simultaneous equations. We briefly outline the general procedure for solving such equations by analyzing the solution of two simultaneous equations more carefully. To that end we consider the two equations $a_1x_1 + b_1x_1 = c_1$ and $a_2x_2 + b_2x_2 = c_2$. Here the two unknowns are x_1 and x_2 and a_1, b_1, a_2, and b_2 are the numerical coefficients; the numbers c_1 and c_2 define the numerical relationships between x_1 and x_2. For a given set of numbers c_1 and c_2, the values of the unknowns depend on certain general relationships among the coefficients a_1, b_1, a_2, and b_2 and the numbers c_1 and c_2 which mathematicians have discovered. These relationships are expressed in terms of what mathematicians call determinants, which are just sums and differences of products of the coefficients taken two or three or four at a time, for example, depending on whether there are two, three or four simultaneous equations.

The development of the theory of simultaneous equations led to the vast subject of linear algebra and, in particular to the algebra of matrices, two branches of algebra which owe a great debt to the British mathematicians Cayley and Sylvester as well as the Irish mathematical physicist William Hamilton. Matrices play an important role not only in linear algebra and transformation theory but also in modern theoretical physics in the form of what physicists call matrix mechanics. Hamilton's mathematical treatment of optics and mechanics later led Erwin Schrödinger to his discovery of the "wave equation" of the electron, the basic mathematical tool of quantum mechanics.

The remarkable generality of algebra—which permits us to apply it to any aggregate of sets provided we can unambiguously define the basic operations of addition and multiplication—is best illustrated by the algebra of propositions, called symbolic logic or Boolean algebra, as expounded by the British mathematician and logician George Boole in his 1854 treatise, "An Investigation of the Laws of Thought." The two basic operations that are used in the algebra of sets are the sum and the product, which are called the "union" or the "logical sum" of the sets being considered, and the "intersection" or the "logical product" of these sets. Here the sum of two sets A and B is a set that consists of all the objects which are in either A or B. This means that the sum $A + A$ must equal A in this algebra and *not* $2A$. The product of two sets A and B, written as AB, is the set which consists of all objects which are in both A and B; if these sets have no object in common the product $AB = 0$. Moreover, the product $AA = A$. With these laws of addition and multiplication, one can develop an algebra of language. Thus $A + B$ means "either A or B" and AB means "both A and B." We can see how the algebra of sets, or Boolean algebra, can be applied to actual physical situations by considering an electric circuit (each branch of which is connected to other branches by switches) as representing a proposition which is true if current flows through the branch but is false if no current flows. The sets of open and closed switches can then represent true or false propositions so that electric circuits can be translated into language or vice versa. Thus the algebra of sets and Boolean algebra in general were the forerunners of the language of today's software that is essential for operating modern electronic computers.

The interrelationships among all branches of mathematics are defined by algebra, which is the mathematical bridge that connects them to each other. The relationship between geometry and algebra is perhaps the most useful of these relationships to mathematics and science. This mathematical bridge is called analytical geometry and was first introduced by Descartes, who saw that a function or a polynomial can be represented graphically by points on a plane, the position of each point being described by two numbers associated with two lines. One number gives the distance (on any arbitrarily chosen scale) of the point from one of the lines and the other number gives the distance from the other line. Thus the function or polynomial is described graphically by points on a plane; this subject is discussed in detail in a later chapter.

CHAPTER 4

From Geometry to Trigonometry

History shows that those heads of empires who have encouraged the cultivation of mathematics, the common source of all the exact sciences, are also those whose reigns have been the most brilliant and whose glory is the most durable.

—MICHEL CHASLES

We may picture mathematics as a huge open-ended maze, with no path leading to a dead end. Rather, each path broadens as we move along it and is connected to other paths by many inviting passageways. Some of these paths merge for a while, run parallel to each other for a stretch and then diverge again. Arithmetic is the main path because both geometry and algebra emerged from it but they immediately began to diverge from arithmetic. Algebra, in a sense, is the continuation of arithmetic by other means, retaining much of the pristine quality of arithmetic, in that it is a product of pure thought with no pretensions to anything concrete and with no compromise for the sake of practical needs. Geometry, on the other hand, is, by its very nature, in its elementary form, concrete because it deals with the nature of mathematical space. The word "space" in a mathematical sense has been extended from its primitive connotation as the mathematics of real space to include fictitious space. In that enlarged sense, geometry is unreal or purely theoretical. Algebra was developed without dependence on or reference to geometry but geometry could not have been developed to its present state without algebra. This is most evident in the branch of geometry, now called trigonometry, which grew out of the study of the relationships between angles and distances. Because these relationships are functional relationships, we treat them algebraically as we do all functional relationships. In this sense, algebra is essential to geometry; this need becomes apparent when we analyze the geometrical difference between angle and distance.

We run into the angle concept as soon as we carry our arithmetic from points on a line to points on a plane. All the positions of points on a line are given by distances from a single arbitrarily chosen point 0 on the line, with the stipulation that these distances on one side of 0 be called positive, marked with a + sign, and distances on the other side of 0 be called negative, marked with a − sign. But this cannot be done for points on the plane because an infinite number of lines can be drawn through the point 0 which is called the origin. Merely giving the distance of a point from 0 does not tell us which point on the plane we are specifying because the same distance applies to an infinitude of points, each lying on a different line. We shall later see that all such points define the circumference of a circle. The angle concept had to be introduced to identify the line on which the given point lay. The angle concept is thus associated with specifying a direction on the plane because geometry itself cannot exist without direction. Trigonometry emerged when the need to express angles in terms of distances became necessary in mathematics. We pointed out in our discussion of geometry that the description of geometric figures requires measurements of two kinds of distinctly different entities—distance and angle. These different kinds of measurements assign numbers to things that are both mathematically and physically different. The measurement of a distance represents the motion of an object or a point along a straight line and this measurement is expressed in distance units such as centimeters, feet, etc. A distance is thus a dimensional entity. Space is three-dimensional because an object can move along three mutually independent directions in space. The measurement of an angle, however, implies a change in direction or a rotation, as exemplified by the rotation of a line from the center of a circle to a point moving along the circumference of the circle. As a measure of rotation, angle is not dimensional; it is a pure number with its own units such as degrees and radians. The degree is defined in terms of one complete rotation, which is then divided into 360 equal parts, each of which is called one degree. Because measuring an amount of turning accurately is more difficult than measuring a distance, the ancient geometers devoted much of their time to finding ways of expressing angle magnitudes in terms of distances. The magnitude of an angle is a pure number whereas distance is a spatial (dimensional) quantity. Angle must be expressed as a ratio of two different distances (one distance divided by another) so that the distance dimensions cancel out in this division leaving a pure number for the angle.

To see how this development led to modern trigonometry, we return to a point moving along the circumference of a circle and follow that radius of the circle whose free outer end is attached to the point so that the radius is turning. We further choose the length of this radius to be our unit of length—one centimeter, one inch, or any other unit of length we please. Mathematicians call such a circle a "unit circle," which is an important mathematical concept because it can be used to define trigonometric concepts in a very simple and obvious way. The unit circle thus leads to a measure of the angle in terms of the length of the arc of the circumference along which the point is moving. This length of arc expressed in our units of length exactly equals the angle through which the rotating radius attached to the point turns, provided the angle is expressed in units called radians. If the radius of a circle is one inch and a point on the circumference moves one inch along the circumference, the radius attached to the point turns through an angle of one radian. This unit of angle (the radian) is expressed in degrees through the number π. If the radius rotates completely around once, it rotates through 360° or 2π radians so that one radian equals very nearly 57.296 degrees. Here we have described how an angle can be measured by measuring a length (the arc of a circle), yet an angle is not a length. In this procedure, it appears to be a length because the length of the radius of the circle we are using is a unit length. An angle is not a length but instead a ratio of two lengths so that an angle itself is dimensionless.

Starting from this definition of angle, mathematicians sought other lengths associated with an angle that can be used to measure the angle. Using the unit circle, mathematicians introduced three such distances associated with the unit circle that measure the angle. These distances are called the cosine, the sine, and the tangent of the angle. But these quantities are not distances; they are ratios of distances so that the cosine, sine, and tangent are pure numbers whose numerical values equal the lengths in the unit circle mentioned above. To define these lengths we consider the unit circle with two perpendicular radii—one horizontal and one vertical— passing through it (two lines that meet at 90° at the center). We start our counterclockwise moving point on the horizontal radius and note that it is then a unit distance from the vertical axis and at zero distance from the horizontal axis. Mathematicians later called these two perpendicular lines coordinate axes, assigning the letter x to the horizontal axis (the abscissa) and the letter y to the vertical axis (the ordinate). These ideas were later extended to locate any point on the plane defined by the axes by assigning two numbers—an x and a y—to a point, called the coordinates of the point, which give the distances of the point from the horizontal and vertical axes, respectively.

Returning now to the point moving on the circumference of the unit circle, we note that the radius attached to it, turning counterclockwise, changes its direction with respect to the horizontal direction (axis) from zero degrees, when the moving point is on the horizontal axis, to 90° when the moving point is on the vertical axis. As this angle increases, the distance of the moving point from the horizontal axis increases from zero to one, while its distance from the vertical axis decreases from 1 to 0. Thus, these two distances are functions of the angle which the rotating radius makes with the horizontal axis. Mathematicians saw these functions as mathematical or geometrical devices for expressing angles in terms of distances; this insight gave birth to the branch of mathematics now known as trigonometry.

The distance of the point from the vertical axis is called the cosine of the angle the rotating radius makes with the horizontal axis, and its distance from the horizontal axis is called the sine of the angle. These names are still used and written as $\cos\theta$ and $\sin\theta$ if the angle is θ. The cosine and sine are not distances; they are the ratios of two distances (one of which, the denominator of the ratio is 1, the radius) and, hence, pure numbers like angles. Both the sine and cosine vary from 0 to 1 but as the cosine increases, the sine decreases and vice versa. Because the rotating radius of the circle and the sine and cosine distances as defined above form a right triangle, these three distances are related to each other by the theorem of Pythagoras: radius^2 = sine^2 + cosine^2 or $\sin^2\theta + \cos^2\theta = 1$. This theorem means that if the sine of any angle θ is known, its cosine can be calculated from the Pythagorean Theorem. Mathematicians later introduced the tangent of the angle θ (written as $\tan\theta$), defining it as the ratio of the sine of the angle to its cosine: $\tan\theta = \sin\theta/\cos\theta$. If any one of these three functions of θ is known, the other two can be calculated.

These trigonometric functions, which are also called circular functions, play enormous roles in science, technology, and mathematics. The tangent is of particular interest owing to its great importance in calculus, which we can infer from the concept of the slope of the rotating radius or its slope with respect to the horizontal axis. If we define this slope as the distance we move vertically divided by the distance we move horizontally as we advance along the radius from the center of the circle to the point on the circle, we see that this distance is $\sin\theta/\cos\theta$ which equals $\tan\theta$.

Trigonometry probably originated in the mathematical speculations of the ancient Egyptians as long ago as 1500 BC when accurate angle measurements were required to build the pyramids. A ratio that may correspond to the cosine of an angle is mentioned in the Egyptian papyrus of Ahmes

(c. 1550 BC). But nothing of a trigonometric nature appears in ancient mathematical literature until the Greek astronomer Hipparchus's analysis of ratios of chords to areas in a circle. Hero of Alexandria and Archimedes certainly used trigonometric concepts to calculate the areas of polygons and to compare these areas to the areas of the circles in which they were inscribed. A good deal of trigonometry also appears in Ptolemy's famous *Almagest*. Because the thirteen volumes of the *Almagest* were required reading for European mathematicians and astronomers following Ptolemy, these volumes probably greatly stimulated the growth of European trigonometry. The concept of the sine of an angle was certainly known to Hindu and Arab astronomers in the Christian era; they wrote the first texts on trigonometry as a pure mathematical discipline independent of geometry or astronomy. The work of these mathematicians filled the gap in mathematics from the time of Ptolemy to the European Renaissance.

Although trigonometry was introduced as a mathematical tool to aid in the analysis of the geometry of triangles, it spread to all branches of mathematics and science, particularly astronomy and physics, where it is used extensively in the study of periodic phenomena including the oscillations of a pendulum and the vibrations of springs. Trigonometry is very useful in solving physical problems because many different physical phenomena can be approximated by or expressed in terms of periodic processes and, thus, in terms of sines and cosines of quantities that need not be physical angles. This is expressed mathematically by saying that functions of all kinds can be represented as sums of sines and cosines. This mathematical technology was introduced in the nineteenth century by the French mathematical physicist Fourier.

Because the sine, cosine, and tangent are functions of a quantity x, which we may call an angle, we now discuss how such trigonometric functions differ from the algebraic functions we discussed in the preceding chapter. Such functions stemmed from polynomials of various degrees (the highest power of x) so we may wonder whether these trigonometric functions can be expressed as algebraic polynomials. They cannot be precisely expressed, however, because they are periodic functions and have the same values for an infinite number of values of x. We recall that these two trigonometric functions describe the distances of a point on the circumference of the unit circle from the horizontal and vertical axes as the point moves counterclockwise along the circumference of the circle. These distances thus repeat over and over again endlessly so that the sine and cosine are functions of the variable x (which may or may not be an angle) which have the same values for an infinite number of x values. These are thus multiple-value functions. We may ask for which values of x are the roots of these functions 0? The answer may be found by following the point as it revolves around the center of the unit circle. Every time the point is on the horizontal axis, its distance from this axis is 0 so that sin$x$ is zero for all values of x (the angle the rotating radius makes with the horizontal axis) for which the rotating axis coincides with the horizontal axis. This happens for the x values (expressed as radians): 0, π, 2π, 3π, 4π, and so on. The sine of x must be a sum of an infinite number of powers of x if it is to have an infinite number of roots (an infinite polynomial, or an infinite "power series," as mathematicians call such a polynomial). This is also true of the cosine but the cosine is 0 for $x = \pi/2$, $3\pi/2$, $5\pi/2$, $7\pi/2$, etc., as can be deduced from the distance of the revolving point from the vertical axis, which is zero for the angles 90°, 270°, etc.

The representation of the trigonometric functions sin$x$ and cos$x$ as infinite polynomials became part of an entire new and sophisticated branch of mathematics called infinite power series; these series are used in all branches of mathematics and physics. In time, with the development of the calculus, mathematicians showed that any function of one variable can be expressed either as a finite polynomial or an infinite series. From our discussion of algebraic equations we know functions that can be expressed as finite polynomials have a finite number of roots but we may not conclude from this fact that every function that can be expressed as an infinite series has an infinite number of roots. We may exhibit a few terms of the infinite series for sin$x$ and cos$x$:

$$\sin x = x - x^3/3 \times 2 + x^5/5 \times 4 \times 3 \times 2 - x^7/7 \times 6 \times 5 \times 4 \times 3 \times 2 + \ldots$$
$$\cos x = 1 - x^2/2 + x^4/4 \times 3 \times 2 - x^6/6 \times 5 \times 4 \times 3 \times 2 + \ldots,$$

where x is a pure number and the radian value of an angle. The denominator in each of these series is called a factorial and is written in shorthand as 3!, 5!, etc.; the factorial is the product of all integers, starting with the number itself and going down to 1. These series have some interesting symmetry features which enable us to deduce all the terms of the series if we know the first two or three terms. Each series is an alternating series; the signs of the coefficients alternate between + and −. The sine series contains only odd powers of x and the cosine series contains only even powers of x (in the cosine series the first term is $1 = x^0$, where 0 may be taken as an even power of x). The denominator of the coefficient of x^5 in the sine series is 5!, etc. From these features of the two series, we see that the fifth term in the sine series is exactly $+x^9/9!$ and the fifth term of the cosine series is $+x^8/8!$ The discovery of these series greatly simplified the preparation of numerical tables of sines, cosines, and tangents. Electronic computers now make it possible to obtain the value of the sine of any angle (any value of x) in a matter of seconds. In general, the sines, cosines, and tangents are irrational numbers; there are certain values of x, however, for which they are fractions. Thus for $x = \pi/6$ (30 degrees) the sine is $\frac{1}{2}$ but the cosine is $\sqrt{3}/2$ and for $x = \pi/3$ (60 degrees) the cosine is $\frac{1}{2}$ and the sine is $\sqrt{3}/2$. Note that the square of the sine plus the square of the cosine always equals 1.

There is an interesting feature of the uniform motion of a point along the circumference of the unit circle which has important applications to motion in general and to periodic motion in particular. We picture a real particle revolving along the circumference and imagine its shadows cast by appropriate beams of light on the horizontal and vertical axes of the circle. As the particle revolves, these two shadows oscillate back and forth but out of phase with each other (by "out of phase" we mean that when one shadow is moving toward the center of the circle the other is moving away from it). The motion of the vertical shadow represents the way the sine varies with time and the motion of the horizontal shadow represents the way the cosine varies with time. These two motions are called simple harmonic motions (SHM); mathematicians show the time dependence of the sine and cosine as follows: $\sin 2\pi\, t/p$ and $\cos 2\pi\, t/p$. Here t is the time, measured from some zero moment, and p is called the period of the motion (the time each shadow requires to complete one full oscillation). The importance of simple harmonic motion in physics, as we have already noted, is that no matter how complex any physical motion may be, it can be expressed as a superposition of a series of simple harmonic motions of different magnitudes and periods. The most important application of this concept in physics is in the mathematical description of the motion of a wave. That wave motion can indeed be described mathematically by sines and cosines can be deduced from a simple physical argument. If we picture our unit circle as moving with constant speed along the horizontal axis, we see that the oscillations of the vertical shadow advance along this axis, thus producing an advancing wave and showing that the sine function does indeed describe a wave.

Trigonometric functions were first introduced to simplify the study of triangles; this subject arises in all kinds of structural, architectural, naviga-

tional, surveying, and physical problems. Surveyors, for example, determine the unknown distance between two points by "solving" a triangle, one
side of which is the distance to be determined and the other two sides of
which are known distances that meet at an angle that can be measured.
Deducing the distance between the two points trigonometrically is called
"solving" the triangle, which is possible because all the properties of a
triangle can be deduced if any three of its properties are known; these three
properties may be the lengths of the three sides of the triangle or the lengths
of two sides and the magnitude of the angle between these two sides or,
finally, the magnitude of any two of the angles of the triangle and the length
of the side between these two angles. Most of the time spent in classes in
elementary trigonometry is used to devise ways of solving the triangle from
knowledge of any one of the triplets listed above. Note that if only the three
angles of a triangle are known, this does not allow one to determine the
triangle (to solve it); the three angles give the shape of the triangle but not
its size.

Because "solving" a triangle means that we can find all three of the
determining parameters of a triangle from the knowledge of any two of
them, some trigonometric equation connecting all three parameters must be
available to us whose solution gives us what we want. Two such equations,
one called the law of sines, and the other, the cosine law, are indeed
available. These laws can be stated simply and can be easily understood;
they relate the lengths of the sides of a triangle to the sizes of the angles. We
consider a triangle whose three angles we label a, b, and c and whose three
sides opposite the respective angles have lengths A, B, and C. The sine
theorem states that $\sin a/A = \sin b/B = \sin c/C$ and the cosine theorem states
that $A^2 = B^2 + C^2 - 2AB\cos a$, which is the extension of the Pythagorean
Theorem to any triangle. These two theorems enable one to solve any
triangle.

Trigonometry has been extremely useful in certain areas of astronomy,
especially in the study and measurements of the apparent positions and
motions of the stars and planets. Because these studies assume that the sky
is the inner surface of an infinite sphere (the celestial sphere) with the earth
at the center, we are concerned only with the positions of stars and planets on
this spherical surface and not at all with their distances from the earth. Each
star or planet is pictured as a point on the interior surface of the celestial
sphere which itself is pictured as revolving from east to west—opposite to
the actual rotation of the earth from west to east. The rate of this rotation

is 15° per hour, which accounts for the 24 hours (solar time) in one day. Relative to the stars, however, the rate of rotation is slightly larger, which accounts for the observed fact that the stars rise four minutes earlier each day. Astronomers have introduced a system of great circles and, perpendicular to them, a system of small circles on the celestial sphere, relative to which they locate stars and planets by giving their angular positions with respect to those two sets of circles. These angles, or celestial coordinates, correspond to the latitude and longitude of a point on the surface of the earth.

In this celestial mapping scheme of stars and planets, two great circles are of special importance—the celestial ecliptic and the celestial equator, both of which stem from the motion of the earth. The celestial ecliptic is the great circle in which the plane of the earth's orbit around the sun cuts the celestial sphere and the celestial equator is the great circle in which the plane of the earth's equator cuts the celestial sphere. These two planes are tilted $23\frac{1}{2}°$ with respect to each other—a very important angle because it accounts for the change of seasons on the earth. The ecliptic and the celestial equator intersect at two points on the celestial sphere—the vernal and autumnal equinoxes.

The axis of the earth's rotation is exactly perpendicular to the plane of the celestial equator. In fact, it defines this plane and pierces the celestial sphere in a point called the North Celestial Pole. The earth rotates on this axis once every 23 hours, 56 minutes; this would be the length of a day if we measured time with respect to the fixed stars rather than with respect to the sun. These two times would be identical if the earth were not revolving around the sun. But the earth does revolve around the sun by about 1° per day, from west to east, in the same sense as it is rotating. As seen from the revolving earth, therefore, the sun appears to move eastward by about 1° per day so that a solar day is about four minutes longer than a sidereal day, that is, the stars rise 4 minutes earlier than the sun.

As the sun moves eastward on the celestial sphere, it traces out the great circle which we have already introduced as the ecliptic. In its apparent annual journey, the sun is $23\frac{1}{2}°$ south of the celestial equator on December 21 (the winter solstice) and $23\frac{1}{2}°$ north of the equator on June 21 (the summer solstice). These solstices are midway between the two equinoxes. As the sun moves away from the winter solstice after December 21, it begins to approach the equator, coinciding with the vernal equinox on March 21, when it is right on the celestial equator. After June 21, when it is at the summer solstice $23\frac{1}{2}°$ north of the equator, the sun begins to move toward the equator again, reaching it on September 21 (the autumnal solstice). The nights and days are equally long at the vernal and autumnal equinoxes. We note that the degree as a unit of angle probably originated among the ancient Babylonians, who had a calendar of twelve lunar months of thirty days each, so that the sun was thought to move eastward by one degree per day. This calendar may have represented the beginning of trigonometry as well as its first practical application.

The study of positional astronomy and, to some extent, the pursuit of surveying led to what we now call spherical trigonometry. We consider a large triangle on the surface of the earth, whose vertices A, B, and C are separated from each other by many miles, to see why spherical trigonometry arose. We cannot draw a plane triangle with these vertices because we cannot draw straight lines connecting these vertices since the earth's surface is spherical. But we can connect them with arcs a, b, and c of great circles whose radii equal the earth's radius. Such arcs on a sphere are equivalent to straight lines on a plane so that the spherical triangle formed by these areas is the counterpart of the plane triangle. The trigonometry of such a triangle is spherical trigonometry. In describing such a triangle, we need not treat the arcs as distances but as angles—the angles they subtend at the center of the earth. The spherical trigonometry of spherical triangles thus deals with the six angles A, B, C, a, b, and c and the law of sines becomes $\sin B/\sin b$ = $\sin A/\sin a$.

Spherical trigonometry is very important in astronomy because the solution of a basic spherical triangle on the celestial sphere can be used to determine one's longitude and latitude, as well as the time of day, from the positions of stars. This procedure, called celestial navigation or nautical astronomy, is of great practical importance in shipping and maritime activities.

Trigonometry is used extensively in astronomical measurements, particularly in the measurement of stellar distances. The distance of a star cannot be measured directly but the shift in its position as the earth moves around the sun can be measured. Because the diameter of the earth's orbit around the sun is 186,000,000 miles, the apparent position of a nearby star, relative to the background distant stars, as observed from the earth, changes by a measurable amount as the earth moves halfway around its orbit. This shift in the star's apparent position every six months is simply the angle that two lines drawn to the star from two diametrically opposite points on the earth's orbit make with each other. It is the angle subtended at the star by the diameter of the earth's orbit; one-half of this angle, called the star's parallax, is extremely small, even for the nearest stars, but its measured value makes it possible for the star's distance to be calculated from a simple algebraic formula. This procedure for measuring a star's distance is called the method of trigonometric parallaxes.

The parallax of the nearest star, Alpha Centaur, at a distance of 4.5 light-years or 27 trillion miles, is $\frac{3}{4}$ second of arc or $\frac{1}{4800}$ degree. Such small angles can only be measured with telescopes and not with the naked eye so that the first stellar parallax was not measured until 1837 when the German mathematician and astronomer, F. W. Bessel, using the best optical instruments available, measured the parallax of the star 61 Cygni. In the sixteenth century, the Danish naked-eye observer, Tycho Brahe, tried to detect the parallactic shift of some of the bright stars; his failure to detect any observable shift coupled with his underestimation of the true distances of the visible stars caused him to reject the Copernican heliocentric theory of the solar system. In 1725, when the Copernican theory had been fully accepted by scientists, a contemporary and friend of Isaac Newton named James Bradley tried unsuccessfully to detect the parallactic shift of the nearby stars. In the process, however, he discovered another important apparent shift in the positions of stars which arises from the motion of the earth. This apparent shift is called the "aberration of light" because the earth's motion changes the direction from which the light from a star seems to emanate. This effect is most pronounced if the true direction of the star is perpendicular to the direction of the earth's motion around the sun at any moment. Because this direction changes continuously, the aberration changes continuously, going from a maximum value of 20″.49 to 0″.

The algebraic relationship between the aberration angle and the speed of the earth, when the aberration is at a maximum, is $\tan\theta = v/c$, where v is the speed of the earth, c is the speed of light, and θ is the aberration angle. By placing $\theta = 20″.49$, Bradley calculated the speed of the earth in its orbit as $v = c\tan\theta = 18.5$ miles per second (29.8 km/sec). This example shows how elementary trigonometry reveals important astronomical quantities. Using this deduced value for the speed of the earth and noting that there are approximately 31,000,000 seconds in a year, we multiply these two numbers to obtain the circumference of the earth's orbit; the radius of the orbit (the mean distance of the earth from the sun) can then be calculated. The earth's distance from the sun actually varies from 96,000,000 miles during the summer in the northern hemisphere to 90,000,000 miles during the winter in the northern hemisphere. The mean distance is half the sum of these two extreme distances—93,000,000 miles. Knowing the distance of the earth from the sun, we can use elementary trigonometry to find the diameter of the sun from a measurement of the apparent size of the angle ($\frac{1}{2}$ degree) which the sun's diameter subtends at the eye of the observer on the earth. If we multiply this angle, expressed in radians, by the earth's distance, we find that the sun's diameter is about 880,000 miles.

Probably the most important and technologically useful application of trigonometry occurs in the field of geometrical optics, which deals with the way light passes from one medium to another, such as air to glass. Since the speed of light varies from one medium to another—decreasing as the density of the medium increases—the direction of a beam of light changes when the beam crosses the interface between two different media. This phenomenon is very important in the design of all kinds of optical systems—camera lenses, telescopes, microscopes, television systems, overhead projectors, optical fibers, eye glasses, etc.—because such systems are used to produce sharp images of objects by focusing the rays of light, coming from any point on the object, that enter the optical system on a single point. As the rays coming from any point on the object spread out, they strike the front lens of the optical system at different angles so that their paths must bend by different amounts as these rays pass through each lens of the system. A sharper image of the origin point of the rays can be produced by the optical system only if all the rays come together at a point after they have left the last surface of the optical system. Very accurate designs of all the lenses in the optical system are required; these designs can be extremely complex if the system contains many lenses. Fortunately, all such designs are based on a very simple trigonometric formula which is known as Snell's law, discovered in the sixteenth century by the Dutch mathematician Willebrod von Roijen Snell.

To describe this law we consider first a flat surface of a transparent material such as glass, and a ray of light passing from the vacuum through this surface into the glass. Since the speed of light in glass is less than its speed in the vacuum, the direction of the path of the ray in the glass is different from its direction in the vacuum. This difference is called refraction. The amount by which the direction of the ray changes depends on the angle that this ray makes with the line perpendicular to the surface, the normal of the surface where the ray penetrates the surface, and on the nature of the glass. If this so-called angle of incidence is represented by i, the amount by which the direction of the path changes depends on $\sin i$. If i equals zero (the entering ray is parallel to the normal) $\sin i$ equals zero and the path of the ray is not altered. As i increases, however, the change in the path's direction increases, reaching a maximum as i approaches $90°$. The change in the ray's path is given by the difference between i and the angle of refraction r that the path makes with the normal to the surface when the ray is in the glass. The relationship between i and r is $\sin i/\sin r = c/v = n$, where v is the speed of the ray in glass, c is the speed of light in the vacuum, and n is called the index of refraction of the glass. We see that n must always be larger than 1 because c, the speed of light in a vacuum, is always larger than v, the speed of light in any medium. For a reflected ray, $v = c$ so that $i = r$, where r is now the angle of reflection. Snell's law describes reflected as well as refracted rays of light and can be used to describe rays of light reflected from mirrors of any shape. This law must be used when designing optical reflecting telescopes and radio telescopes.

In designing a complex system of lenses, the optical engineer must trace various rays of light through all the lenses in the system, applying Snell's law at each lens surface, whether the ray is leaving or entering that surface. By combining enough lenses with surfaces of different shapes and with different kinds of glass, the optical engineer can design a lens system that produces sharp images. The images produced by any lens system are not perfectly sharp; the diffuseness of such an image is produced by aberrations of the lens system. Reducing these aberrations to a minimum is the principle goal of the lens designer. Although obtaining the best optical design for a lens requires applying Snell's law in its correct trigonometric form to many different rays that pass through a system, one can obtain a good approximation by considering rays that enter the front surface of the first lens close to its center at a small angle i. The angle of refraction r is then also small so that $\sin i \approx i$ and $\sin r \approx r$, and Snell's law is approximately $i/r = n$. We can then describe the complete path of a ray of light through the system algebraically and express the various aberrations algebraically so that obtaining the first approximation to a good optical system becomes an algebraic problem. The seventeenth-century French mathematician Pierre de Fermat discovered that when rays of light pass through a series of different media (e.g., air, glass, water) in going from some initial point to a final point, they move along paths so that their travel time is a minimum. Fermat called this discovery the principle of least time.

One of the most important uses of trigonometry is in the description of the rotation of a body or a system of bodies. When a rigid disk is rotating uniformly around a fixed axis—that is, perpendicular to the disk and passing through its center—the position of any point on the disk can be described by giving its distance from the axis and giving the angle that is made by the line from the axis to the point with a fixed horizontal direction. As the disk rotates, this angle increases at a constant rate; this rate is called the angular velocity of the disk. The German astronomer Johannes Kepler discovered in the seventeenth century that if he described the position of a planet with respect to the sun by its distance r from the sun (its radius vector) and the angle θ this radius vector makes with the largest diameter of the planet's orbit (a fixed line), he could then describe the planet's motion with respect to the sun in the form of three laws—Kepler's three laws of planetary motion. An important feature of these laws is that they assume a simple trigonometric form if they are expressed in terms of r and θ. Later, Newton showed mathematically that Kepler's laws are a direct mathematical consequence of Newton's law of gravity. Kepler was led to the ellipse as the correct orbit of a planet around the sun by some 36 years of arithmetic juggling of Tycho Brahe's observational data. Newton's laws of motion and gravity lead to a simple trigonometric formula for the orbit of a planet; this formula contains only two variables—the distance r of the planet from the sun and the angle θ that r makes with a fixed direction which we mentioned above. This deduced formula not only emphasizes the usefulness of Newton's laws, but also leads to a much more general property of the laws than one might infer from Kepler's laws. The derivation of Kepler's laws from Newton's laws shows that the relative orbits of any two bodies interacting gravitationally may be any one of three possible types of orbits: ellipses, parabolas, or hyperbolas. Owing to the generality of Newton's laws we may apply them to the dynamics of two stars that are gravitationally bound to each other to form a double-star or a binary system.

The remarkable interrelationships that exist among diverse and disparate concepts that we cannot infer from first considerations of such concepts are often revealed by mathematics. This is perhaps the most important role that mathematics plays in the theoretical researches of physicists and astronomers. In other words, mathematics, in this case, trigonometry, shows that the orbits of planets, comets, stars, and satellites are all related to each other and to the figures one obtains by cutting through a right circular cone.

The mathematics of the orbits of planets reveals two other remarkable features about orbits which one cannot infer merely by observing the orbits, regardless of the accuracy of the observations. The orbits have a range of sizes and shapes which are incorporated in the mathematical formula for an orbit by two parameters (measurable quantities): the average distance of the planet from the sun and the eccentricity of the orbit—meaning the extent of its departure from circularity. This in itself is not very astounding since we expect a mathematical formula for an orbit to say something about the orbit's size and shape, which define mathematically what physicists call the energy and angular momentum of the planet—two very important dynamical quantities. We return to these concepts later when we consider the mathematical harmony that governs the universe.

One of the most beautiful features of trigonometry was discovered by the eighteenth-century Swiss mathematician Leonhard Euler (1707–1783) who expressed it in the form of what is now called Euler's equation. This discovery stemmed from the mathematical relationship between trigonometry and complex numbers. In our discussion of arithmetic in Chapter 1 we pointed out that the number $\sqrt{-1}$ cannot be placed among the natural numbers consisting of all negative and positive integers, fractions, and irrational numbers—the real numbers—nor represented by any point on a line. To accommodate all numbers containing $\sqrt{-1}$ (called i) such as $3 + 5i$, the number system was enlarged to include all numbers $a + bi$ (complex numbers), where a and b are any two real numbers. Gauss accomplished this by introducing the "complex plane," each point of which represents a complex number. This plane is defined by any two interacting perpendicular lines, along one of which, the real axis, the points represent all positive and negative real numbers, and along the other, the imaginary axis, the points represent all the negative and positive imaginary numbers. Each point on this plane then represents a complex number—the sum of a real number (a point) on the real axis and an imaginary number (a point) on the imaginary axis—such as $3 + 7i$. This sum means that the point in the complex plane representing this complex number is three units of distance to the right of the imaginary axis and seven units of length above the real axis. We now consider the angle θ that the line from the origin 0, the point of intersection of the real and imaginary axes, makes with the real axis. We know from our definitions of the cosine and sine of an angle that the number 3 of the complex number above is related to the cosine of θ (but not equal to it as the cosine must always be less than 1) and the number 7 is related to the sine of θ in the same way that the cosine of θ is related to 3. Every complex number must be related to the complex sum $\cos\theta + i\sin\theta$. This represents the trigonometric formula for any complex number on the unit circle in the complex plane whose center is at 0. All such numbers are represented by points which are at a unit distance from 0 but differ only in their directions from 0.

Euler discovered a relationship between $\cos\theta + i\sin\theta$ and a very important transcendental number called e (the exponential), the base of what mathematicians call the natural logarithms. Mathematicians discovered that e enters into the calculation of compound interest if the compounding is done continuously, instant by instant, rather than in discrete intervals. A simple but unrealistic example will help to illustrate this point: Imagine receiving simple interest at 100% per year so that after one year the amount of the principal doubles. Suppose now that this interest is compounded continuously. At the end of one year your principal would more than double; it would be multiplied by e which is somewhat larger than 2.7. Because e is a transcendental number, it cannot be written down exactly as 2 plus a fraction, but it can be approximated to any desired degree by the infinite series

$$e = 1 + 1 + \frac{1}{2} + \frac{1}{2 \times 3} + \frac{1}{2 \times 3 \times 4} + \frac{1}{2 \times 3 \times 4 \times 5} + \cdots$$

Euler introduced the number e (also called Euler's number, as defined by this infinite series) to mathematics in 1748. The more terms we include in this infinite sum, the closer we get to an exact value of e. Because we can never obtain the exact value, however, we simply write it as e. Euler's equation, which is quite surprising when we first meet it, shows that e raised to the power $i\theta$ is exactly $\cos\theta + i\sin\theta$: $e^{i\theta} = \cos\theta + i\sin\theta$. This means that $e^{\pi i/2} = i$, $e^{i\pi} = -1$, and $e^{2\pi i} = 1$. The reason this expression surprises us is that we may find it difficult to understand how the infinite sum for e described above raised to the power $2\pi i$ can possibly be 1. This is but one of the many oddities one meets when dealing with complex numbers.

Euler's equation teaches us that although all the rules of arithmetic apply to complex numbers, we often obtain results working with these numbers that appear magical. Owing to these hidden properties of complex numbers, a whole branch of mathematics called the "theory of functions of a complex variable" is devoted to the study of the relationships among such numbers. Although complex numbers appear to be unrelated to reality, they play a very important role in the mathematical description of physical phenomena that range from the propagation of all kinds of waves to the behavior of electrons.

That Euler made such important contributions to the field of complex numbers is not surprising given his varied interests and background. The son of a Lutheran clergyman, Euler was born in Basel, Switzerland. Euler's father wanted his son to follow in his footsteps but these plans were forever changed when young Leonhard began studying with the distinguished mathematician Johann Bernoulli and discovered his own talent for mathematical calculations. Leonhard's father, recognizing his son's aptitude and himself having studied with Jacob Bernoulli, encouraged Leonhard's studies and even instructed him in certain basic areas of mathematics. As Leonhard passed through adolescence, his interests broadened and he became adept in disciplines ranging from astronomy and physics to medicine and theology. By the time he reached his twentieth birthday, his intellectual gifts had become well known among many of Europe's leading mathematicians. That much of Euler's notoriety at this point in his career was due to his long-term association with the Bernoullis did not detract from his own fledgling talents as a mathematician. But he had already acquired something of an international reputation after receiving an honorable mention from the Paris Academy of Sciences for an essay on the masting of ships. Although Euler's paper was technically correct, it did not read as though it had been written by one familiar with the ways of the sea. This remoteness from nautical reality was doubtless due to the fact that Euler had never seen a ship, having never ventured beyond the borders of his landlocked, mountainous Switzerland. But his surprisingly high finish in the Academy's essay contest was a precursor of things to come. In the ensuing years, Euler would win the Academy's biennial prize twelve times.

In 1727, Euler learned of a vacancy at the St. Petersburg Academy in Russia in medicine and applied for the position. Both Nikolaus and Daniel Bernoulli, the sons of his former teacher Jean Bernoulli, had already served as professors of mathematics in St. Petersburg. The considerable weight of Daniel Bernoulli's recommendation prompted the academy to offer Euler the job even though it lay outside his chosen field. Overjoyed at finally being able to earn a living as a teacher, Euler eagerly packed his bags and left for Russia. He arrived in St. Petersburg on the same day that Catherine I, the widow of Peter the Great, who had established the Academy as part of his ongoing efforts to help Russia catch up with Europe, died. Her death left the very future of the Academy in question because her successors were much less sanguine about the benefits of inviting foreign scholars to sit on the faculty at the Academy. But the international prestige of the Academy

Leonhard Euler (1707–1783) (Courtesy AIP Emilio Segrè Visual Archives)

and its intellectual luminaries was such that even the most vocal opponents of its foreign membership were unable to convince the new government of the merits of purging its ranks of all foreign visitors.

Euler taught medicine and physiology for several years. By 1730 he had succeeded to the chair of natural philosophy. The tragic drowning of Nikolaus Bernoulli in 1726, coupled with Daniel Bernoulli's decision to leave Russia to return to Basel to teach mathematics in 1733, left Euler as the senior mathematician at the Academy—at the ripe old age of twenty-six. Now that he had established himself professionally in the field of study dearest to his heart, Euler began to live and breathe mathematics. Indeed, it was not very long after receiving his appointment as the senior mathematician in St. Petersburg that Euler began to contribute a steady stream of important and often profound papers to the Academy's research journal. Indeed, the journal's editors could have very well named the journal after Euler in view of the many contributions he made to its pages. But this productivity was not merely an outburst of youthful enthusiasm, destined to burn out in a few years. According to the eighteenth and nineteenth-century French mathematician François Arago, Euler possessed an almost super-human ability to do important mathematical research, with the ability to make laborious calculations effortlessly "just as men breathe, as eagles sustain themselves in the air." Indeed, the loss of sight in his right eye in 1735, reportedly from overwork, had little apparent effect on the torrent of papers and books produced by Euler's pen. According to Carl B. Boyer, "[Euler] is supposed to have said that his pencil seemed to surpass him in intelligence, so easily did his memoirs flow; and he published more than 500 books and papers during his lifetime." Euler's original papers continued to appear with mind-numbing regularity in the Academy's journals well into the nineteenth century (half a century after his death), the legacy of an annual output of mathematical papers averaging about 800 pages each year. It has been estimated that a collection of all his mathematical works would fill the equivalent of three sets of encyclopedias. In any event, he was without question the most prolific mathematician in history, and the man whom his colleagues called "analysis incarnate."

As Euler's mathematical career in Russia flourished, he married Catharina Gsell; she was a woman whose endurance rivaled that of her husband in that she ultimately gave birth to thirteen children, only five of whom survived beyond their first few years of life. Euler seems to have taken great joy in his offspring; he would often write mathematical papers while holding a baby in one arm. Euler also thought nothing of dashing off a paper while waiting for supper to be served or while sitting out in the garden sipping tea. He would then toss the finished paper on top of the pile to be submitted to the printer. But because the papers were not always published in chronological order owing to Euler's somewhat haphazard filing system, some of his later works were published before his earlier works, thereby creating a crazy-quilt trail of mathematical work in which the intellectual lineage was often impossible to discern.

But Euler did not simply confine himself to abstract mathematical papers. Indeed, his sponsorship by the Academy might have been jeopardized had he limited himself to mathematical research. According to Eric Temple Bell, Euler gave the Russian government its money's worth: "Euler wrote the elementary mathematical textbooks for the Russian schools, supervised the government department of geography, helped to reform the weights and measures, and devised practical means for testing scales." But no matter how deeply involved he became in these arguably pedestrian activities, Euler managed to churn out a steady stream of mathematical papers. He also authored an important treatise on mechanics, which applied Newton's calculus to mechanics and thus placed that discipline on a firm scientific footing for the first time.

By 1740, Euler had begun to tire of Russia and longed to return to Central Europe. His wish was granted the following year when Frederick the Great invited Euler to join the Berlin Academy. Euler moved to Germany and became a prominent fixture of Frederick's court for the next quarter century. But so great was the esteem with which Euler was held in Russia that he continued to receive a pension from St. Petersburg during his entire stay in Prussia. Euler repaid the favor to some extent by continuing to favor the St. Petersburg Academy with his mathematical papers. He also became a regular contributor to the journals of the Prussian Academy.

Euler's relationship with Frederick was never warm. Frederick saw the plain-mannered Euler, despite his evident brilliance, as something of a country bumpkin. Euler, for his part, did his best to win the emperor's favor, bringing his considerable talents to bear on a wide variety of problems in areas ranging from navigation to the nation's coinage system. Amply compensated by the crown, Euler and his eighteen dependents lived in a house in Berlin; he also owned a farm near Charlottenburg. But Euler was never very comfortable with the rituals of the court. Frederick, for his part, found Euler's simple dress to be somewhat disconcerting. Moreover, Euler, unlike some of the other distinguished luminaries in residence at that time, such as Voltaire, did not fawn over the king. In Frederick's eyes, this was a significant flaw and he began to search for a new mathematician. But finding someone to replace Euler was not easy, particularly when visitors to the court such as the French mathematician Jean d'Alembert would bluntly tell Frederick that no mathematician alive could fill Euler's shoes. Frederick, recognizing the prestige Euler brought to his court, refrained from sacking him but Frederick's hostility toward his court mathematician became more apparent over the years and eventually led Euler to return to Russia in 1766 at the invitation of Catherine the Great.

Euler was adored by Catherine and repaid her lavish favors by helping her get rid of the French philosopher Denis Diderot, who had been invited to the Russian court only to make a nuisance of himself by trying to convert the members of Catherine's entourage to atheism. Euler, knowing that Diderot was an ignoramus about even the most basic mathematics, devised a clever plan that he believed would solve the problem without forcing Catherine to throw the troublesome Diderot out of the country. According to Augustus de Morgan's *Budget of Paradoxes*, "Diderot was informed that a learned mathematician was in possession of an algebraical demonstration of the existence of God, and would give it before all the Court, if he desired to hear it. Diderot gladly consented. . . . Euler advanced toward Diderot, and said gravely, and in a tone of perfect conviction: 'Sir, $a + b^n/n = x$, hence God exists; reply!' " The equation appeared to be impressive enough to Diderot but he was at a loss as to how he should respond to Euler's "proof." As the courtiers broke out in howls of laughter, Diderot, turning a deep shade of red, saw that he had been humiliated. He asked Catherine if she would consent to his return to France. Permission was granted and the troublesome Diderot never set foot in Catherine's court again.

Despite this pleasant beginning, Euler's second stay in Russia was not without sadness. Soon after his arrival at St. Petersburg, Euler's sight began to fade in his remaining good eye. His contemporaries such as Lagrange and D'Alembert seemed to have viewed his deteriorating eyesight as a death sentence but Euler refused to wallow in self-pity. Instead, he began writing his mathematical equations on a slate with a piece of chalk and dictating his papers to a secretary. Freed from having to write his papers in his own hand, Euler's output actually increased, even as the last bits of vision faded from his eye. Because Euler had the ability to perform complex calculations in his head, his poor eyesight had little effect on the quality of his work. Indeed,

Euler was able to dispense with reference books altogether, having completely memorized all of the significant mathematical formulas known at that time. He also amused himself by memorizing classic works of poetry such as Virgil's *Aeneid*.

Euler also suffered several personal tragedies while in Catherine's court. In 1771, a devastating fire swept through St. Petersburg and destroyed his house. Euler himself would have perished had his faithful servant, Peter Grimm, not carried his master out to safety. Fortunately for mathematics, nearly all of Euler's manuscripts were saved. Undaunted, Euler resumed his work and, thanks to the generosity of Catherine, was soon living in a new home. Five years later, however, his wife died, plunging Euler into a deep depression. He was comforted by his wife's half-sister, Salome, whom he later married in 1777. Buoyed by the upturn in his personal fortunes, he agreed to a risky operation on his left eye to restore his eyesight. Miraculously, the operation was a success and the world of light and color returned to Euler. Like a small child wandering among the presents under a Christmas tree, Euler drank in the sights of a world that had been shrouded in darkness for many years. But his eye became infected and, after suffering in excruciating pain for several days, the darkness soon returned for good.

By the time Euler suffered a stroke and died on September 18, 1783, he had created an incredible body of work in mathematics. Not only had he helped to clarify and unify vast areas of mathematics but he had also completely changed the way in which mathematicians approached their subject, even though he did tend to be rather impulsive about engaging in calculations without necessarily having a grasp of the physical features of the problem at hand. But for Euler the universe of mathematics instead of the universe of stars and galaxies was often the only subject of interest to him; he preferred to leave it to others to worry about the practical moorings of his work. We have already seen that Euler was no fluff-headed academic, having greatly assisted both Russia and Prussia in a variety of practical problems. Nevertheless, he often saw calculations as an end to themselves and was not always successful in resisting the impulse to perform complex mathematical operations for their own sake. Despite this weakness, however, his legacy to mathematics and science remains secure. Euler must surely be ranked as one of the top mathematicians of all time whose unceasing efforts not only helped to overhaul mathematics itself but also unify much of its chaotic landscape.

One of the most remarkable and useful relationships in elementary mathematics was discovered by the French mathematician Abraham de Moivre about the time that Euler introduced the number e and the formula $e^{i\theta} = \cos\theta + i\sin\theta$, but whether or not this formula led de Moivre to his discovery or influenced him is not clear. In any case, de Moivre's discovery was presented in the form of a very striking algebraic equation involving sines and cosines.

To grasp the full significance of de Moivre's theorem we must first point out that if we raise a sum of two quantities $x + y$ to any power—$(x + y)^2$, for example—we must have a sum of terms, each of which is x or y raised to a power of 2 or products of x and y such as xy raised to powers that add up to 2. Thus, the binomial $(x + y)^2$ equals $x^2 + xy + yx + y^2$, and the binomial $(x + y)^3$ equals $x^3 + 3x^2y + 3xy^2 + y^3$. The sum of all the coefficients equals 2 raised to the power to which the binomial is raised (in the first case 2^2 or 4 and in the second case 2^3 or 8). This statement is true regardless of the power (exponent) so that the binomial $(x + y)^n$, where n is any number, must be a sum of all products of powers of x and y with the sums of these powers (exponents) adding up to n, and the sum of all the numerical coefficients in this sum must equal n.

De Moivre's theorem shows that this result need not be so if the binomial is $\cos\theta + i\sin\theta$. In fact, de Moivre discovered the formula $(\cos\theta + i\sin\theta)^n = \cos n\theta + i\sin n\theta$, so that the number of terms of the right-hand side of the equation is always 2, but the magnitude of the angle to which the sine and cosine apply is n times the angle to which the sine and cosine apply on the left-hand side of de Moivre's formula. This formula is important and useful in trigonometry because if we know the sine and cosine of any angle, we can find the sine and cosine of any multiple of that angle, regardless of the nature of n as a pure number. Two examples illustrate this point: $(\cos\theta + i\sin\theta)^2 = \cos 2\theta + i\sin 2\theta$ and $(\cos\theta + i\sin\theta)^3 = \cos 3\theta + i\sin 3\theta$. In the first example, we square the left-hand side and obtain $\cos^2\theta + 2i\sin\theta\cos\theta - \sin^2\theta = \cos 2\theta + i\sin 2\theta$ or $(\cos^2\theta - \sin^2\theta) + 2i\sin\theta\cos\theta = \cos 2\theta + i\sin 2\theta$. We now apply the rule that in any equation involving complex numbers the real part of the left-hand side of the equation must equal the real part of the right-hand side and the imaginary parts of the two sides of the equation must be equal. Thus $\cos^2\theta - \sin^2\theta = \cos 2\theta$ and $2\sin\theta\cos\theta = \sin 2\theta$. In the same way, the second example tells us that $\cos 3\theta = 4\cos^3\theta - 3\cos\theta$ and $\sin 3\theta = -4\sin^3\theta + 3\sin\theta$. These are examples of what mathematicians call trigonometric identities; they hold true regardless of the value of the angle θ.

As trigonometry grew, trigonometric identities also developed. These identities are formal relationships among the sine, cosine, and tangent like de Moivre's formulas that can be manipulated without violating any mathematical principles to produce new relationships that look quite different from the original relationships. These have no particular usefulness except as exercises in trigonometry to test how well students have mastered the subject. All trigonometry is contained in the definition of the sine (or cosine) of an angle because all trigonometric relationships can be deduced from that single definition. In spite of its simple and lowly origin, however, we shall see that trigonometry plays an important simplifying role in all branches of mathematics and physics.

CHAPTER 5

Analytic Geometry

The Geometrization of Arithmetic and Algebra

I have resolved to quit only abstract geometry, that is to say, the consideration of questions which serve only to exercise the mind, and this, in order to study another kind of geometry, which has for its object the explanation of the phenomena of nature.

—RENÉ DESCARTES

Until the seventeenth century, mathematics evolved along what appeared to be disparate and essentially independent lines but this was forever altered owing to the mathematical contributions of three mathematicians—Fermat, Descartes, and Leibnitz—and three scientists—Kepler, Galileo, and Newton—whose theoretical discoveries greatly influenced and contributed to the convergence of the various branches of mathematics into a single discipline. Fermat and Descartes saw that algebra and arithmetic could be made more understandable if they were presented graphically. This insight led to one of the most important concepts in mathematics, science, and even in the social sciences—the coordinate system. Leibnitz's contributions extended the concept of the coordinate system to all functional relationships.

The contributions of Kepler, Galileo, and Newton to analytic geometry stemmed from the very nature of research in astronomy and physics. Kepler's work was purely empirical in the sense that his final conclusion that the orbits of the planets are ellipses was based on an exhaustive numerical analysis of Tycho Brahe's very accurate naked-eye data, collected over a span of many years. These dealt with the motion and positions of Mars in its orbit around the sun. Brahe's observations, taken with respect to the earth, were in the form of numbers, which Kepler then had to translate into a Martian orbit with respect to the sun. This problem required long hours of tedious arithmetic and some trigonometry. In his resolution of this problem, Kepler always represented the planet's orbit with respect to the sun graphically. This procedure seemed very natural since it merely required Kepler to place a dot on a piece of paper to represent the sun and two other dots to represent the earth and Mars. Lines connecting these dots formed a triangle, one vertex of which, the sun, remained fixed, while the other two vertices, earth and Mars, moved about, changing the shape of the triangle continuously. The length of one side of the triangle—the side from the sun to the earth—remained constant but the lengths of the other two sides changed continuously. Kepler's problem entailed expressing the length of the side of the triangle from the sun to Mars, in terms of the fixed distance from the sun to earth and the varying distance from the earth to Mars. This procedure led Kepler to a geometrical figure for the orbit of Mars, which Kepler recognized as an ellipse.

In a very bold extrapolation, Kepler extended the concept of the elliptical orbit to all the planets, including the earth, but this involved nothing more than geometrizing arithmetic. The idea of expressing the orbits algebraically—as functions connecting two variables of the orbit—did not occur until Fermat, in 1629, and Descartes, in 1637, introduced their concepts of a coordinate system, by which time Kepler had already completed his arithmetic analysis of Tycho Brahe's data. Here we have a striking example of how the evolution of mathematics and science supplement each other when the need for this kind of associated evolution of ideas arises. The remarkable aspect of this process is that the mathematical concepts evolve as pure thought with no immediately apparent relationship to the physical world. In time, however, these pure ideas acquire a concreteness and physical reality that transcends their mathematical roots. Mathematics thus becomes physics and physics, in turn, stimulates mathematics.

Galileo was the first experimental physicist, who insisted on experimental verification of Aristotelian principles, particularly in the area of the motion of bodies. His experiments, which demonstrated the falsity of Aristotelian physics, established the principles of dynamics, which later became the foundations of Newton's laws of motion and, ultimately, modern mechanics. Galileo's primary experiments on motion dealt with freely falling bodies, which led him to the very important concept of acceleration, specifically the acceleration of gravity. Because measuring the time rate of change of the velocity (acceleration) of a freely falling body was very difficult for him with his crude clocks, he introduced the idea of studying small spheres rolling down inclined planes whose slopes he could alter. He was thus able to reduce the speed of the rolling sphere to any desired amount and measure the rate of change in its speed. To deduce his law of freely falling bodies and the role of gravity, Galileo had to relate the acceleration of gravity for a freely falling body to the acceleration along the incline of the sphere rolling down the inclined plane. He saw at once that the two accelerations are related by the sine of the angle of incline of the plane; the acceleration along the incline equals the freely falling acceleration multiplied by the sine of the angle of the incline. Galileo had deduced an algebraic formula relating the length of the inclined plane to the acceleration of the rolling ball along the plane and the time it took the ball to roll from the top to the bottom of the plane. However, instead of using this formula, Galileo constructed a graph on which he plotted the distance the ball rolled along the inclined plane against the rolling time. In that way he obtained a distance–time graph which showed that the distance is proportional to the square of the time with a proportionality factor that is exactly one-half the acceleration of the ball along the incline. Multiplying this quantity by twice the sine of the angle of the incline gave him the acceleration of gravity.

The importance of the use of the graph to express experimental results is that it enables the experimental physicist to present his findings visually without reference to any preconceived mathematical formula. Moreover, both the experimental and theoretical physicist can see at a glance the essential functional relationships among the various parameters that describe any particular phenomenon. This is important because the physicist can see from the graph whether or not a proposed physical theory that explains some particular phenomenon is correct and because the graph itself can suggest a theory (a mathematical formula) if none has been proposed. Kepler's discovery of the elliptical orbits of the planets around the sun is an interesting example of this.

Having set himself the task of proving the correctness of the Copernican heliocentric model of the solar system, Kepler decided to work on the orbit of Mars, using Tycho Brahe's data, as we have already noted. Instead of just contemplating the numerical results he had deduced from Tycho's data, Kepler plotted the calculated Martian orbit with respect to the sun, hoping to obtain a circle. He very nearly succeeded but the plotted orbit was not quite a circle. Most investigators might have been satisfied with such a result but not Kepler. He was forced to give up the idea of circular orbits but he then considered the idea of an ellipse and everything fit.

Unlike Kepler, Isaac Newton was a master of all phases of the mathematics known at the time and he used arithmetic, algebra, geometry, and trigonometry as needed, to demonstrate his own mathematical theorems and his deductions in physics. One of his important contributions to algebra was his binomial theorem which gives the formula for the expansion of any power of the sum of two terms such as $(x + y)^n$ into a sum of individual products such as x^3y^{n-3}. This theorem is useful in all branches of mathematics and science. In formulating his laws of motion and of gravity mathematically, Newton set the pattern of research for future scientists, establishing what was first called mathematical physics, and, later, theoretical physics. This was the beginning of the dichotomy in physics research (theory and experiment) which has grown to such an extent that theoretical and experimental physicists are as familiar with each other as two very close kinsfolk can be. The experimental physicist generally presents his results graphically and the theoretician then formulates a theory mathematically based on the results of the experimentalist; with his theory already formulated mathematically, the theoretician checks his theory against the graphical data of the experimentalist.

When Descartes first introduced his concept of the graphical representation of algebraic functional relationships, the Galilean–Newtonian classical laws of physics were unknown, so Descartes was not prompted by the desire to express these laws graphically. Indeed, he went so far as to reject the concept of atoms and argued that all forms of matter were but various manifestations of vortices. He introduced analytic geometry as a way of visualizing algebraic formulas, not to aid the development of physics. This visualization required relating algebraic equations to spatial configurations. Because the simplest of such configurations are produced by connecting points on a plane, Descartes saw that he could describe algebraically geometrical configurations on a plane if he could describe the position of a point on a plane algebraically. This realization led him to the concept of the coordinate system, which has become perhaps the single most important theoretical construct in the evolution of theoretical physics as well as mathematics. Descartes introduced the coordinate system merely as a device to locate points on a plane but it quickly took on a life of its own, in a sense, and it became part of the laws of mathematics and physics. The coordinate system ultimately became a kind of acid test of the validity of a physical law or theorem.

The number system (complex numbers included) and, therefore, arithmetic, can be described by points on a line, but geometry, in its simplest form (plane geometry), requires the extension of arithmetic to points on the plane. In other words, geometry requires a two-dimensional manifold so that describing the position of any point on this manifold requires two numbers or, put differently, two coordinates. Descartes began by laying off two mutually perpendicular lines on the plane—one horizontal and the other vertical—and defining the point 0 of intersection of these two lines as the "origin" of the coordinate system. He called the two intersecting lines the coordinate axes of his coordinate system, which, since then, has been called a "rectangular" (right angle) Cartesian coordinate system. The position of any point on the plane is specified in this kind of coordinate system by assigning two numbers to the point, one of which is its perpendicular distance, on an arbitrary scale, from one of the perpendicular axes and the other of which is its perpendicular distance from the other axis. Descartes designated the horizontal axis as the x axis or the abscissa and the vertical axis as the y axis or the ordinate. The position of a point in such a coordinate system is then specified by its two coordinates x and y, written as (x,y).

From a geometrical point of view, these are only distances, but this coordinate representation is not limited to spatial geometrical configurations and can be used to express graphically the relationship between any two quantities that are functionally related to each other. The relationship between the pressure in a gas (the push of a gas on a unit area of its container) and the volume in which it is confined is a useful physical example of this concept. One plots the pressure (in suitable units) along the ordinate axis and the volume (in units such as cubic centimeters) along the abscissa. Robert Boyle, a contemporary of Newton, showed that if the temperature of the gas is kept constant, the pressure increases in a very definite way as the volume decreases. The pressure varies inversely with the volume. Each point on this pressure–volume graph represents a definite pressure and volume of the gas; if we connect all such points we obtain a curve which is a visual representation of Boyle's law of gases, as his discovery is called. Note that such a curve has great empirical value because it enables us to find the pressure of a gas for any volume. We can cite many other examples of useful graphs in areas as diverse as sociology, economics, chemistry, and astronomy.

Considered by many to be the first modern mathematician, René Descartes was born on March 31, 1596 near Tours, France. His father, a councillor in the local parliament at Rennes, divided his time between his official duties at Rennes and the management of the family estate at La Haye. Even though René's mother, Jeanne Brochard, died a few days after giving birth to him, young René had a pleasant childhood. Indeed, his father appears to have taken it upon himself to try to make up for the loss of their mother to his three children, going so far as to hire a caring nurse and personally monitoring their educational progress. René himself received his early education at a Jesuit school in La Fleche. Although he was already beginning to show signs of brilliance, René did not shrink from the disciplined regimen of the school. Indeed, he always spoke very highly of his experience with the Jesuits even though he was continually plagued by poor health. It was his recurrent illnesses that prompted the school rector, Father Charlet, to grant René permission to lie in bed in the mornings until he felt like getting up. Not only did this practice bolster his health but it also seemed to stimulate his interest in mathematics.

Descartes received a rigorous classical education at La Fleche, concentrating on Greek and Latin. But he also made a number of lifelong friends, including Marin Mersenne, who later became famous as an amateur mathematician and scientist in his own right. But Descartes' rebellious mind could not accept at face value the Jesuits' dogmatic teachings in philosophy and ethics. Indeed, he had already developed a strong distrust of unquestioned authority and this skepticism began to spread to long-held tenets of even the most basic areas of study, such as theology. After leaving the school, Descartes spent several aimless years in Paris, enjoying the pleasures of the flesh with youthful abandon. He also became quite an accomplished gambler. But this rollicking life left a stale taste. Surely he was destined for better things! Perhaps it was the classical training of the Jesuits that prompted him to step back and take stock of his life. It is curious that Descartes chose to purge himself by enlisting in the military in 1617, joining the army of Prince Maurice at Orange. Although Descartes disliked the daily routine of army life, he did find that it offered ample leisure time which he could devote to mathematics.

Because Prince Maurice's army saw little action, Descartes left the camp and eventually reenlisted in the service of the Elector of Bavaria, which was then engaged in a bloody war with Bohemia. Descartes was a good soldier who was not averse to taking risks on the battlefield. What is surprising is that he survived his military career at all. But Descartes' own interest in a military career soon changed. On November 10, 1619, Descartes experienced three vivid dreams that changed his life. Although some historians have wondered whether the dreams were inspired by his heavy consumption of wine, Descartes himself seemed to believe that the dreams were heaven-sent. Although the dreams contained a kaleidoscope of colorful images, the message that Descartes drew from them was that the key to understanding nature was to apply algebra to geometry and thereby use mathematics to unlock the secrets of the universe. Thus, analytic geometry was born and the modern era of mathematics began. Because nearly twenty years elapsed before Descartes actually published his ideas, however, the birthday party had only a single celebrant.

Descartes saw limited action on the battlefield during the next two years before returning to Paris in 1622, where he attempted unsuccessfully to secure a commission in the army. After a bloody campaign with the Duke of Savoy, Descartes left the military and spent the next three years contemplating life and nature in Paris. He did not hide from humanity like a hermit but instead spent his days on the streets of the City of Lights, and passed many of his evenings with some of France's most beautiful women. Descartes himself was the consummate gentleman, always attired in fashionable clothes. Indeed, his appearance gave little indication that he had any interest in any academic subject—let alone mathematics. But appearances can be deceiving and the dapper Descartes continued the mathematical investigations that had first attracted his attention while he was still a schoolboy at La Fleche. But once again, Descartes rebelled against the carousel of parties and dances and gratuitous love affairs, joining the King of France on the battlefield at La Rochelle. Following the successful conclusion of hostilities, Descartes journeyed to Holland, where he remained for the next twenty years. Here he tried to settle down for good.

He engaged in lengthy correspondences with many of the leading scientists and philosophers in Europe and began conducting his own intellectual investigations. He did not stay in any one place for very long, however, preferring to move from village to village, leaving it to his old friend, Father Marsenne, to forward his mail to him. But Descartes' physical wanderings around Holland only seemed to increase the diversity of subjects that he studied including medicine, optics, astronomy, meteorology, and chemistry. During this time, he also put the finishing touches on his treatise, *Le Monde*, which was to provide a scientific (and heliocentric) account of God's creation of the world. But news that Galileo had been turned over to the Inquisition and forced to declare that the Copernican system was heresy convinced Descartes to delay publication of *Le Monde* until his death. Descartes was genuinely fearful for his own life as his treatise had gone far beyond the works of Copernicus and Galileo in expounding a rationalist universe that did not require the incessant intervention of a divine entity. But Descartes' own strong feelings for the Church and the papacy itself dissuaded him from concluding that the religious authorities were mindless boobs who should be ignored. Faced with two irreconcilable positions, Descartes decided that the only solution was to delay publication until all of the interested parties had died.

Descartes might never have published his work in analytical geometry had it not been for the pleas of his many admirers. His masterpiece in mathematics, known by its abbreviated title as *La Méthode*, appeared in 1637 and immediately established the forty-one-year-old Descartes as one of the greatest mathematicians in Europe. According to W. W. Rouse Ball, "the great advance made by Descartes was that he saw that a point in a plane could be completely determined if its distances, say x and y, from two fixed lines drawn at right angles in the plane were given, with the convention familiar to us as to the interpretation of positive and negative values; and that though an equation $f(x, y) = 0$ was indeterminate and could be satisfied by an infinite number of values of x and y, yet these values of x and y determined the co-ordinates of a number of points which form a curve, of which the equation $f(x, y) = 0$ expresses some geometrical property, that is, a property true of the curve at every point on it." Although Descartes confined his investigations to plane geometry, his method showed how any point in space could be precisely located by the use of three coordinates. This graphical approach to geometry was revolutionary and, despite sniping remarks by some of Descartes' detractors that it involved little more than a reworking of the methods of Archimedes, this approach was of tremendous importance to the future growth of all mathematics.

Four years later, Descartes' *Meditationes de Prima Philosophia* appeared, laying out his philosophy of skepticism, which was of great importance to philosophy but of little relevance to mathematics because it recounted Descartes' interminable efforts to examine the same timeless, unanswerable questions that had confounded philosophers for thousands of years. Descartes then turned his attention to the physical sciences in the 1644 publication of his *Principia Philosophiae* and offered his own cosmology in which he suggested that all empty space was actually a continuous plenum. Although much of Descartes' theory of the universe was incorrect, he did provide a detailed analysis of the laws of motion as well as his own theory of vortices. So great was Descartes' influence in Europe that he was awarded a pension by the French government in 1647 in recognition of his scientific achievements.

Descartes' towering reputation attracted the attention of Princess Elizabeth of Bohemia who was living in exile at the Hague. Fluent in six languages and familiar with the leading classical writers, Elizabeth was

René Descartes (1596–1650) (Courtesy AIP Emilio Segrè Visual Archives)

something of an intellectual heavyweight in her own right; her voracious appetite for knowledge could not be satisfied. Filled with hope that Descartes might have the answers to all her questions about truth and wisdom, Elizabeth summoned him to the Hague and literally begged the reluctant philosopher to become her tutor. Descartes initially resisted Elizabeth's pleas but he eventually agreed to give her lessons. Soon thereafter, he began instructing his eager student in analytical geometry, philosophy, and physics. There is no evidence that Descartes, ever the proper gentleman, had a love affair with his infatuated student but his own raucous past precludes ruling out such a possibility altogether.

After Elizabeth left Holland, Descartes took up residence in Egmond, dividing his time between his mathematical investigations and puttering about in his garden. He also continued to stay abreast of the leading developments in the sciences. Despite his worldwide fame, he now lived a somewhat monastic life, even though the tranquility he sought continued to be interrupted by a continuous stream of visitors. Some might view his daily routine as being tedious or dull, particularly in view of the rambunctiousness of his earlier years, yet Descartes was reasonably content. But this satisfactory state of affairs was not to last, because Queen Christine of Sweden decided that Descartes must come to her kingdom.

According to Eric Temple Bell, "this somewhat masculine young woman [Queen Christine] was then nineteen, already a capable ruler, reputedly a good classicist, a wiry athlete with the physical endurance of Satan himself, a ruthless huntress, an expert horsewoman who thought nothing of ten hours in the saddle without once getting off, and finally a tough morsel of femininity who was as hardened to cold as a Swedish lumberjack." She was surrounded by fawning courtiers who told her that her judgment was infallible and that her knowledge was complete. Able to get by on several hours of sleep each night, Christine was a restless, headstrong young woman who seems—like Elizabeth before her—to have seized upon Descartes as the one who could quench her thirst for knowledge. Unfortunately for Descartes, he was unable to resist Christine's requests and so he left Holland for good in the fall of 1649.

Any illusions Descartes might have had about maintaining an easygoing lifestyle in which ample time would be available for mathematical and philosophical investigations were shattered soon after his arrival in Stockholm. Queen Christine, never one to wait for good things to come to her, immediately instituted a daily schedule which required Descartes to show up at the palace each morning at 5:00 AM. This schedule took a tremendous psychic toll of Descartes, who detested climbing out of bed before the morning sun had risen high above the horizon. The royal carriage invariably arrived at his home each day before he had roused himself from his sleep. Despite his disdain for his new job, however, Descartes could not bring himself to break off his teaching engagement. Perhaps it was his reverence for royalty or even the fear that Queen Christine would simply follow him back to wherever he went that caused him to continue this intolerable regimen. Sadly for Descartes, he soon fell ill and, after a short illness, died on February 11, 1650. Christine, for her part, was overcome with grief and refused to part with Descartes' remains for the next seventeen years; they were finally returned to France in 1667 where they were entombed in what is now the Pantheon.

Having discussed Descartes' life, we now turn to his greatest discovery—analytic geometry. But before we consider specific examples of the application of analytic geometry to physical phenomena, such as the gas law, we apply it to pure algebra where algebraic quantities that are functionally related to each other are pure numbers having no concrete significance. We can represent such functional algebraic relationships between two quantities y and x in the form of an equation in which y equals a polynomial in x; this functional relationship may be as simple or complex as we please, depending on how few or many terms in increasing powers of x we have in the polynomial. We can now plot this functional relation in our coordinate system by assigning various numerical values to x in the polynomial and calculating the associated values for y. By connecting all the points that these (x, y) numerical doublets represent, we obtain a curve which is a visual expression of the functional relationship. This graph is important, both from the purely mathematical and the empirical point of view because it shows us, at a glance, the functional relationship over a complete domain of x values. Although the graph may have been plotted only for a discrete set of x values, it defines the functional relationship for all intermediate values of x.

A few general observations and remarks about the functional relationships and the nature of the graphs that represent them in a given coordinate system will clarify any obscure features. Because the functional relationship between y and x becomes increasingly more complex as we increase the number of x terms, we obtain the simplest relationship, and therefore the simplest graph if we have no x terms at all. We may write this relation as $y = c$, where c is a definite number, such as 4. This expression states that

y does not depend on x at all; it has the same value c for all values of x. But the y value of a point in our coordinate system is the distance of the point above the x axis. This is a straight line parallel to the x axis and, in our example, is four units of distance above this axis. Although the graph in this coordinate system is expressed in terms of distances along the x and y coordinate axes, the actual functional relationship may have nothing at all to do with distances. The distances are then merely scale factors where an inch on the graph along the y axis may mean one physical entity and an inch along the x axis may mean an entirely different physical entity.

If we now add a term involving x to our functional relationship between y and x, we obtain $y = c + ax$, where a is a constant unchanging number such as 3. The functional relationship is now $y = 4 + 3x$. As x changes, y also changes so that the functional relationship is represented graphically by a straight line and we then say, in analytic geometry, that the functional relationship is the equation of a straight line or that y is a linear or first-degree function of x.

This concept is all quite elementary but it marked the beginning of a vast new branch of mathematics—the theory of algebraic equations or, more generally, the theory of functions of real variables. The mathematician is not at all concerned with what these functions and variables represent physically or concretely; he is interested primarily in deriving general theorems or relationships that are consistent with the mathematical constraints imposed on him. To the scientist, particularly the mathematical or theoretical physicist, the only important aspect of the deductions of the mathematician is their usefulness in furthering the physicist's research. The mathematical physicist himself may, of course, be charmed by the beauty of the mathematics and actually contribute substantially to its development. Some of the outstanding physicist–mathematicians include Newton, William R. Hamilton, Karl Friedrich Gauss, Hermann Weyl, and John Von Neuman, but owing to the rapid growth in the complexity of both mathematics and physics, this dualism is becoming increasingly more difficult to pursue.

Returning now to our simple linear function we see that we can change it dramatically by adding a term bx^2 so that the functional relationship becomes $y = c + ax + bx^2$, where c, a, and b may be any positive or negative numbers. We see at once that the term bx^2 alters the shape of the graphs from a straight line to a curve of some sort, which must either bend toward the x axis or away from it, depending on whether b is negative or positive. If the curve bends toward the x axis, it must continue moving toward it until it cuts the x axis and passes right through it; it then goes on never to turn up again. We have thus learned how to produce a functional relationship represented by a curve that has a single dip in it by adding a term such as bx^2 to the sum of x terms to which y is set equal.

Returning to the straight line $y = c + ax$ and following it along in our coordinate system we see that it must cut the x axis (the abscissa) at some point and that y equals 0 at that point. We then say that the function y has a "root" at that point (at the value of x). In this simple example, the root of our linear function can be found very easily by setting $y = 0 = c + ax$, or $x = -c/a$. Because the straight line can intersect the x axis only once, the function $y = c + ax$ has only one root; the straight line starts from infinity on the plane, cuts the x axis and then goes off to infinity, never to return. But by adding bx^2 (where b is any number) to our function we can make it turn toward the x axis and cut it again, making the function y equal to zero again. We thus obtain a function that is zero twice (for two values of x); it has two roots. The graph of the function then continues moving away from the x axis, never to cut it again. But if we add a term such as dx^3 (where d is any number) to the function we make the graph turn toward the x axis and cut it again, which means that the function has three roots. This result shows us the general relation between a function y and the graph of that function if the function can be expressed as a sum of increasing powers of the variable x. The graph (the plotted curve in our x, y coordinate system, that is in the x, y plane) must cut the x axis a number of times equal to the highest power of x.

The graph reveals two other interesting features of the function. As the plotted curve of the function in the coordinate system wiggles up and down with respect to the x axis, it creates hills and dales, like a road in a fixed direction. Mathematicians call the hills "maxima" and the dales "minima" of the function. From our description above we see that the number of hills and dales equals the highest power of x. Thus if the function contains the final term kx^5 (where k is any number) its graph must contain a total of four hills and dales (two maxima and two minima). The graph also teaches us that the curve of the function must approach and recede from the x axis from the same side of the x axis (from above it or from below it) if the maximum power of x in the sum of x powers is even. If this power is odd, however, the curve approaches and recedes from the x axis on opposite sides of this axis. These points illustrate how useful the graph of a function can be in revealing mathematical properties of a function that are not immediately apparent from the algebra alone.

The general features of the graph of a polynomial function of a variable are determined by the powers of x in the polynomial. For any specific polynomial, however, the numerical values of the coefficients c, a, b, d, etc., determine the exact shape of the curve (the heights of the hills, the depths of the dales, and where the curve cuts the x axis—the values of the roots). The following simple example illustrates the care we must exercise in interpreting the functional relationship in terms of the general discussion given above: $y = 1 + x + x^2$. According to our general discussions, this function only has two roots; the curve in our coordinate system describing this function should cut the x axis twice, but we see at once that no value of x, positive or negative, can make the function zero; it remains positive for all values of x. But we see that it does have a dale (a minimum) for $x = -\frac{1}{2}$, which is $\frac{3}{4}$ of a unit of length above the x axis. This means that the curve approaches the x axis, dipping down to a point $\frac{3}{4}$ of a unit of length above the x axis, then turns upward and recedes from the x axis. But where are the two roots of this function? They are not on the x axis because they are complex numbers. Indeed, we see that the two values of x, $\frac{1}{2} + (i\sqrt{3}/2)$ and $\frac{1}{2} - (i\sqrt{3}/2)$, make the function zero. The graph thus teaches us that if the curve of the function wiggles up and down with hills and dales, but does not cut the x axis in our chosen frame of reference (coordinate system) each such hill or dale represents a pair of complex roots. These roots are identical except that for one of these roots the real and imaginary parts are separated by a plus sign while for the other they are separated by a minus sign. Such pairs of numbers are called complex conjugates.

The most dramatic application of analytic geometry to mathematics, physics, and astronomy was to the algebraic representation of conic sections—the shape of the perimeter of a section obtained by cutting a right circular cone with a plane. A right circular cone may be obtained by rotating two intersecting lines in space around the line that bisects their angle of intersection. One thus obtains two cones attached to each other and mirror images of each other at the point where the two rotated intersecting lines (called generators of the cone) meet. Depending on the angle of intersection of the two cone generators, we obtain very narrow or broad cones, but the general mathematical principles that govern the relationship between the geometrical shape of the section obtained on cutting the cone and the algebraic equation that describes or defines the shape is the same. This is an excellent example of the way mathematics generalizes relationships and enlarges our understanding of such things as geometrical figures.

Mathematicians discovered that all conic sections can be represented by the general quadratic equation in x and y:

$$ax^2 + by^2 + cyx + dx + ey + f = 0,$$

where the coefficients a, b, c, d, e, and f are real numbers, whereas x and y are variables. This is a special form of the general functional relationship $f(x,y) = 0$. The conic section which this quadratic equation describes depends on the numerical choice of the coefficients a, b, c, etc. By making the appropriate choice of these coefficients, we obtain the equation of a circle, an ellipse, a parabola, or a hyperbola. This shows the power of analytic geometry to relate geometry to algebra.

With the discovery of conic sections and their application to planetary orbits by Kepler, mathematics began its irreversible penetration into physics with tremendous benefits to both physics and mathematics. As we shall see, mathematics permitted physicists to perform ideal or, as Einstein later described them, "gedanken" (thought) experiments using only pencil and paper; this was the beginning of "mathematical" or theoretical physics. When Kepler proposed the ellipse as the generic orbit of planets, he did so empirically or pragmatically; he had no theoretical basis for such a proposal yet it worked. Only after Newton discovered the laws of motion and gravity and expressed them algebraically did mathematics become an indispensable "branch" or adjunct of physics. This marriage of physics and mathematics was a happy union for mathematicians because it provided another vast intellectual domain for them to explore. Yet theoretical or mathematical physicists rarely thought of themselves as mathematicians even though they contributed significantly to the development of certain branches of mathematics. But mathematicians unabashedly called themselves mathematical physicists and, to some extent, still do so. Measured by their contributions to theoretical physics, those mathematicians are entitled to classify themselves as "part-time" physicists.

Kepler was a much greater astronomer than mathematician, whereas Descartes was a much better mathematician than physicist. In fact, Descartes' physics of the structure of matter, based on the mathematical theory of vortices, is all wrong. Newton, of course, was both a great mathematician and a great physicist, whereas Leibnitz was primarily a mathematician who contributed only peripherally to physics. But Newton did far more than just apply mathematics to physics; his formulation of the laws of motion and the law of gravity algebraically was a watershed in the evolution of physics and mathematics because it demonstrated the power of mathematical reasoning, when combined with basic physical principles, to reveal and elucidate the most profound mysteries of nature. Newton thus opened the floodgates of the pent-up ideas that until then had been purely speculative. Mathematics, as practiced by Newton, permitted one to formulate these speculations in a mathematical form which endowed them with vestiges of reality and made them part of the laws of nature. It is no wonder that the post-Newtonian era was marked by the contributions to science of some of the greatest mathematicians of all time.

Returning to conic sections, we note a few applications of these principles to some physical problems. In the seventeenth century, as we noted earlier, the physicist Robert Boyle, a contemporary of Newton, discovered the relationship between the volume V of a gas and its pressure P if the temperature of the gas is kept constant as its volume is changed. The plot of the relationship in a Cartesian coordinate system, in which the ordinate is the pressure P and the abscissa is the volume V, is exactly an equilateral hyperbola, which is expressed algebraically as $PV = C$, where C is a constant. This means that the pressure varies inversely with the volume. Each such curve is called an isothermal because the temperature of the gas is the same for every point on the curve. This curve thus gives us a picture of the changes in a gas if the temperature of the gas remains constant as the gas expands or contracts. For each different value of the temperature the isothermal curve is shifted, but all such curves are shifted parallel to each other. Each point on the P–V (pressure–volume) diagram represents a definite state of the gas, but the state of the gas cannot go from one point on a definite isothermal curve to any arbitrary point along another isothermal curve; only if the two points lie on the same equilateral hyperbola can the state of the gas change from one point to the other isothermally. We can, of course, connect any two points in this P–V coordinate system by an infinite number of paths and picture the state of the gas changing from point to point along any one of these curves. During any such change, the pressure, volume, and temperature will all change. This example shows us the usefulness of analytic geometry applied to physical problems.

This example of the application of a graph to the equation of state of a gas shows the great versatility of analytic geometry and how much it has evolved since Descartes first introduced it as a way of expressing geometrical entities algebraically. Because geometry deals with the spatial relationships among points, for example, distances, both coordinate axes were assumed to be distances when coordinate systems were first introduced, but this need not be so in the graphical representation of physical phenomena.

We now describe another application of a conic section—the parabola—to an important problem in telescope design. The history of astronomical telescopes begins with Galileo, who constructed the first astronomical telescope based on the design of the sixteenth century Dutch spectacle maker Hans Lippershey. The lenses in the earliest telescopes were constructed according to Galileo's design, but had serious flaws, called aberrations, which produced fuzzy images. The most important of these aberrations are spherical aberrations and chromatic aberrations. Spherical aberration causes the rays of light from a point that pass through the edge of a lens and those that pass through the center of the lens to be focused at different points. Chromatic aberration arises because the lens acts a bit like a prism, breaking up the light into its different colors, which are then focused differently, thus producing a colored border around an image. Newton, who contributed greatly to optics, turned his attention to telescope design and saw at once that using a concave mirror instead of a lens to form an image eliminates chromatic aberration because all rays of light, regardless of their color, are reflected in exactly the same way by a mirror. But if the concave reflecting surface of the mirror is perfectly spherical, spherical aberration is still present. If the shape of the mirror is parabolic (shape of a parabola), however, then spherical aberration is eliminated. Today all large optical telescopes are parabolic reflectors.

Soon after the introduction of two-dimensional analytic geometry (a coordinate system on a plane), three-dimensional analytic geometry (a coordinate system in space) was developed. In its simplest form the three-dimensional coordinate system is based on three mutually perpendicular axes, x, y, and z, passing through a single point 0 (the origin) in space. Each point in space is then located by assigning to it three numbers which give its distances from the three coordinate axes. Because only three mutually perpendicular lines can be drawn through any point in space, real space is three dimensional, but this did not prevent mathematicians from inventing purely abstract "spaces" of any number of dimensions. Three-dimensional analytical geometry led to another branch of mathematics, vector analysis, which has played a very important role in simplifying the mathematics of physics.

Descartes' introduction of the coordinate system, which led to the merging of algebra and geometry, was the first step in the growth of a mathematical discipline that influenced every branch of mathematics and physics in a way that Descartes and his immediate disciples could not have foreseen. This was particularly true of the impact that the coordinate system concept has had on the formulation of the laws of physics. To see how this discipline evolved from the primitive concept of the coordinate system introduced by Descartes, we consider what mathematicians call the transformation of coordinates. This transformation means that we may change our coordinate system in any manner we please and study how such a transformation changes the algebra that describes the intrinsic geometry of the mathematical relations being studied.

We consider here two kinds of changes that we can impose on a coordinate system—a change in its shape and a change in its position and orientation. By a change in its shape, the mathematician means either using curved or straight lines for the axes, but having them intersect at an angle different from 90°. Such a coordinate system is called an oblique Cartesian coordinate system. In fact, all straight-line coordinate systems are called Cartesian coordinate systems. The difference between a rectangular Cartesian and an oblique Cartesian coordinate system can best be illustrated by considering these coordinate systems laid off on a plane. The rectangular coordinate system may be pictured as two sets of equally spaced intersecting (at 90°) parallel lines, one set of lines parallel to the x axis and the other set parallel to the y axis. These intersecting sets of parallel lines divide the plane into equal squares. The rectangular Cartesian coordinate system is thus equivalent to covering the plane with a square mesh or grid. This leads to the concept of area which can be measured by counting the number of unit squares (one unit of length on each side) which cover the area defined by any four points on the plane. The square mesh or grid thus becomes a multiplication table.

An oblique Cartesian coordinate system can be created if we cover the plane with two sets of parallel lines that cut each other at an angle greater or less than 90°, thus giving a diamond mesh or grid. This grid covers the plane as well as the square grid but the algebraic relationships describing various geometric shapes or curves on a plane change their forms if we change from a square mesh to a diamond mesh. Yet this change forces us to ask whether anything remains unaltered (invariant) when we transform from one coordinate system to another.

Another type of coordinate system which has a shape different from that of the Cartesian coordinate system is the famous polar coordinate system, which consists of a system of straight lines radiating from a point 0 on the plane and a system of concentric circles having 0 (the origin) as their common center. In this coordinate system the position of any point on the plane is given by the circle it lies on (the distance of the point from 0) and the straight line it lies on (the angle this line makes with a fixed straight line such as, for example, the x axis). The great nineteenth-century mathematician Karl Friedrich Gauss extended the idea of coordinate systems that do not use straight lines by proposing two different sets of curved lines of any sort (e.g., parabolas or hyperbolas) that cut each other. This led to the most general way of representing geometry algebraically. A coordinate system of this general sort is called a Gaussian coordinate system; it plays a very important role in the theory of relativity which uses four-dimensional coordinate systems and incorporates the very important concept that the laws of physics must be expressed in the most general ways possible and must, therefore, be formulated in the most general kind of coordinate system.

We now turn to the discussion of the second type of transformation of a coordinate system—a change in the position and the orientation of a coordinate system. This is most easily understood and explained if we limit ourselves to rectangular coordinate systems on a plane. Because changing the position of a coordinate system simply means shifting the origin of the coordinate system from one point to another, such a transformation is called a translation of coordinates; it is the simplest and most basic of all transformations but is of great importance in physics. To see how any transformation alters the algebraic relationships or equations that describe geometric figures or curves on our plane, we need only determine how the transformation changes the coordinates of a point. To that end we consider shifting the origin 0 of our Cartesian coordinate system to a new origin $0'$, a distance h, to the right of 0 along the x axis. The x coordinates of all points are then reduced by the amount h, while all y coordinates remain the same. If we label the new x coordinates x' and the new y coordinates y', this simple transformation is expressed algebraically as $x' = x - h$ and $y' = y$. If we shift the origin of our coordinate system a distance k along the y axis, each x coordinate remains the same, but each y coordinate is reduced by the amount k. This translation is written algebraically as $x' = x$ and $y' = y - k$, where x' and y' are the coordinates in our translated coordinate system. The most general translation is one in which the origin 0 is shifted to any point $0'$ on the plane which is at a distance k to the right of 0 and k distance units above

0. If x' and y' are the coordinates of any point in the shifted coordinate system, the algebraic representation of this general translation is $x' = x - h$ and $y' = y - k$. These expressions are fairly simple but they form the basis for all transformations, regardless of their complexity.

Certain questions arose in connection with coordinate transformations which not only greatly enriched mathematics and physics but led to new branches of mathematics, the most important of which is called the theory of groups. For now, however, we consider the algebraic relationships and geometric properties of figures and curves that are unaltered when we transform our coordinate system in any way. Such entities, whether geometric or algebraic, are called "invariants," to which a whole branch of mathematics—the theory of invariants—is devoted. Invariants are also very important because they must correspond to intrinsic aspects of nature and therefore must represent the laws of nature in some way or other. A translation of a coordinate system cannot change geometric figures, alter distances between points, change angles between lines, alter the shape of a curve, or the intrinsic functional relationship between the independent variable x and the dependent variable y which is defined as a function of x.

The translations of a coordinate system constitute the simplest of what mathematicians call a group. Because group theory has played (and still plays) a very important role in mathematics and physics, it is useful to consider it and see how it stemmed from the translations of coordinates. To this end, we picture a translation of a coordinate system or a physical operation which lifts the coordinate system bodily and shifts it to a new position. The same result may be achieved by moving to a different point and setting up an identical coordinate system at this new point. If we represent this operation symbolically as T_1, we may then picture many such operations (in this case, an infinite number) and collect them into an ensemble which we label as a group provided we apply certain restrictions to them. The first of these restrictions is that this ensemble or aggregate must contain the unit or identity translation T, which leaves the coordinate system unchanged. The second restriction is that for every translation T in the group, the inverse translation $1/T$ or T^{-1} must be present. Finally, the elements of the group must obey a certain product rule, namely that the product of any two elements of the group must itself be an element of the group.

These rules may be presented symbolically to show the great importance of the group and illustrate its algebra. In limiting ourselves to a translation, which we write as T, the group operation is $TS = S'$, which tells us that if we apply the operation (translation) T to a coordinate system S, we obtain another coordinate system S'. If T_i is the identity transformation, then $T_iS = S$. The inverse translation is $S = T^{-1}S'$ and the product of two translations T_1T_2 is itself a translation: $T_1T_2 = T_3$. A vast branch of mathematics has evolved from these simple rules which is associated with the names of many great mathematicians.

To pursue the concept of the coordinate transformation a bit further, we consider those features of a graph or the algebraic representation of a geometry which a transformation does not change. These are the invariants of the geometry and represent the intrinsic entities that remain the same in all coordinate systems. The most important of these entities for mathematics and physics is the formula for the distance between two points—the Pythagorean theorem.

Thus far we have been discussing only the translation of a coordinate system. Now we consider briefly the rotation of a coordinate system in which the origin of the system itself is fixed but the x and y axes are rotated through a certain angle. This rotation changes the algebraic formulas of all the curves. A simple example illustrates the basic features of these changes. An equation such as $y = 4$ represents a straight line parallel to the x axis and four units of distance above it. If we rotate our axes, the line will then cut the new x axis (called x') so that the equation of the line becomes $y' = ax' + b$ (where a and b are numbers that depend on the magnitude of the angle through which we rotate our coordinate system) in the rotated coordinate system x', y'. All of the algebraic changes that a rotation of a coordinate system produce can be deduced from the changes that the rotation produces in the coordinates of a point. If x, y are the coordinates of a point in the unrotated coordinate system and x', y' are the coordinates of the same point in the rotated system, then x' depends on both x and y and the angle of rotation and y' also depends on x and y, and the angle of rotation.

This brief discussion of coordinate transformations does not cover adequately the great importance of such transformations for mathematics or for physics. The laws of physics are expressed algebraically as functional relationships between two different physical entities, as, for example, the dependence of the force of gravity between two masses on the distance between them. Having some sieve or mathematical device that can separate laws from false mathematical statements about the universe is extremely important. The transformation of a coordinate system performs this task so that the laws of nature are those algebraic functional relationships among physical entities that do not change when we transform our coordinate system in any way we may desire. In fact, this concept of invariance of a coordinate transformation is the basis of the theory of relativity. This concept led to the discovery that mathematical laws must possess a certain mathematical symmetry. Thus symmetry in nature and invariance go together. The concept of invariance led to still more profound interpretations of basic laws; they express the conservation of certain basic entities such as energy and momentum. All of these concepts will become clear when, in later chapters, we develop the relationships between mathematics and science more fully.

CHAPTER 6

The Birth of the Calculus

As God calculates, so the world is made.

—GOTTFRIED WILHELM LEIBNITZ

The history of ideas is replete with examples of remarkable breakthroughs that have drastically altered the direction, methodology, and the conceptual basis of a discipline or a philosophy. Such breakthroughs in the past seem to have come at just the right time to fulfill important intellectual needs. The calculus is a prime example of such a fortuitous coincidence in both mathematics and science. Without the calculus, much of theoretical physics, chemistry, and astronomy would have never evolved. The remarkable aspect of the birth of calculus is that it occurred almost simultaneously in theoretical physics and in pure mathematics, which led to one of the most famous controversies in the history of mathematics, and to the most heated intellectual rivalry between two nations.

Because the relationship of the origin of the calculus to the evolution of theoretical physics is somewhat easier to trace than its relationship to the evolution of pure mathematics, we consider the calculus–physics connection first. This combination stemmed almost entirely from the work of Isaac Newton (1642–1727), who initiated theoretical physics when he expressed the laws of motion and the law of gravity algebraically. Until Newton's great discoveries, it never occurred to "scientists" that mathematics could be used to express basic principles about nature or the universe itself. Geometry, of course, had been applied to represent positional relationships among objects in the universe because space itself is geometrical; it was therefore quite natural to apply Euclidean geometry to describe spatial relationships, as Kepler did in deducing his three laws of planetary motion. Kepler went beyond pure geometry in his description of planetary motion. In proposing the ellipse as the natural orbit for planets, Kepler introduced trigonometry as well as geometry into science, and, in expounding his second and third laws, he introduced algebra as a powerful analytical tool to describe the motions of the planets.

One may wonder then why Kepler's discoveries—unlike those of Newton—were not the "launching pad" of theoretical physics. Kepler's discoveries, though much more profound and significant than those of Copernicus, and couched in fairly sophisticated mathematics, did not expound or otherwise incorporate fundamental physical principles. In other words, Kepler's laws did not present a mathematical methodology for solving problems of any kind. These laws were promulgated only to describe the orbits and motions of the planets. Kepler deduced his laws empirically from the observations of Tycho Brahe, not mathematically from basic physical laws. Newton's discoveries, on the other hand, were purely theoretical and entirely mathematical in form so that mathematics, in a sense, became part of physics. Mathematics became the pathway to the presentation of natural laws in precise mathematically manageable forms, and the tool for manipulating these laws to reveal features and aspects of nature that are not immediately evident from the simple expressions of the laws themselves. The reason mathematics has this power to extend one's knowledge beyond the bare mathematical statement of the law itself is that one may manipulate the law mathematically in any way one may desire as long as no mathematical or physical principle is violated. These manipulations alone can (and do) reveal new features and new avenues of theoretical and experimental research. They also made it possible for Newton to discover the calculus.

Newton's laws of motion and gravity—which are both simple algebraic formulas—form the basis of classical physics and led Newton to the calculus. The difference between what is often called Greek science and modern science is contained in Newton's second law of motion, which opened the gates of science to mathematics and mathematicians. Newton's second law is important because it establishes the mathematical relationship between the motion of a body and the force acting on the body, which Aristotle and later Aristotelian physicists did not understand. Indeed, the nature of motion and its relationship to force was misunderstood by the Aristotelians, who wrongly believed that forces are required just to keep bodies moving at constant speeds.

The first step in separating fact from fiction in describing motion was taken by Galileo in the seventeenth century when his experiments on moving bodies clearly demonstrated that two or more bodies left alone (no forces acting on them) maintain their states of motion forever. This result introduced the concept of inertia into physics. Inertia is that property of a body (we now call it mass) which resists any change in the body's motion whether it is at rest or moving at constant speed in a straight line. Yet nobody knew what to do with this important discovery. When Galileo died in 1643, his discoveries about motion had not yet been expressed as laws in any mathematical form. Some twenty years later, however, Newton formulated his three laws of motion and thus clothed the concept of inertia in mathematical garb.

Before discussing Newton's laws of motion and his law of gravity in greater detail, we recall the circumstances of his life and the environment in which he made his pivotal contributions to science. Isaac Newton was born on Christmas Day, 1642, in the hamlet of Woolsthorpe, in Lincoln County, England. The son of a farmer, also named Isaac Newton, who had died before the baby was born, and Hannah Ayscough, little Isaac was so underweight at birth that he was not expected to live through his very first day. Indeed, he was so puny that he reportedly could have fit completely inside a quart jug. Despite this unpromising beginning, however, young Isaac survived. Soon afterward, Isaac's mother—still in mourning for her deceased husband—became engaged to the Reverend Barnabas Smith, a resident of the neighboring North Witham parish. The two were married when Isaac was three years old. The new Mrs. Smith moved to her husband's home, leaving Isaac in the care of his grandmother. Although Newton's mother had three more children with Mr. Smith, she never invited young Isaac to live with her. We can only speculate as to how this rejection by his own mother affected Newton psychologically, particularly when he reached adulthood. But it seems clear that the first symptoms of this estrangement saw the boy become much more withdrawn and distrustful of others. Perhaps it also contributed to his unwillingness—even after he became the most famous scientist in the world—to acknowledge freely the valuable contributions of his colleagues, most notably Leibnitz and Huygens. As a boy, however, Isaac was a loner. His poor health prevented him from engaging in rough and tumble sports with the other children, so he amused himself by designing and building a variety of clever, sometimes ingenious, contrivances and devices including mechanical toys with moving parts, sundials, and clocks. He also took an unusual interest in the world around him, filling up numerous notebooks with his observations of the nearby lands like some modern-day Aristotle attempting to catalog all the flora and fauna of nature.

But for his maternal uncle, the Reverend William Ayscough, Newton might have passed the rest of his life on the family farm and never made his remarkable discoveries in physics and mathematics. Ayscough had first noticed Newton's academic promise when his nephew was attending the Grantham Grammar School. Although young Isaac had initially shown little interest in his subjects, the encouragement of one of his teachers had stoked the boy's competitive fires, causing him to focus on his subjects with the same intensity of concentration that he would devote to his later work in mathematics and physics. Newton soon surpassed his schoolmates and convinced both the headmaster and his uncle that he was meant for better things than cleaning stables and working in the fields—namely, a university education at Cambridge.

Before leaving for Cambridge in 1661, Newton became involved in perhaps the only love affair of his life. The object of his affections was the daughter of the village apothecary. The two became engaged in the spring of 1661 but their romance cooled after Newton enrolled at Cambridge when distance and time conspired to snuff out the passion between them. Despite the loss of the only woman whom he would ever love, Newton retained a fondness for her that continued until her death, even though the incessant demands of his work tended to dilute his admittedly feeble romantic inclinations.

Upon arriving at Cambridge, Newton immersed himself in his studies and found a mentor in Isaac Barrow, the Lucasian professor of mathematics. Barrow saw glimmers of promise in his student but he was not convinced that Newton was more talented than all of the other hardworking students at Cambridge. Barrow's lectures in mathematics did inspire Newton to think about the problems that would eventually lead him to the discovery of the calculus. Although Newton, like everybody else at Cambridge, studied the classics, he preferred mathematics and physics, finding them more challenging than the literary works of antiquity. Newton's own preferences were doubtless bolstered by his own aptitude for mechanics but he did not make any important discoveries or conduct significant research as an undergraduate. He did spend some time at the taverns, drinking and playing cards with some of his classmates. Newton also acquired an interest in astrology, a curious hobby for the man who would do more than any other to create the image of the rational, mechanistic universe.

This tranquil if uneventful time in Newton's life was interrupted by an outbreak of the bubonic plague in 1664, forcing the closing of the University. Newton returned to Woolsthorpe for the next two years, while the plague

Isaac Newton (1642–1727) (Permission of National Portrait Gallery; Courtesy AIP Emilio Segrè Visual Archives)

ravaged much of England. But Newton's self-imposed isolation provided him with the time to contemplate nature and think about the forces that govern the universe. In those two years, Newton made several of the most important scientific and mathematical discoveries of all time—the binomial theorem, the calculus (which he called "fluxions"), and the law of universal gravitation. He also performed several experiments with glass lenses which enabled him to prove that white light is really a composite of many different colored bands of light; this conclusion led him to propose a corpuscular theory of light. Although Newton undoubtedly realized that his work was revolutionary, he did not attempt to publish his findings right away. Indeed, a 1665 manuscript is one of the few pieces of evidence of the work that Newton did on the calculus at Woolsthorpe; its author made no attempt to establish his priority until many years later when the supporters of Newton's rival, Leibnitz, began claiming that the German had been the first to discover the calculus.

When Newton returned to Cambridge in 1667, he was elected a fellow of Trinity College. He spent the next two years lecturing to students and assisting Barrow with his teaching duties. In 1669, Barrow, having learned of Newton's work at Woolsthorpe, resigned his chair in favor of his young student, then only twenty-seven years of age. The faculty was aghast at the prospect of having such a young man hold one of the most prestigious professorships in the world and would not hear of Barrow's request. But it soon became clear to those members of the faculty who could actually understand Newton's ideas that he was a truly gifted individual who had made several revolutionary discoveries. Thus mollified, the faculty adopted Barrow's recommendation and Newton assumed the chair in the fall of 1669.

The next two decades of Newton's life were taken up by teaching and administrative duties. Newton developed his corpuscular theory of light, defending it vigorously against the proponents of the undulatory, wave theory of light, originally offered by Christian Huygens. Newton also built a reflecting telescope and made extensive observations of the moons of Jupiter. In 1672, Newton was elected to the Royal Society, based on his work with his telescopes and his theory of optics. The Society knew next to nothing about the calculus or his law of universal gravitation because Newton had limited the circulation of the few papers he had written on those subjects to a handful of acquaintances.

Over the next few years, Newton became more deeply interested in chemistry, particularly the idea that base metals could be transmuted into gold. He also learned firsthand about the politics of science, engaging in lengthy correspondences from 1672 to 1675 to defend his theory of light against the often vicious attacks mounted by Robert Hooke and several other advocates of the undulatory theory. Newton's initial reaction was to respond in detail to each of the objections raised by his critics. When it became clear that many of their comments were motivated more by spite than by interest or curiosity, however, Newton became more reluctant to continue responding to his detractors. Indeed, he became rather sullen and introspective, resolving never again to become embroiled in such petty controversies.

Only in 1684 did Newton begin to prepare a treatise on his theory of gravitation; this was due to the dogged insistence of Edmund Halley—the astronomer who is best remembered for the discovery of the comet that bears his name—that Newton's work should be made available to all scientists. For the next two years, Newton devoted almost all of his waking hours to working out the mathematical intricacies of his masterpiece, the *Principia Mathematica*. When it was finally completed, the *Principia* described in laborious detail how a complete dynamical system based on Newton's law of gravitation could be formulated to explain the motions of the heavenly bodies. In it, Newton demonstrated how Kepler's laws could be derived from the law of gravitation. Newton also showed how the orbit of any single heavenly body, such as the moon, is affected by the gravitational pulls of neighboring bodies, such as the sun and the earth. He also explained such things as the orbits of the comets, the flattening of the earth at its poles, the precession of the equinoxes, and the rising and falling of the tides. Indeed, Newton's *Principia* brought humanity's understanding of the universe from the darkness of ignorance and theologically inspired speculation to the light of pervasive universal principles that can explain the falling of an apple on earth or the movement of a planet around the furthest sun.

So awe-inspiring was the sweep of the *Principia* that most of Newton's colleagues were unable to comprehend fully its true worth. According to Eric Temple Bell, "it is to the credit of Newton's contemporaries that they recognized at least dimly the magnitude of what had been done, although but few of them could follow the reasoning by which the stupendous miracle of unification had been achieved, and made of the author of the *Principia* a demigod." But it was not too many years before the sheer brilliance of Newton's vision won over Britain's scientific establishment. Acceptance on the continent was somewhat delayed by the lingering vestiges of Descartes' theory of vortices but that timeworn concept was also destined to succumb to the mighty thrust of Newtonian physics.

Newton spent the better part of three years preparing the *Principia* for publication, even though Halley himself had generously paid the cost of the publication and reviewed all of the proofs. Once the task was completed in 1687 and his place in history as perhaps the greatest physicist and mathematician of all time was firmly established, Newton seemed to be physically and mentally exhausted. Indeed, he claimed that he no longer wished to pursue physics. Soon afterward, Newton was drawn into the treacherous swirl of national politics when he agreed to accompany several representatives of Cambridge University to London to protest an attempt by King James II to force Cambridge to grant a degree to a priest who refused to take the oath of allegiance. Newton was instrumental in convincing the university officials not to sign a compromise agreement that he believed would have disgraced the institution. Newton's political courage did not go unrecognized as he was elected to represent the university in Parliament in 1689 after James's abdication. Any expectations held by the other members of Parliament that Newton would demonstrate a political acumen to rival that of his scientific abilities were soon dashed. During the year in which he sat in Parliament, Newton never participated in a single debate. Indeed, the one time he did speak—an event which caused a hush to fall over the room—was to ask that a window be opened to let in the breeze.

Upon his return to Cambridge in 1690, Newton seemed less interested in mathematics and science than ever before. He continued to correspond with other scientists but his research was halfhearted. He still retained his ability to concentrate on a problem with unmatched intensity until he had solved it, but his great intuition for physics and mathematics seemed to be eroding. Perhaps it was to be expected that Newton should slow down; after all, he was now nearly fifty years old and the sciences are supposed to be a young man's game. Any attempts Newton might have made to rouse himself from his slumber were further delayed by a debilitating illness that nearly immobilized him for two years. Although he apparently made a complete recovery by 1695, Newton was even less enthusiastic about original research and, indeed, did little new physics.

But it was Newton's appointment as warden of the mint in 1696 that ended his scientific career for good. Although some of his earlier works, most notably his investigations into the nature of light would be subsequently published (*Optics*, 1704), Newton's decision to become a civil servant and attack the problems of the nation's coinage rather than the gravitational perturbations of the solar system marked the end of his work at Cambridge. Two years after being appointed Master of the Mint in 1699, Newton resigned the Lucasian chair at Cambridge. In 1703, he was elected president of the Royal Society. He became Sir Isaac Newton in 1705 when he was knighted by Queen Anne. Although Newton had "retired" from science, however, the subsequent decade saw him become embroiled in the famous controversy with Leibnitz over the discovery of the calculus. He also wrote immense but virtually worthless tracts on theology and alchemy, revealing his obsession with the mystical side of nature.

Despite his flirtations with astrology and alchemy, however, Newton was a very religious man. He was honest and straightforward even though he was sometimes provoked into petty arguments by some of his more virulent adversaries, such as Hooke. Newton was also quick to acknowledge his debt to his predecessors, particularly Galileo and Kepler, whom he considered to be the first true scientists. Although he was doubtlessly aware of the unprecedented ways in which his work changed the way humanity viewed the universe, he was reluctant to make extravagant claims on his own behalf. This modesty was reflected in the following remark: "I do not know what I may appear to the world; but to myself I seem to have been only like a boy, playing on the sea-shore, and diverting myself, in now and then finding a smoother pebble or a prettier shell than ordinary, whilst the great ocean of truth lay all undiscovered before me." Newton was also a very generous man, who could always be counted on to give generously to the poor.

Like many other persons of brilliant intellect, Newton was also absent-minded. According to W. W. Rouse Ball, "once when riding home from Grantham [Newton] dismounted to lead his horse up a steep hill; when he turned at the top to remount, he found that he had the bridle in his hand, while his horse had slipped it and gone away." Newton was also known to invite guests over for supper but then forget all about them after becoming immersed in solving a problem.

Although Newton stopped making new scientific discoveries in his middle years, he remained the preeminent mathematician in the world. Two mathematical problems posed by Johann Bernoulli and Leibnitz in 1696 stumped the finest mathematicians in Europe for six months. Newton, having learned of the problems after an exhausting day at the mint, solved them both while taking his supper. Although he submitted his results to the Royal Society anonymously, the author's identity was clear to all who read the answer. The same thing occurred in 1716 when Leibnitz posed a problem that he fervently hoped would humble Newton. But Newton solved Leibnitz's problem in a few hours.

By the time Newton died on March 20, 1727, he was the most highly honored scientist of all time. His name was synonymous with scientific achievement and he had almost single-handedly elevated British science to a position of supremacy (even though his English successors would soon see their achievements fall behind those of their French counterparts). Newton also brought order and predictability to the universe and established a scientific legacy which, despite more recent modifications necessitated by Einstein's theory of relativity, has continued to dominate much of physics to the present day.

Of particular importance to Newton's scientific legacy were his three laws of motion. Because only the second of Newton's three laws of motion is expressed as a mathematical equation, it is both the simplest and most profound of his three laws. From this simple equation, one can deduce a number of important concepts which have dominated all branches of physics. Because the law is so simple, it is puzzling that some twenty centuries elapsed from the time of Aristotle before the law was discovered. Newton was the first scientist to recognize the importance of the dominant role of acceleration—as opposed to velocity—in the motions of bodies.

As acceleration is the time rate of change of the velocity of a body (the change per unit time), Newton saw that the basic law of motion of a body must relate the acceleration of the body to whatever quantity it is that changes the velocity of the body. Galileo had shown that the velocity of a body cannot change by itself; it requires an agent of change, which Newton called force, and so Newton's law of motion relates force to acceleration in a very simple and elementary way. But a third quantity, which is a physical property of the moving body must enter the law as well. This is the mass (inertia) of the body. Just how these three physical entities are linked together algebraically to express a law is fairly easy to deduce from very general physical considerations and from our own experiences with motion— walking, running, driving automobiles, playing physical games, etc.

Our own experience shows us that setting an automobile in motion requires a greater force than setting a small cart in motion, because the automobile is more massive than the cart. Thus the force required to move the automobile depends directly on the mass of the body being moved. Newton might have written the first step in the derivation of the law of motion as $F = m(?)$, where F is force, m is mass, and the question mark (?) stands for some unknown factor that must multiply m to complete the law as an algebraic equation or formula that correctly describes the motion of a body of mass m under the action of a force F. The only factor available to us is the time rate at which the velocity of the body changes (the acceleration) under the action of the force. If we call this acceleration a, then the factor (?) is a and we obtain one of the great laws in nature $F = ma$. This law is deceptively simple in appearance and illustrates in a very elegant manner the power of mathematics to reveal new aspects of natural phenomena. That the force required to alter the state of motion of a body depends on the mass of the body and on how rapidly its motion changes (acceleration) is clear from our direct experience, but its great simplicity and generality suggest its correctness even if we had no direct experience with motion. Why did Aristotle fail to see this simple relationship, given the fact that he knew about force and acceleration? Aristotle probably did not associate these concepts with the concept of inertia so that he had no reason to relate force to the product of the mass of a body and its acceleration. Indeed, the concept of mass, as distinct from weight, two different physical entities, was not understood by the Greeks so that the equation $F = ma$ escaped their attention.

A detailed study of this equation shows how manipulating it algebraically reveals new truths and leads to new physical concepts that have greatly expanded our understanding of natural phenomena. This law is of a very general nature; it does not refer to any particular force but applies to any force in nature. The law cannot be used to describe the motion of a body unless the nature of F is specified. Either the magnitude of the force must be given numerically for each case of the motion of the body or the force must be expressed algebraically so that its magnitude can be calculated for any special set of conditions.

An equation which expresses a law of nature must always relate physically identical things to each other. Since the left-hand side of the equation $F = ma$ is a force, the product ma (mass times acceleration) must also be a force, regardless of the physical situation in which this product appears. But this equation raises a conceptual problem which is cloaked in an aura of mystery. We all understand the meaning of force because it is related to our muscular responses and we also comprehend the nature of acceleration because we have learned about it through our daily experience with motion. Yet what is mass? Herein lies the mystery expressed by the law $F = ma$. In all the years that scientists have studied the properties of matter in all its forms, they have not been able to answer this question satisfactorily. Fortunately, we do not need to know the answer to this question to solve the various problems that are presented by the motions of bodies. Although we do not know the nature of mass, we know its two most important properties for motion: It is a source of and responsive to gravity and it possesses inertia—resistance to any change in its state of motion. But now, in the algebraic equation that expresses the law of motion, we have a way of measuring the mass of a body. This illustrates the great usefulness of mathematics in science. To see how the algebra leads to the measurement of mass, we rewrite the equation as $m = F/a$, which expresses mass as the quotient of two measurable quantities, force F and acceleration a. This equation tells us that the mass of a body is the number we get when we divide the force acting on a body by the acceleration imparted to the body by the force. Before we can carry out this simple arithmetic, we must introduce units for force and units for acceleration, which, when divided into each other, give units of mass.

The equation of motion can be rewritten in still another form which emphasizes the role of the acceleration, namely $a = F/m$, which tells us at once that if the same force is applied to two different bodies, the one with the smaller mass acquires the larger acceleration. These deductions from this simple law reveal almost all we have to know to solve even complex dynamical problems, but a few other features of the law are notable. This law contains all three of Newton's laws of motion. If $F = 0$ (no force acting on a body), for example, the right-hand side ma is also 0 so that the acceleration a is also 0 and the body must be moving at constant velocity. This is Newton's first law of motion, the law of inertia. We can also deduce Newton's third law of motion from $F = ma$. This law may be better understood by picturing a person who is standing still. Being at rest, he experiences no acceleration. If he suddenly moves to his left or his right, however, his state of motion changes from rest to a velocity to his left or to his right, which means that he was accelerated. But the law $F = ma$ tells us that he must have been pushed to the left or right. Yet this answer seems queer because the only body that could have pushed him was the earth itself. But the earth did push him because the man first pushed the earth. This is known as the law of action and reaction because the man pushes the surface of the earth with his foot and the surface itself reacts by pushing the man's foot with exactly the same force in the opposite direction. We have thus deduced Newton's third law of motion from a simple algebraic equation.

The importance of mathematics to physics is that once we have discovered the mathematical form of a law, any mathematical deduction we make from that law is as valid as the law itself. Mathematics thus leads us to new discoveries which are not immediately evident in the law itself. A simple example illustrates the "truth-discovering" power which mathematics gives to us. We apply a constant force F for a time t to a body of mass m so that it moves a distance d while the force is acting on it. The motion, of course, is governed by the law $F = ma$. We now multiply both sides of the equation by the distance d that the body moves: $F \times d = ma \times d$, that is mad. $F \times d$ is the work done on the body by the force F. What is the right-hand side, the product mad? If v is the final speed of the body, and it started from rest, then $v/2$ is its average speed, so that $d = (v/2)(t)$, and $a = v/t$. The equation of motion now becomes $Fd = m(v/t) \times v/2(t) = (\frac{1}{2})mv^2$, which is called the kinetic energy of the body. Elementary algebra has thus led us to the new but very important concept of energy and the equation of motion now tells us that the work done on a body equals the kinetic energy the body acquires. This is essentially a statement of the principle of the conservation of energy, a concept that permeates all of science and our daily activities.

There is one other feature of the laws of motion which ultimately led to a new branch of mathematicas called vector analysis. As an algebraic equation, $F = ma$ has no spatial attributes; it may refer to any three quantities, one of which equals the product of the other two. But if these quantities are related to the motion of a body, space and direction are involved. Force and acceleration are directional quantities because we can exert a force in any one of the three independent directions that describe or define space. To be explicit we describe the motion of a particle in a three-dimensional rectangular Cartesian coordinate system, defined by three mutually perpendicular directions x, y, and z so that the motion of a particle must be represented by a vector, which combines both the magnitude (quantity) of the motion and its spatial direction in a single entity, a vector. The acceleration of a body is thus a vector. Because the force acting on the body that produces its acceleration is also a vector, Newton's law of motion is really a vector equation because it relates two vectors to each other and tells us not only that the force acting on a body equals the product of its mass and its acceleration but also that the force acting on the body and the acceleration are in the same direction.

When we discuss vector mathematics in more detail, we will point out the simplification and brevity produced by vectors in the mathematical presentation of the laws of nature. Here we note that Newton's second law of motion—owing to its vector property—is really a combination of three equations of motion which we obtain immediately if we describe the motion of a body in a three-dimensional Cartesian coordinate system. If x, y, and z represent the three orthogonal (mutually perpendicular) Cartesian coordinate axes, we may consider the motion of the body as a vectorial sum of the separate motions along the x, y, and z axes, which are independent of each other. If the components of the force F in the x, y, and z directions are F_x, F_y, and F_z and the independent accelerations are a_x, a_y, and a_z along the x, y, and z axes, respectively, then Newton's law $F = ma$ becomes the three equations $F_x = ma_x$, $F_y = ma_y$, $F_z = ma_z$. The practical importance of this concept is that the study of the motion is enormously simplified by breaking the motion up into three independent motions which do not affect each other. Newton did not know the vector concept, which emerged roughly two centuries later, but he did know that force and motion can be expressed as sums of independent components. All of these ideas follow from the concept of the coordinate system and the dimensionality of space.

Newton's law of motion alone cannot be used to describe the motion of a body because it is merely a general mathematical statement about the relationship between force and motion; it applies to the motion produced by any kind of force acting on a body. We cannot apply the law to a moving body until we know the mathematical nature of the force acting on the body. The law of motion then becomes a manageable algebraic equation which we can solve. Newton's great genius lay not only in his discovery of the laws of motion but also in his discovery of the mathematical expression for the gravitational force. This enabled him to apply his law of motion to determining the orbits of the planets on the assumption that the planets move in response to the gravitational pull of the sun. Newton did not write down the law of gravity for the pull of the sun on a planet directly, but rather the formula for the pull of one mass on another under the assumption that each mass is concentrated in a point. This assumption is, of course, an idealization because point masses do not exist, but it greatly simplified Newton's task because it eliminated the need to deal with the dependence of the force of gravity on the sizes and shapes of the two interacting bodies. Therefore, Newton's reasoning which led him to his law of gravity is easy to follow.

A force formula of any kind must depend on two different features of the interacting particles. The extrinsic feature is the distance between the two particles and the intrinsic feature is some physical property of the particles that generates the force. The appearance of the interparticle distance in the formula for the gravitational force between the two particles tells us that Newton chose particles of zero size (mass points) because only for such particles is the distance well defined. For real bodies we cannot speak of the distance between them in a unique way because an infinite number of different such distances exists, depending on which two points in the bodies we choose. Because the gravitational force exerted by a mass point on any other mass point spreads out equally in all directions like light from a point source, it weakens with the square of the distance between the two mass points. If r is the distance between the two mass points, the algebraic formula for the gravitational force between them must have r^2 in the denominator. For this reason Newton's law of gravity is called an inverse square law of force.

The intrinsic feature of the law of gravity is a physical property of the gravitationally interacting particles. This is some algebraic combination of the particle masses because masses generate gravity. But only the product of the masses will do because the formula must be completely symmetrical between the two masses, as required by the law of action and reaction: the pull of body 1 on body 2 is exactly equal and opposite to the pull of 2 on 1 (the numbers 2 and 1 are just identification symbols). If m_1 is the mass of body 1 and m_2 is the mass of body 2, then the only algebraic combination of these masses that has the required symmetry demanded by the correct law is the product m_1m_2 of these masses. Newton had the physical insight and genius to see this and to state his famous law of gravity accordingly: the gravitational force between two bodies is proportional to the product of their masses and inversely proportional to the square of the distance between them. The word "proportional" means that the force does not equal the algebraic expression m_1m_2/r^2, rather it varies as this expression varies. Indeed, this expression, by itself, cannot be placed equal to the gravitational force because it does not have the characteristic of a force which involves mass, space, and time. The formula m_1m_2/r^2 contains mass and space but not time. Newton multiplied the expression m_1m_2/r^2 by a universal constant G which itself is just the right combination of mass, space, and time to make the total expression Gm_1m_2/r^2 a force. Newton thus wrote his law of force as $F = Gm_1m_2/r^2$, where G is a constant that was measured a century later by the British physicist Henry Cavendish. Its numerical value is 6.7 divided by 100 million, written as $6.7/10^8$ and its dimensional formula, expressed in mass, length, and time units is length^3 per mass per time^2 or L^3/MT^2. G is not a pure number like π but a dimensional quantity that is the same everywhere and for all time.

Newton's next step in this development was to combine this law of gravity with his law of motion and to deduce from it the orbit of a planet around the sun. This led him to the calculus.

To follow Newton's reasoning we consider the specific example of a planet moving around the sun, kept in its orbit by the sun's gravitational pull.

Kepler deduced the orbit of Mars empirically by plotting its distance from the sun, for equal time intervals, using Tycho Brahe's observational data, and then drawing a smooth curve through all the plotted points. He thus obtained an ellipse and made the bold and brilliant generalization that all the planets move in elliptical orbits around the sun. Newton set himself the task of deducing the elliptical planetary orbits mathematically from his law of gravity, but he saw that this was far more complicated and difficult than solving a simple algebraic equation like his law of motion.

The law of motion does give us the planet's acceleration for any distance of the planet from the sun if we picture a planet now at a certain moment in its orbit, moving with a definite velocity and at a definite distance from the sun. His laws of motion and gravity gave Newton the planet's acceleration at that moment, but from that alone he could not possibly deduce all future positions of the planet—its orbit—because the planet's distance from the sun and its velocity change continuously from moment to moment so that one cannot predict the planet's positions over a finite time interval.

Newton's great mathematical genius led him to the concept of the infinitesimal time interval. One can solve the problem of the planet's orbit—not by trying to obtain a picture of the complete orbit all at once—but rather by breaking up a finite time interval in the planet's motion into an infinite number of infinitesimal intervals and then summing all of these intervals for one complete period of the planet to obtain the planet's orbit. Following Newton's reasoning we picture a planet at a certain moment t at the distance r from the sun moving at the speed v in a definite direction, a velocity v. We consider the position and velocity of the planet at an infinitesimal time dt later (the d in front of the t merely identifies the time interval as infinitesimal; it plays the same role when placed in front of any quality so that a d placed in front of an r, means an infinitesimal change in the distance r). After the time interval dt the distance of the planet is $r + vdt$ and the velocity of the plane is $v + adt$, where a is the planet's acceleration at the moment t. Because the acceleration a can be obtained from the distance r by applying Newton's law of gravity, the procedure described above gives us a new distance r and a new velocity v for the planet's position at the time $t + dt$. The procedure can now be repeated until an orbit is obtained. This is nothing more than arithmetic or algebra of infinitesimals; we now consider it in somewhat more detail to reveal the essential features of the calculus as Newton developed it. We therefore simplify things by considering the motion of a particle in a straight line.

If the accelerated particle is moving in a straight line with a definite speed v at some moment t, and moves an infinitesimal distance ds in the infinitesimal time dt, then $ds = vdt$. This statement holds only if the speed v does not change in the time dt, even though the particle is accelerated. Newton saw that this statement is true if dt is infinitesimal—that is, allowed to go to 0. This led Newton to the definition of the instantaneous velocity of the particle at the time t as $v = ds/dt$, the ratio of two infinitesimals. The quantity ds is now (in modern calculus) called the derivative of s with respect to t. Newton called it the fluxion of s and represented it by a dot: $ds/dt = \dot{s} = v$. Because the velocity of the particle is constantly changing if it is accelerated (like planets), we may, following Newton, say that if the velocity of the particle changes by the amount dv (it may become larger or smaller or change direction) in the time dt, then $dv = adt$ where a is the instantaneous acceleration of the particle at the time t. The instantaneous acceleration was thus defined by Newton as $a = dv/dt$, written as $\dot{v}$ or d^2s/dt^2 or $\ddot{s}$. The quantity d^2s/dt^2 is called the second derivative of s with respect to t. Newton's second law of motion is now written as $F = m(d^2s/dt^2) = m(dv/dt)$; such an equation is called a differential equation. The mathematics of modern physics is essentially the mathematics of differential equations which is, in turn, the algebra of infinitesimals.

Although we have discussed Newton's invention of the differential calculus because of its very close relationship to physics, no story of mathematics would be complete without an account of Gottfried Wilhelm Leibnitz's (1646–1716) great contribution to the calculus. His work was more mathematical and, hence, more general, than that of Newton. During his productive years, Leibnitz was Newton's contemporary but his interests lay more in philosophy than in either mathematics or science. He was an expert mathematician and a very good scientist but he did not pursue mathematics as a tool to master physics. As an expert algebraist, he was naturally drawn to the function concept and to how the dependence of one quantity, say y, on another quantity, say x, is altered if x changes by a very tiny amount, say dx (an infinitesimal). This type of analysis led him to the algebra of infinitesimals and, hence, to the differential calculus.

Leibnitz was born in Leipzig on July 1, 1646, four years after the birth of Newton. His father, a professor of moral philosophy, came from a family which had served the Saxony government with distinction for several generations. Young Gottfried was probably exposed to the works of the leading political theorists in his earliest years. Even though his father died in 1652, Gottfried maintained a passion for learning. Over the next few years, he taught himself both Latin and Greek. Gottfried also became deeply interested in the idea of revamping the structure of classical logic, a task which consumed much of its own intellectual life and has remained incomplete to the present day.

Perhaps because of his wide interests, Gottfried began to study law at the University of Leipzig. He also found the leisure time to acquaint himself thoroughly with the treatises of Plato and Aristotle as well as the more recent works of leading natural philosophers such as Descartes, Galileo, and Kepler. So intrigued was Leibnitz by this emerging vision of a universe governed by simple mathematical laws (Kepler's three laws of planetary motion) that he attended a number of lectures in mathematics at the university.

That Leibnitz was not destined for a career in the law became apparent in 1666 when he was ready to receive his doctorate in law from the university. The faculty refused to confer the degree on Leibnitz, declaring that he was too young to receive the degree. Leibnitz himself was enraged by the faculty's decision, which was prompted in large part by the jealousies of several boorish professors. But Leibnitz lived in an age in which people did not automatically resort to the courts to right their wrongs; instead he resigned himself to the irreversibility of the decision and left the university. He then traveled to Nuremberg where he submitted an essay on teaching law to the faculty at the University of Altdorf. The faculty there was so impressed with Leibnitz's essay that they voted to grant him his doctorate of laws and also offered him a position as professor of law. Somewhat stunned at the turnabout of his own fortunes, Leibnitz nevertheless declined the offer, perhaps believing that the works of Archimedes and Kepler would play a more important role in his career than those of Grotius and Hammurabi.

Although Leibnitz had received a rudimentary education in mathematics while at Leipzig, it was not until he met the Dutch physicist, Christian Huygens, in Paris in 1672, that he first learned about the intricacies of geometry. Leibnitz and Huygens seemed to get along very well, and Huygens went so far as to give his new friend some lessons in basic mathematics. His appetite for mathematics whetted, Leibnitz began what became a lifelong study of the subject, making it a habit to jot down thoughts about what eventually led him to the calculus. He also obtained employment with the Elector of Mainz as a lawyer but soon demonstrated his aptitude for double-talk and intrigue and was promoted to the diplomatic ranks. Leibnitz's ascension to the international arena was soon followed by his completion of a plan designed to divert the imperial ambitions of Louis XIV of France, which had already begun to nibble away at parts of Alsace. Leibnitz proposed that Germany assist France in taking over Egypt in exchange for France's promise to leave Germany undisturbed. The French never acted on the proposal but it brought Leibnitz to the attention of the other German principalities.

Leibnitz's travels took him to London in 1673 where he met many of the members of the Royal Society. Leibnitz's prior invention of a calculating machine that could do addition, subtraction, multiplication, division, and square roots so impressed the Society that he was elected a foreign member. He also began to formulate some of his ideas about the calculus at this time although several more years would pass before Leibnitz had constructed a complete system.

Following the death of the Elector of Mainz in 1673, Leibnitz obtained a position as an advisor with the Duke of Brunswick. Along with his duties as a diplomat, Leibnitz also agreed to complete the Brunswick family genealogy, a tortuous trail of infidelity and inbreeding that would confound even the most talented of detectives. But the true nature of this genetic quagmire was not immediately apparent to Leibnitz, so he cheerfully agreed to take on what he imagined would be a pleasant little excursion among his employer's ancestors.

Over the next few years, Leibnitz completed the basic work on the calculus and published his results in 1677, more than a decade after Newton's discovery of the calculus at Woolsthorpe. The appearance of Leibnitz's paper eventually prompted Newton to publish his own version of the calculus, which had previously been circulated only among a few mathematicians and scientists. The resulting controversy over whether the English Newton or the German Leibnitz had been the first to discover the calculus touched off a highly nationalistic debate that was fueled by the reluctance of either Newton or Leibnitz to calm their angry supporters. Indeed, Leibnitz tossed a few coals on the fire by publishing an anonymous essay that savagely attacked Newton's work. Newton, for his part, initially resisted becoming involved in the debate over priority, but he eventually jumped into the fray and lobbed his own verbal cannon balls at Leibnitz's work. Fortunately, history saw fit to split the difference: Newton was recognized as the first to develop the calculus while Leibnitz was acknowledged to have developed the superior notation. Indeed, English mathematics languished for a century after Newton's death owing to the cumbersomeness of his "fluxions" while mathematicians on the continent (such as Laplace and Lagrange) raced ahead using Leibnitz's more elegant notations and constructed an entire mathematical framework for the Newtonian world system.

Leibnitz's philosophical speculations, particularly his argument that the entire universe consists of monads (discrete individual entities), consumed much of the remainder of his life. He also developed the field of combinatorial analysis—a process by which shared but often obscure characteristics among objects or things that appear to be completely dissimilar to each other might be identified comparatively quickly. His contributions to mathematics, both in the calculus and in combinatorial analysis, ensured that he would be remembered as one of its most prolific and versatile contributors.

And what of the Brunswick family history? Sadly for Leibnitz, this imposing tangle of half-truths and myths continued to hang over his head like the proverbial millstone until his death. The Brunswicks continued to insist that he complete the chronology of their illustrious lineage; one of history's most notorious examples of a great man's talent being squandered on a completely worthless task. But Leibnitz's skill and lack of scruples as a diplomat did serve him well in his ambitious but unsuccessful attempt to link the Brunswick name to every throne in Europe, even when the actual evidence for these ties had to be manufactured by Leibnitz himself.

Leibnitz's affinity for politics also led him to attempt to reconcile the Protestant and Catholic Churches. Unfortunately, his attempts to repair the schism that had existed since the time of Martin Luther were doomed to failure. The conferences that he set up for the representatives of the two organizations were disastrous shouting matches which ended with both sides more distrustful and suspicious of each other than ever before. Leibnitz was more successful with scientific organizations; he was instrumental in the founding of the Berlin Academy of Sciences, which would remain one of the premier institutions of its kind until it was gutted by the Nazis in the 1930s.

Throughout his life, Leibnitz retained a restless intellect that ranged across wide areas of knowledge ranging from mathematics and politics to economics and religion. Unlike Newton, he did not periodically tire of his areas of expertise but seemed to press forward on all fronts, even though progress in particular areas was not always immediately forthcoming. By the time of his death at Hanover on November 14, 1716, Leibnitz, the greatest polymath of his era, had accomplished more in his seventy years than all but a handful of history's greatest mathematicians and philosophers. But he passed away in relative obscurity. Unlike Newton, who was laid to rest in Westminster Abbey with all the honors that the British could give to him, Leibnitz's bodily remains were buried in a simple grave with only a few persons in attendance. But Leibnitz's death did little to quell the continuing controversy of his role in the development of the calculus.

We can best follow Leibnitz's reasoning in developing his version of the calculus by starting from the function concept and writing $y = f(x)$ which is the symbolic statement that y is a function of x. This simply means that any change in x implies a change in y. The nature of this relationship or dependence may be anything we please but we impose the condition that it is continuous—that a small change in x does not produce a large change in y. With this idea in mind we may now accept Leibnitz's idea that if x increases or decreases by the infinitesimal amount dx, then $f(x)$ becomes $f(x + dx)$ and the change in $f(x)$ is $f(x + dx) - f(x)$, which we may call dy. Leibnitz recognized that the important quantity is not the total change in $f(x)$ but the rate of change with respect to x which is just $[f(x + dx) - f(x]/dx$ or dy/dx. Leibnitz called this ratio of the two infinitesimals dy and dx the derivative of the function $f(x)$ with respect to x. Because dy and dx are infinitesimals and, therefore, essentially zero, it may appear that the derivative is nothing more than $0/0$ and therefore meaningless, because $0/0$ is forbidden in algebra. Yet one can demonstrate that the ratio dy/dx, if properly treated, is finite and has a precise and well-defined meaning and value. Leibnitz deduced general rules for finding the derivatives of functions of all kinds such as sines, cosines, and exponentials. This is probably the single greatest intellectual tool ever devised because science and technology could not have grown anywhere nearly as rapidly as they did without the calculus.

Leibnitz's approach to the development of the differential calculus was essentially algebraic or analytic, but a visual or graphical approach reveals certain important features of the calculus which are not immediately apparent in the algebra of calculus. The function $y = f(x)$ may be plotted in a rectangular Cartesian coordinate system, with y (the value of the function for a given value of x) measured along the vertical axis and x measured along the horizontal axis. Each pair of values y, x is represented by a point in the x, y plane and the curve passing through all of these points is a graphical representation of the function $f(x)$. The derivative of $f(x)$ may be obtained in a simple geometric way. We imagine walking along the curve as though it were a road and note that the slope of the curve changes from point to point as we move along, sometimes rising, sometimes falling, and still at certain points neither rising nor falling. The slope of the curve (its steepness) at a given point is obtained graphically by drawing a straight line which just touches the curve at the chosen point. This line is called the tangent of the curve, or of the function, at the point. This line cuts the x axis at a definite angle; the tangent of this angle, as defined in our chapter on trigonometry, is the numerical value of the derivative of the function with respect to x and, hence, its slope at the point being considered. We can thus find the values of the derivative or slope of $f(x)$ with respect to x graphically for all values of x.

This graphical approach to the calculus immediately reveals its practical importance in the study of functions and the plotting of curves. If we again imagine ourselves walking along the curve as though it were a road, its slope or steepness constantly changes, rising and falling, as we move along the curve. After rising to a hill, the road begins to dip so that the slope goes from positive (rising curve) to negative (falling curve). This means that at the very top of the hill, the slope is zero because the tangent is parallel to the x axis. But the slope is the derivative of the function, which means that the derivative is zero when the curve is at its highest value. This is also true if the curve dips down to a lowest point where the slope is also zero. This phase of the differential calculus, the theory of maxima and minima, developed soon after Newton and Leibnitz had completed their pioneering work, became very useful in the study of algebraic curves and also in many physical problems which require some knowledge about the maximum and minimum values of certain physical entities.

With the development of the differential calculus, physicists applied it with increasing frequency to the formulation of physical laws and the solution of physical problems. This was particularly true of the theory of maxima and minima which may be illustrated by a few examples from physics. The energy of a system of bodies, for example the energy of the planets in the solar system, or of a single body, like the bob of a swinging pendulum, tends to a minimum as the system settles down to a state of equilibrium. One can therefore find this state in terms of a defining parameter by expressing the energy in terms of this parameter and differentiating the energy with respect to this parameter. The state of equilibrium is then found by setting this derivative equal to zero. Another example is Fermat's discovery that the path of a ray of light between two points is that one, among all possible paths connecting the points, along which the time of passage of the light is a minimum. This is known as Fermat's principle of least time. The calculus can be used to find such a minimum and therefore the path of a ray of light. This principle developed into a powerful mathematical theory of optics. Using these same general principles, the eighteenth-century French mathematician Maupertuis introduced a minimal principle to describe the motion of a particle without using Newton's second law of motion. To this end he defined the action of a particle moving an infinitesimal distance dx as the product of this infinitesimal and the mass times the velocity of the particle. He then showed that the path along which a particle moves between two points is the one, among all such paths, along which the total action is a minimum. This is known as Maupertuis's principle of least action and is one of the most useful principles in physics.

Finally, we consider the application of the calculus to the development of the branch of physics called statistical mechanics. In dealing with the behavior of a system consisting of a very large number of particles—e.g., the molecules in a gas confined to a cylinder at a definite temperature and occupying a definite volume—the physicist, studying such a gas, is not concerned with the behavior of each individual molecule but with the gross behavior or properties of the entire ensemble. Because the molecules in a gas move randomly and with a range of different speeds, the physicist calculates the average velocities of molecules by introducing probabilities for various possible velocity distributions. The distributions with the maximum probabilities then determine the physical state of the system from which the properties of the system can be deduced. Finding the maximum probability is just a matter of setting the derivative of the probability with respect to the number of particles in any given state equal to zero. These few examples illustrate the importance of the differential calculus to physics.

Not only did the differential calculus greatly influence and spur the development of physics, but it also did as much for the development of pure mathematics. It thus altered entirely the evolution of geometry, leading geometers to what is now called differential geometry, which played a very important role in Einstein's synthesis of the general theory of relativity. Until the discovery of the calculus, geometers dealt only with Euclidean (plane) geometry; the geometry of curved surfaces such as spherical surfaces was not developed independently. The important difference between plane geometry and the geometry of curved surfaces is that the geometry on a plane is everywhere the same, as illustrated by the Pythagorean theorem for the distance between two points. It is the same, expressed in terms of the Cartesian coordinates of the two points, regardless of the distance between the points. But this is not true for any two arbitrarily chosen points on a curved surface because the contour of the curved surface changes continuously from point to point. One cannot even introduce a coordinate system of any kind that can be used over the entire surface. Owing to this very important stricture, nineteenth-century mathematicians broke the curved surface down into infinitesimal domains in each of which a different coordinate system, which could vary from domain from domain, was introduced. This led to what we now call differential geometry. Although differential geometry was first introduced for two-dimensional surfaces such as those of an ordinary sphere or an ellipsoid, it was quickly extended to surfaces of any number of dimensions or n-dimensional geometry. As such it is very important in the general theory of relativity and other branches of theoretical physics.

The differential calculus led to the development of two other very important branches of mathematics: infinite series and differential equations. The theory of infinite series grew out of the mathematicians' need and desire to express in all possible detail the way a given function of a variable changes with infinitesimal changes of the variable. To illustrate this point graphically we picture the curve that represents the function $y = f(x)$ graphically as plotted in a Cartesian coordinate system. By comparing this function to a road along which we are walking, we note that at each point along this road the slope has a definite value, which is the rate of change of the function, or its derivative with respect to x. But this slope itself may change from point to point so that the function is represented not only by its rate of change at each point but also by the rate of change of the rate of change. Thus a complete representation of the function requires knowledge of the rate of change of the slope; this is called the second derivative of the function with respect to x and is written as $d^2f(x)/dx^2$. But this may still not represent the function completely because third, and higher derivatives may be involved. The function may be represented by a sum of such higher derivatives, each multiplied by a different power of x. This led mathematicians to the representation of functions by sums of an infinite number of terms, and so the theory of the infinite power series was born, which is one of the most useful outgrowths of the differential calculus.

The differential calculus is but half of the calculus; the other half is the integral calculus, which is intimately related to differential equations. The integral calculus, or integration, as a mathematical operation is the inverse of the differential calculus. We recall that the essential operation of the differential calculus is differentiating a function with respect to the variable; a given function can, in general, always be differentiated, to give another definite function of the variable. Nearly every function, with some rare exceptions, has a definite derivative, but the inverse is not true. A given function may be too complex to be the derivative of another function. We may see the reason for this by considering first a very simple function of x such as $2x$. This is just the derivative of x^2 with respect to x. We can now multiply $2x$ by some complicated function of x such as $(1 + x)^x$. The new function $2x(1 + x)^x$, if a derivative at all, must be the derivative of a much more complicated function than x^2. Thus by multiplying our initial function $2x$ by increasingly more complex factors we can obtain functions that are too complex to be derivatives of other functions. Integrating a function is thus generally extremely difficult, if not impossible. Mathematicians have therefore constructed tables of integrals of a variety of functions that are important in pure mathematics and science—particularly in physics. The theory of the integral has evolved into its own branch of mathematics to which some of the greatest mathematicians have contributed.

A particularly important outgrowth of the integral calculus is the differential equation, whose solution is the integral we have just discussed. From what we have already said about the integral, we may conclude that a differential equation may be so complex that its solution cannot be expressed in terms of simple functions. In general, differential equations have no closed solutions; they can be solved only approximately. But approximate solutions are sufficient for most practical purposes. A very simple example of a differential equation is $dy/dx = 2x$, whose solution is just $y = x^2 + a$, where a is any constant. Only in the most simple cases can the solution of a differential equation be obtained by inspection.

Differential equations are the language of the scientists who can relate two or more widely differing branches of science. Thus the same general type of differential equation may appear in atomic theory, optics, astronomy, cosmology, and biology. The reason for the ubiquity of the differential equation in science is that using it permits the scientist to express functional relationships over infinitesimal domains of variables (such as space and time) on which the functions depend. The solutions of such equations reveal functional relationships that are valid over wide domains. In a sense, a differential equation is a microscopic, piecemeal guide to the construction of a macroscopic function. We consider a typical differential equation which we write as $dy/dx = f(x)$, where the right-hand side is any arbitrary function of x, and y is the solution of this differential equation that we seek. We write this equation as $dy = f(x)dx$ and note that all it tells us is the infinitesimal amount dy by which y changes if the dependent variable x changes by the infinitesimal amount dx. If we know the value of y for any given value of x, the differential equation gives us its values for $x - dx$, and $x + dx$. In other words, the differential equation enables us to trace the function y from point to point. It breaks up the function into an infinitude of infinitesimal bits, which, when added together, give the function y which is the solution of the differential equation. Because differential equations describe infinitesimal relationships among interdependent entities, they are ideally suited to the needs of the physicist who expresses the laws of nature in terms of infinitesimals. The laws of gravity and of motion, as stated previously, from which the astronomer deduces the orbits of planets, are written as differential equations.

Before we leave the calculus, we point out another of its branches which is very important to mathematics and science; this branch was generated by the concept of the partial derivative. To present this concept and explain it, we consider a function that depends on more than one variable. As a specific example, we consider a function $F(x,y,z)$ which depends on the three spatial coordinates of a particle that is moving from point to point. If x alone changes (the particle moves only along the x axis), then F changes only with respect to the change in x; its rate of change with respect to x is written as dF/dx. This is called the partial derivative of F with respect to x. If the particle is moving randomly so that x, y, and z are all changing simultaneously, then the infinitesimal change, dF, in F, equals $(dF/dx)(dx) + (dF/dy)(dy) + (dF/dz)(dz)$. This analysis led to partial differential equations, which are the principal language of modern physics.

CHAPTER 7

Post-Newtonian Mathematics

A mathematician who is not also something of a poet will never be a complete mathematician.

—KARL WEIERSTRASS

By the beginning of the fourth decade of the eighteenth century, both Newton and Leibnitz had died. Yet their passing did not deal a sudden blow to science because Newton had already abandoned science and mathematics in his later years while Leibnitz had returned to philosophy, his first love. The mathematics that developed in the seven decades following Newton's death may be characterized as the beginning of the golden age of mathematics, which reached its peak in the nineteenth century. Post-Newtonian mathematics was dominated by two great figures, Lagrange and Euler. Their contributions differed, however, in the sense that Euler's work was primarily in the area of pure mathematics, whereas Lagrange contributed as much to mathematical physics as he did to pure mathematics. Indeed, Lagrange laid the foundation of what was originally called mathematical physics but is now called theoretical physics, which goes far beyond Lagrange's original work.

When Lagrange began his work in dynamics, Newton's laws of motion and his law of gravity were universally accepted, but certain dynamical concepts such as the energy and the momentum of a body were not fully understood. Lagrange clarified these concepts and used the energy concept to express Newton's laws of motion in the form of partial differential equations, which can be used to describe the motions of groups of particles and force fields as well as the motion of a single particle. The energy concept, associated with the motion of a single particle, was known to both Newton and Leibnitz. Called the kinetic energy of the particle (KE), it is defined as one-half the mass of the particle times the square of its speed, $\frac{1}{2}mv^2$, if the particle is moving freely with no forces acting on it. With the introduction of the gravitational field, as defined by Newton, the energy concept had to be extended to include the contribution of the gravitational field to the total energy of the particle. This contribution is called the potential energy (PE) of the particle in the field. The total energy (TE) is then written as the sum of these two different kinds of energy: TE = KE + PE. The potential energy is always taken as negative. Lagrange, however, introduced the difference between these two energies, KE − PE, now called the Lagrangian of the particle, and showed that Newton's laws of motion, in their most general mathematical form for systems of particles, can be deduced from this entity, which is called the Lagrangian L of the system.

To see how the Lagrangian is related to Newton's laws of motion, we recall the contribution of the late seventeenth-century French mathematician Maupertuis to mathematical dynamics—the principle of least action. This principle states that a freely moving particle chooses that path connecting two points along which its action is a minimum. The problem which one confronts in applying this principle is determining the action of the system being studied. No problem arises when considering a single particle because its action is only its momentum times its infinitesimal displacement. But for a system of particles and force fields, the action may be a very complex function of space and time. For such systems one can deduce the action from the Lagrangian of the system which can be calculated from the kinetic and potential energies of the system. The action of the system is then obtained over a definite time interval by multiplying the Lagrangian by an infinitesimal time interval and summing this product from moment to moment over the entire time span. The space-time path over which the system evolves from some initial moment to some final moment is the one for which the sum of the infinitesimal products just defined is at a minimum.

Obtaining the correct dynamical history (the space-time path) of the system in this way seems forbidding because it appears to entail comparing the actions over an infinite number of paths to find the correct one—the path of minimum action. But Euler and Lagrange showed independently that the correct path is given as the solution of a second-order partial differential equation, the Euler–Lagrange equation, which is probably the most famous and useful equation in mathematics and physics. In mathematics it is particularly important in n-dimensional non-Euclidean geometry. The most important property of any geometry is the formula for the length of the infinitesimal interval between neighboring points expressed in terms of the coordinates of those points in any arbitrary coordinate system. In general, one is interested in the equation or formula for the shortest such interval which is the Pythagorean formula for Euclidean geometry (plane or flat geometry). But for a curved surface, one obtains a different equation for this interval (a geodesic) which is one form of the Euler–Lagrange equation. This mathematical concept has nothing to do with dynamics—the area of inquiry for which the Euler–Lagrange equation was first deduced—but Einstein showed that this geodesic is the basic equation that one must use in describing motions in general relativity. The calculus of variations is thus the bridge that connects pure geometry with the most sophisticated branch of physics. Lagrange and Euler knew nothing about the theory of relativity or the quantum theory, but their work in mathematical dynamics, particularly their concepts of the Lagrangian and action, are of fundamental importance in both of these remarkable physical theories which enable scientists to probe the universe from quarks to galaxies.

Euler, who considered himself a pure mathematician rather than a mathematical physicist, devoted most of his time to exploring algebra, the theory of algebraic equations, and analytic geometry. He published numerous important papers in these areas; his mathematical researches are summarized in his famous book *The Introduction to the Analysis of Infinities*. His work with the calculus led him to the study of infinite series.

A function of a single variable can be represented as a sum of powers of this variable. Thus the function $y = 1/1-x$ can be written as the infinite series $y = 1 + x + x^2 + x^3 + \ldots$, but this infinite series gives the correct value for y for only a restricted set of x values that lie between 0 and 1, with 0 included, but with 1 excluded. For x equal to 0, the function $y = 1/1-x$ equals 1 and the infinite power series, $y = 1 + x + x^2 + x^3 + \ldots$, also equals 1. But if $x = 1$, both the function and the infinite series are meaningless, hence, they can be compared only for values of x equal to or larger than 0 but less than 1. Mathematicians then say that the series converges for values of x between 0 and 1 and diverges for values of x equal to or larger than 1, whereas the function $y = 1/1-x$ has a finite, negative value for x larger than 1. Note that for x less than 0 (x negative) the function has positive values less than 1, and the infinite series has the same positive values. The function, however, has positive values for all negative values of x smaller than -1, but the infinite series does not. This brief analysis of the function and its infinite series representation shows that the series is a true representation of the function only in the domain of x between -1 and $+1$ where the series has finite values. For values of x larger than 1 and less than -1, the series

becomes infinite and so cannot represent the function. Euler fully understood this and saw that finding the domain of convergence of a series is of
great importance in the study of the representation of functions by infinite
series. He therefore initiated the study of the convergence of infinite series,
which became the principal field of mathematical research of the great
nineteenth-century French mathematician Augustin Cauchy, the founder of
the general theory of functions.

Euler's interest in the theory of infinite series and their representations
of functions led him to the discovery of one of the most remarkable equations
in mathematics—the famous Euler equation. Euler was led to this equation
in his study of the infinite series that represent the sine (sin) and cosine
(cos) as functions of the variable x. Euler also discovered another series that
is important in the calculation of compound interest. If x is the interest rate
for a given finite period, for example, a year, and the interest is compounded
continuously (an infinite number of infinitesimal time intervals) during this
period, then the principal P becomes $P(1 + x + x^2/2 + x^3/2 \times 3 + x^4/2 \times 3 \times 4 + . . .)$. If the interest x were 100 percent (that is, x is 1 for the
given period), the principal P would become $P(1 + 1 + \frac{1}{2} + 1/2 \times 3 + 1/2 \times 3 \times 4 + . . .)$. Euler evaluated the infinite series in the parentheses
and showed that it equals approximately 2.718282. . . . He called the exact
number given by the series e, which is also written as

$$e = \lim_{n \to \infty} \left(1 + \frac{1}{n} \right)^n$$

which is the number we get if we evaluate $(1 + 1/n)^n$ as n becomes infinite.
The number e is one of the most important numerical constants in pure
mathematics and physics, appearing in such diverse areas of mathematics as
probability theory and the theory of numbers. It also appears in almost every
branch of physics.

For the interest x, where x may be any value, the principal P as already
noted, becomes $P(1 + x + x^2/2 + x^3/2 \times 3. . .)$, and the value of the infinite
series in the parenthesis is e^x, the famous exponential function. Euler noted
that if x is an imaginary number, the series for e^x breaks up into a sum of two
series; viz., the cosine series and i times the sine series. Thus, if $x = iu$,
$e^{iu} = \cos u + i \sin u$, so that $e^{2\pi i} = 1$. It was the beginning of the study of
complex numbers which was carried on by the great nineteenth-century
mathematician Gauss. Euler introduced i as the symbol for $\sqrt{-1}$ and this
nomenclature is now used universally. A complex number Z which is the

sum of a real number and an imaginary number is always written as $x + iy$, where x and y are any two real numbers.

Euler was primarily a pure mathematician who occasionally contributed to mathematical physics whereas Lagrange is best remembered for his work in theoretical physics, primarily in dynamics and gravity. Lagrange's greatest efforts were directed to the solution of the n-body gravitational problem, which he failed to solve and which is still unsolved today even though it has attracted the efforts of the greatest mathematicians in the ensuing years. The n-body problem is the following: Given n gravitationally interacting bodies with arbitrary masses and velocities distributed randomly throughout space, find the mathematical expressions for the orbits of these bodies. Newton's laws of motion and his law of gravity permit one to write down one equation of motion for each body so that the problem entails solving n simultaneous partial differential equations involving the time and spatial coordinates of all of the bodies. Because each body has three spatial coordinates, all the equations of motion involve $3n + 1$ variables which includes time as one variable.

Newton solved the two-body problem. Despite its apparent simplicity, however, the three-body problem has never been solved in its most general form, but Lagrange, giving all his attention to it, obtained two different solutions of the famous restricted three-body problem. It is called the "restricted" three-body problem because the three bodies are constrained to move in the same plane. Lagrange found two distinct solutions: the equilateral triangle solution and the straight-line solution. In the equilateral triangle solution, the three bodies move in such a way that they form a rotating equilateral triangle whose dimensions remain fixed. In the straight-line solution, the three bodies lie on a rotating straight line with the distances between the bodies remaining fixed as the straight line rotates uniformly about an axis that intersects the straight line at the center of mass of the three bodies. The motion of certain asteroids near Jupiter is an excellent example of Lagrange's equilateral triangle solution of the three-body problem. The asteroids constitute an ensemble of thousands of bodies moving in orbits around the sun between Mars and Jupiter. Ceres, the largest of these asteroids, which was first discovered by G. Piazzi in 1801, has a radius of 350 kilometers and a mass of 6×10^{23} grams. The orbits of all the asteroids are greatly affected by Jupiter's gravity but about twelve of them are at the apex of an equilateral triangle, with Jupiter and the sun occupying the other two vertices of this triangle, which rotates in accordance with Lagrange's solution of the restricted three-body problem. Seven of the asteroids occupy the vertex of the equilateral triangle that precedes Jupiter in its orbit around the sun and five asteroids are at the vertex that follows Jupiter.

Lagrange, primarily concerned with the stability of the solar system when he discovered the equilateral triangle three-body solution, pointed out that the vertices of such a triangle, now called Lagrangian points, are points of stability of the three-body system; a body initially at rest at one of these points with respect to the other two Lagrangian points (vertices) remains so unless disturbed by an external force. If the body is displaced from this point momentarily, it oscillates about this point but ultimately returns to it at rest. The gravitational and centrifugal forces at such points just balance each other so that equilibrium and stability exist. Lagrange was particularly interested in a very special case of the restricted three-body problem: the case in which the mass of one of the three bodies is so small with respect to that of either of the other two bodies that it may be neglected by the theoretician. In that case, as Lagrange showed, five different Lagrange points exist; if the infinitesimal mass particle is placed at any of these points at rest, it remains at rest. By treating the solar system as consisting essentially of only two massive bodies (the sun and Jupiter), Lagrange showed that the motions of the other planets are highly stable.

Joseph Louis Lagrange (1736–1813) is widely regarded as the finest mathematician of the eighteenth century and perhaps the most humble genius ever produced by France. He was born the youngest of eleven children but was the only child to survive beyond infancy. His father, once a prominent official in the Sardinian government, was married to the daughter of a wealthy physician. Although Joseph's early years were spent in a luxurious lifestyle, his father's inept business investments taxed the family finances severely and it was not too many years before the considerable legacy that would have otherwise passed to Joseph had been squandered.

As a boy, Joseph studied the classics. His early training in mathematics included the works of the early Greek geometers such as Euclid and Archimedes. He did not find these works terribly difficult to master, but they did not excite his imagination. Lagrange's passion for mathematics was ignited only after he happened to come across a paper by Edmund Halley which discussed Newton's calculus. With the zeal of a religious fanatic, Lagrange read everything he could find about the calculus and soon mastered its most subtle intricacies. But Lagrange did not stop with the calculus; his fervor extended to other branches of mathematics such as geometry. Word of his skills as a mathematician also spread and Lagrange was appointed professor of mathematics at Turin at the age of nineteen.

Although Lagrange's youth may have bothered his students—all of whom were considerably older than he—he soon gained their respect with his unassuming manner and his brilliant lectures. He also formed a society which eventually became the Turin Academy of Sciences; this organization published a journal, *Miscellanea Taurinensia*, which served as a forum for many of Lagrange's earliest mathematical papers. But Lagrange first became known to other mathematicians outside of Turin when he was only nineteen years old after he sent a letter to Euler which contained the solution of a famous problem that had stymied mathematicians for many years; its solution required Lagrange to invent the calculus of variations. Euler, then recognized as the leading mathematician in the world, saw Lagrange's genius at once and graciously withheld from publication a paper he had written on the same subject so that Lagrange might be given adequate time to develop fully the new calculus. Lagrange's successful completion of this daunting task over the next four years placed him alongside Euler and the Bernoullis as one of the leading mathematicians of that era.

During Lagrange's tenure at Turin, he churned out numerous important papers on such things as the theory of sound, vibrating strings, dynamics, the calculus of variations, and the three-body problem. His mathematical versatility made clear his genius but his unceasing labors exacted a frightful toll on his health. According to W. W. Rouse Ball, his doctors were able to persuade Lagrange to relax the tempo of his work somewhat but "although his health was temporarily restored his nervous system never quite recovered its tone, and henceforth he constantly suffered from attacks of profound melancholy."

While still a young man, Lagrange was elected as a foreign member of the Berlin Academy. His election, despite his obvious mathematical genius, however, had been secured only after Euler intervened on his behalf. This election was merely a precursor to Lagrange's appointment in 1766 as court mathematician to Frederick the Great. Lagrange lived in Germany for twenty years; during that time he completed his most important mathematical work, including his masterpiece, *Mécanique Analytique*, which showed how the entire field of mechanics can be deduced from a single fundamental principle—the principle of virtual work. He also became a regular contributor of mathematical papers to the Berlin, Paris, and Turin Academies. Among the many subjects featured in the papers that he submitted to the

Joseph Louis Lagrange (1736–1813) (Courtesy AIP Emilio Segrè Visual Archives; E. Scott Barr Collection)

Paris Academy of Sciences were the motions of fluids, infinite series, and perturbations in the orbits of comets. Lagrange saved his best work for Berlin and his benefactor, Frederick, however, authoring papers on the theory of numbers, analytic geometry, differential equations, the stability of the planetary orbits, and the theory of the potential.

When Lagrange arrived in Berlin, Frederick welcomed him as "the greatest mathematician in Europe." The warmth of Frederick's greeting was not echoed by most of the other members of the Berlin Academy; many resented the fact that a foreigner—particularly, one of mixed French-Italian blood—had so successfully curried the emperor's favor. But Lagrange was patient and his gentle personality and unfailing courtesy to the other members gradually won their grudging admiration. Unlike Euler, who enthusiastically listened to even the most inane philosophical argument, Lagrange was more circumspect and offered his opinions about subjects beyond his areas of expertise only with the greatest caution. Lagrange's reticence was also prompted by his own aversion to controversy, a character trait he shared with Newton. Lagrange soon became the emperor's favorite mathematician and was periodically subjected to long-winded discourses by Frederick on matters ranging from court manners to marital bliss.

Partly at Frederick's urgings, Lagrange decided to take a wife. Driven less by feelings of love than by the desire to have a more "regular" lifestyle, Lagrange married a young girl who was a distant relative and, much to Lagrange's horror, an extravagant shopper. Even though the marriage turned out to be a happy choice for them both, Lagrange himself always had a curiously detached attitude toward his personal life. In a letter to d'Alembert, for example, Lagrange declared that he "never had a taste for marriage" and that his failure to inform d'Alembert sooner of the nuptial was "because the whole thing seemed to me so inconsequential in itself that it was not worth the trouble of informing you of it." Despite these seemingly mechanical statements, Lagrange was fond of his young wife. Her untimely death soon after their marriage dealt a serious blow to him and caused him to bury himself further in his mathematics to ease the pain of his loss. He focused his greatest efforts on arithmetic problems, particularly the solution of Fermat's unproved last theorem. But Lagrange solved only some of them and his attitude toward arithmetic in particular and mathematics in general became increasingly morose.

The death of Frederick in 1786 unleashed much of the xenophobic sentiment that had always percolated beneath the surface of German society.

Fortunately for Lagrange, he was offered a position with the French Academy in Paris. Weary but relieved, Lagrange packed his bags and returned to Paris, where he was given a hero's welcome by the royal family. He was given a comfortable apartment in the Louvre and he became a good friend of Marie Antoinette. But the move to France did not cure Lagrange of the depression that had dogged him for so many years. He did little work in mathematics, preferring to follow Newton's example and study metaphysics, theology, and chemistry. Although Lagrange never indulged in impossible dreams such as transmuting lead to gold, he did wander far away from the mathematical pastures that had nourished him throughout his professional life.

Angry mobs stormed the Bastille on July 14, 1789, two years after Lagrange's arrival in France, thus ushering in the French Revolution. Lagrange himself was never in any real danger because of both his mathematical reputation and lack of strong political convictions. Indeed, he seemed to welcome the invigorating shock of the revolt as a way to jolt the ineffective and bumbling French government into acting on the wide variety of pressing social problems tearing that country's fabric. But Lagrange was not sympathetic to the calls by the revolutionaries that the human race be renewed, particularly after it became clear that murder was the instrument by which this regeneration was to be accomplished. Lagrange was especially sickened by the beheading of his benefactors, Louis and Marie. But Lagrange himself was left untouched by Robespierre and his supporters. He was even given a pension by the new government and appointed to a succession of government posts. Lagrange helped to introduce a system of weights and measures and, after being appointed to the newly founded École Polytechnique in 1797, began to teach engineering for the first time. Many of his students later played a vital role in helping to transport Napoleon's troops across the expanses of Europe and Africa.

Lagrange's persistent despondency about his own life was finally brought to an end by his marriage to the daughter of his friend, the astronomer Lemonnier. The girl, who was nearly forty years younger than Lagrange, had fallen in love with the sad man who always had a kind word for others. Using her considerable charm, she finally persuaded Lagrange to take her as his wife. Lagrange, stunned by the turnaround in his personal life, went along with what he may have initially supposed would be a marriage of convenience. But Lagrange found that marriage suited him superbly and helped to relieve him of his recurring depression. According to

Eric Temple Bell, Lagrange's wife "reawakened his desire to live," prompting Lagrange to accompany her to social functions where he would never have gone by himself. Indeed, he soon found that he could not bear to be alone without her.

At the same time, Lagrange continued to be treated as something akin to royalty by the very same persons who had executed Louis XVI. He was France's greatest living scientist and a symbol of the nation's scientific prowess. Napoleon himself praised Lagrange as "the lofty pyramid of the mathematical sciences." Although these honors tended to sound like post-mortem plaudits, Lagrange was still very much alive and would devote the remaining fifteen years of his life to numerous mathematical pursuits, particularly a doomed effort to purge the differential calculus of infinitesimals. He also returned to a long-forgotten area—the stability of planetary orbits—and began to undertake a detailed revision of his treatise on mechanics. But he was not much more than halfway through the composition of the Second Edition when he died on April 10, 1813, at the age of seventy-six.

Lagrange's work in pure mathematics was as important as his work in gravitational dynamics. Indeed, his work in mathematics was extremely important in the development of modern mathematics, particularly his investigations into the solution of algebraic equations. In intermediate algebra we are introduced to quadratic equations and quickly learn the formula for the two roots (solutions) of the general quadratic equations, but the formulas for the three roots of a cubic equation are exceedingly complicated. Lagrange investigated the solutions of the cubic and higher-degree equations and discovered general theories which simplified the analysis of such equations. This work was carried on later by the brilliant French mathematician Galois, who introduced the techniques of group theory to the solution of algebraic equations. Lagrange's greatest work, however, is contained in his *Mécanique Analytique*.

Although Euler and Lagrange dominated post-Newtonian mathematics, significant contributions were made by their contemporaries, such as Marquis Pierre Simon de Laplace. He is best remembered for his nebular hypothesis of the origin of the solar system, which is generally accepted as the basic model for the formation of the sun and planets from a rotating cloud of dust and gas. Laplace assumed that the precursor or primordial solar nebula emitted heat as it rotated, gradually cooling off and shrinking, with the formation of the sun and the planets as a consequence of this process. The

German philosopher Immanuel Kant had proposed a similar model earlier than Laplace, hence, today we speak of the Kant–Laplace hypothesis. But the emphasis must be on Laplace's work because Kant merely proposed the model without doing any of the mathematics whereas Laplace performed the necessary mathematical analysis to make the model acceptable. Although the basic ideas of the Laplace nebular hypothesis are accepted today, important changes in the theory have been introduced to account for the distances of the planets from the sun, as expressed approximately by Bode's law (an empirical numerical relationship among the distances of the planets) and for the occurrence of two distinctly different groups of planets: the four inner or terrestrial planets—Mercury, Venus, Earth, and Mars—which have few satellites and little hydrogen, and the four larger outer planets—Jupiter, Saturn, Uranus, and Neptune—which have many satellites and consist primarily of hydrogen.

Pierre Simon de Laplace (1749–1827) was born in the village of Beaumont-en-Ange in Normandy to parents who made their living as farm laborers. Little is known about Laplace's youth because Laplace himself later took every measure he could to hide the truth about his peasant upbringing. Laplace went so far as to distance himself from both his family members and the neighbors who had first taken notice of his talent and helped to pay for his schooling. This attempt to upgrade his autobiography began when Laplace attended a military academy as an adolescent and discovered his affinity for mathematics. Laplace's talents soon became evident to the school's instructors and to the cocky Laplace himself. In his eighteenth year, feeling that he had learned all that he could at Beaumont, Laplace left his hometown for good, traveling east to Paris. He had outgrown the village that had been his entire world and now sought fame and fortune in the French capital.

Upon reaching Paris, Laplace managed to secure an audience with Jean d'Alembert, one of the top mathematicians of the day, after submitting a letter on the principles of mechanics. D'Alembert himself was captivated by Laplace's knowledgeable exposition, and after appealing to his colleagues, managed to secure Laplace an appointment as a professor of mathematics at the Military School in Paris. Scrutinized by professional mathematicians for the first time, Laplace was stimulated to begin what became a lifelong task—applying Newtonian physics to the entire solar system. He also began to indulge in an unfortunate habit of incorporating the works of others into his own researches without giving the proper credit. Newton's famous statement about having been able to see further than other men because he had stood on the shoulders of giants such as Kepler and Galileo probably could not have been uttered with a straight face by Laplace. According to Eric Temple Bell, "Laplace stole outrageously, left and right, wherever he could lay his hands on anything of his contemporaries and predecessors which he could use. From Lagrange, for example, he lifted the fundamental concept of the potential; from Legendre he took whatever he needed in the way of analysis; and finally, in his masterpiece, the *Traité de Mécanique Céleste*, he deliberately omits references to the work of others incorporated in his own, with the intention of leaving posterity to infer that he alone created the mathematical theory of the heavens." Indeed, Newton was the only scientist whom Laplace saw fit to give his due. But that acknowledgment was probably unavoidable since his treatise dealt with Newton's law of gravitation.

Laplace's first breakthrough was to show that the mean distances of the planets from the sun do not vary significantly over time. But this finding, by itself, did not prove that the solar system itself is stable and that the planets will continue spinning around the sun forever. This was only a first step because the question of stability could not be conclusively answered until the gravitational forces exerted by the sun and all of the planets on each other had been completely analyzed. But Laplace's first step did convince the French Academy of Sciences to elect him as a member in 1773. His results also prompted him to begin the investigations that consumed the next half century of his life and culminated in the publication of his *Traité de Mécanique Céleste*, an imposing five-volume monument to Laplace's mathematical researches. This treatise's accessibility was limited somewhat by the failure of Laplace in some instances to include the elaborate mathematical reasoning that was necessary to justify his own conclusions. Brilliant man that he was, Laplace thought that the conclusions were patently obvious even though he was often unable to recall how he had arrived at them.

Laplace, as noted above, is also associated with the nebular hypothesis— that the solar system evolved from a condensing cloud of gases—originally proposed by the German philosopher Immanuel Kant. Kant did not approach the problem in a rigorously scientific manner and Laplace himself did not construct an adequate mathematical model to provide convincing evidence for this theory. But the nebular hypothesis—despite its imperfections—has continued to influence the thinking of astronomers to the present.

Pierre Simon de Laplace (1749–1827) (Courtesy AIP Emilio Segrè Visual Archives)

In 1785, Laplace became a full member of the Academy. He also served as an examiner at the Military School and made the acquaintance of a young man studying there named Napoleon Bonaparte. Napoleon himself must have been impressed by Laplace because he later became the mathematician's leading benefactor after seizing power. Before Napoleon's ascension, however, France endured the bloody chaos of the Revolution. Like Lagrange, Laplace managed to avoid the guillotine by lending his considerable talents to directing the manufacture of armaments. But Laplace was much more of a politician than Lagrange, coveting the power and influence that came with being associated with the nation's leaders. Indeed, Lagrange himself cared little for such things and doubtless took a dim view of Laplace's maneuverings for influence. But Laplace's instinctive talent for disregarding views once held dear when the circumstances required it was underscored when he won Napoleon's favor, obtaining France's highest honors such as the Grand Cross of the Legion of Honor and the Order of the Reunion. Napoleon also appointed Laplace to be Minister of the Interior, a disastrous tenure that lasted a mere six weeks. Of Laplace's shortcomings as an administrator, Napoleon is reported to have said: "A mathematician of the first rank, Laplace quickly revealed himself as only a mediocre administrator; from his first work we saw that we had been deceived. Laplace saw no question from its true point of view; he sought subtleties everywhere, had only doubtful ideas, and finally carried the spirit of the infinitely small into administration."

In 1812, Laplace published his *Théorie Analytique des Probabilités*, perhaps the most important work on probability theory ever written. Laplace himself viewed his book as little more than common sense expressed in a mathematical form. But his approach to the subject was wide-ranging and he attacked problems relating to probability theory on all levels. As with all of his other books and papers, Laplace approached his subject by treating mathematics as merely the tool with which a better understanding of nature can be obtained. Lagrange, by contrast, viewed mathematics as the end, an infinitely branching literary work with profound philosophical implications. As a result, Laplace's works have never been accorded as high a place in the pantheon of mathematics as have those of Lagrange. But Laplace himself was probably more highly regarded by his fellow scientists because he used mathematics to attack problems relating to the dynamics of the universe itself while Lagrange mostly confined himself to abstract mathematical questions that did not appear to be as applicable to the real world.

Much has been said about how Laplace toadied up to Napoleon and then, following the restoration of the monarchy, to Louis XVIII. Indeed, Louis rewarded Laplace for his foresight by appointing him to be president of the committee that was established to reorganize the École Polytechnique. But to be fair, Laplace did stand up for his principles on occasion, even to Napoleon. One noteworthy example occurred when Napoleon had wondered aloud why the *Traité de Mécanique Céleste* did not mention God as the creator of the universe. Without hesitating, Laplace responded: "Sir, I have no need for that hypothesis."

Although Laplace did appropriate the works of other mathematicians without giving them due credit, he could be very kind to his own students. Jean-Baptiste Biot, then a young mathematician, read a mathematical paper to the French Academy on a day in which Laplace was sitting in the audience. After completing his presentation, Biot was approached by Laplace, who showed him an old manuscript that Laplace had written many years before; it contained the same points made by Biot in his paper. Amused by the startled look in the young man's face, Laplace told Biot not to worry about his priority but instead to submit the paper so that Biot himself might receive full credit for the discovery.

Laplace is famous for two other concepts—one in physics and the other in pure mathematics. In his analysis of the gravitational field surrounding a sphere of mass M and radius R, Laplace noted that the square of the speed of escape from the surface of such a sphere equals $2GM/R$, where G is Newton's universal gravitational constant. He then reasoned that if this quantity equals the speed of light, c, that is $2GM/R = c^2$, then light itself cannot escape from such a sphere which is then invisible. This is essentially the concept of the black hole whose detailed properties can be deduced only from Einstein's general theory of relativity.

In pure mathematics, Laplace is famous for his discovery of the Laplace transform, which is a very simple and elegant mathematical technique for solving what are known as integral equations. These appear in many different branches of physics, such as those which deal with the flow of radiation through a gas, the diffusion of neutrons, and in electrical circuit theory. Although Laplace introduced his transform idea in the context of his theory of probability, we can best understand the concept of the Laplace transform by considering a very important problem in stellar structure. The radiation emitted from the surface of a star such as the sun consists of radiation from many layers below the surface. As the radiation from the deep interior flows toward the surface, it is altered by the intervening stellar layers. The Laplace transform permits us to find the individual contributions of these various layers from the observed radiation emitted from the surface.

Laplace's great contributions to mathematics and physics, as noted above, are contained in his two superb books *Théorie Analytique des Probabilités* and *Traité de Mécanique Celeste*. In the first of these books,

Laplace summarized almost all the mathematics known at that time with special emphasis on the theory of games, geometrical probabilities, the theory of least squares (the basis of the calculus of errors), the solutions of differential equations, and Laplace transforms, which the great nineteenth-century electrical engineer Heaviside later developed into operational calculus. In his celestial mechanics, Laplace completed the work begun by Newton and extended Lagrange's planetary work. Laplace is best known by physicists, however, for the Laplace differential equation which is the basis of what is known today as potential theory. This theory describes gravity as a force field, represented at each point of space by a potential function. The importance of this equation is that it enables one to study the gravitational interaction among many bodies without introducing their masses specifically. These ideas were later expanded to encompass the electromagnetic field and the study of electrostatic interactions among electric charges. It was also applied to the study of acoustics and hydrodynamics. Indeed, the Laplace equation is an excellent example of the universality of mathematical formulas in their applications to physics.

Lagrange and Laplace were at the peak of a pyramid of eighteenth century French mathematicians who established a tradition of excellence in mathematics that has continued up to the present. Among the lesser figures in this pyramid were Clairaut, d'Alembert, Legendre, and de Moivre. D'Alembert, for example, wrote extensively on all branches of mathematics, but his most brilliant work dealt with the mathematics of vibratory motions—vibrations of strings and acoustical vibrations which led him to the development of the theory and solutions of partial differential equations. He also contributed to the theory of probabilities. He is, however, most famous for his theoretical work in the dynamics of solid bodies; in this field he introduced the famous d'Alembert principle which reduces the dynamics of a rigid body to statics. This principle, also known as the principle of virtual work, states that if the particles that constitute a body are displaced by an infinitesimal amount, the work done in this displacement (the virtual work) must be zero if the body is at rest or moving with constant speed in a fixed direction. As formulated by d'Alembert, this principle states that the effective force on a system of bodies equals the sum of all the individual forces acting on the bodies. His best known contributions to mathematics and physics are contained in this *Mémoir sur le calcul integral* and his *Traité de dynamique*. In his later years he devoted his time to philosophy and contributed articles in science, philosophy, and mathematics to the great encyclopedia begun by Diderot, of which d'Alembert was the first editor.

Here we have concentrated on the great post-Newtonian mathematicians, but some Newtonian figures and Newtonian contemporaries also made important contributions to the calculus. Foremost among these was Fermat, whose work on mathematics we have already mentioned without pointing out his early contributions to the calculus. Fermat developed a method of finding the maximum or minimum value of an algebraic expression—a function of a variable x—by considering infinitesimal changes in the variable x and allowing the changes produced in the algebraic function to go to zero. In this approach, he anticipated the calculus and would probably have discovered it if he had devoted all of his time to mathematics instead of being a part-time mathematician whose primary activities were concerned with law. Fermat is not famous for his contributions to the development of the calculus but for his work in the theory of numbers and for what is called Fermat's last theorem. Although we have already mentioned his last theorem, Fermat wrote the following passage about it in the margin of a page in his copy of the book *Arithmetica* by Diophantos: "To divide a cube into two other cubes, a fourth power, or, in general, any power whatever into two powers of the same denomination above the second is impossible, and I surely have found an admirable proof of this, but the margin is too narrow to contain it." Recently, Dr. Andrew Wiles of Princeton University gave a proof of Fermat's last theorem in a lecture at Cambridge University. Dr. Wiles's proof, however, has not yet gained wide acceptance by the mathematical community. Fermat's research in the theory of numbers stimulated other mathematicians of that period to work in number theory so that Fermat is properly called the "father of the theory of numbers."

Like Lagrange, Fermat pursued the study of mathematics for the sake of mathematics, even though he made a number of important discoveries in areas of applied science, such as optics. He was less concerned than Laplace or Newton with the ways in which mathematics can be used as a tool to increase our knowledge of the universe. Because Fermat lived before Newton had made his great discoveries, however, the power of Newtonian mechanics to reveal the subtleties of nature had not yet been demonstrated. As such, applied mathematics was still relatively primitive, particularly when it was used to study physical phenomena even though the works of Kepler and Galileo had hinted of its potential value.

Pierre de Fermat (1601–1665) was born in the village of Beaumont-de-Lomagne, France. His early years were spent in relative comfort as his father, a merchant, was able to provide the family with most of the luxuries that the money of that time could purchase. That young Pierre might pursue a mathematical career was not preordained, but the union of his father and his mother, the daughter of a judge, seemed to give some impetus to his pursuing a career that would require some skill with numbers. Little else is known about his early years or the extent of formal education that he received. It is probable that Pierre was taught arithmetic and the classics at home and then acquired his knowledge of mathematics through sheer dint of effort. But his career in mathematics seems to have been given a jump start when he became the councillor for the local parliament at Toulouse at the age of thirty. Although Fermat was an extremely diligent worker who discharged his duties competently and efficiently, the job also provided him with ample leisure time to think about mathematics. He also devoted himself to "restoring" lost works of antiquity, such as Apollonius's *Plane Loci*, by culling contextual clues from other classical treatises. Although Fermat was a lawyer by training and not a professional mathematician, his efforts to reconstruct Apollonius's great work may, according to Carl B. Boyer, have also led him to the discovery of a fundamental principle of analytic geometry in 1636, a finding which he failed to publicize.

His reluctance to publish his mathematical papers was unfortunate because it did raise some questions about the dates of Fermat's discoveries in analytic geometry and the theory of numbers. It was only after his death that the true scope of Fermat's achievements began to be fully appreciated by mathematicians as his boxes of private papers were examined. Fermat himself would have scoffed at the idea that he should be remembered as a great mathematician even though he was undoubtedly aware of his own talents. Indeed, he viewed his own life as simple and uneventful, seeing himself as a lawyer who indulged in his passion for mathematics during his leisure hours. Because he pursued mathematics for the sheer pleasure of solving a particular problem or discovering a new theorem, Fermat was not very concerned about whether posterity would amply credit him for his brilliant work. Fermat viewed mathematics as an enjoyable hobby as opposed to an occupation; this nonchalance about priority is not commonplace among professional mathematicians.

Fermat's own private life was not the grist of gossip mongers. He married a distant cousin, Louise de Long, in 1631. The couple had a son and two daughters, none of whom seemed to inherit their father's aptitude for mathematics. By all accounts, Fermat was a doting father and a devoted husband. Certainly he did not place material gain above the happiness of his family. His career as a public servant did not bring great riches but it did permit him to provide his family with a comfortable living.

His placid temperament was mistaken by some for passivity, particularly when he became involved in debates with Descartes over problems in analytic geometry as well as questions relating to its invention. Descartes, ever the soldier, openly criticized Fermat for what Descartes argued were errors in Fermat's mathematics. Fermat, for his part, cheerfully deflected Descartes' verbal hand grenades, often getting the better of Descartes. Yet, the two men did respect each other and did not allow these debates to poison their professional relationship.

Fermat's occupation might have seemed, at first glance, to be something of a disadvantage. But Fermat's legal training imbued in him a skepticism about unfounded philosophical speculations that others, such as Descartes, could not resist. One can only wonder how Fermat responded when he first read about Descartes' preposterous theory of vortices or his argument that empty space is actually a continuous plenum. Unlike Descartes, Fermat was not interested in speculating about the physical universe itself. The abstract, pristine world of mathematics was the only universe that Fermat sought to explore. According to Eric Temple Bell, "Fermat seems never to have been tempted, as both Descartes and Pascal were, by the insidious seductiveness of philosophizing about God, man, and the universe as a whole; so, after having disposed of his part in the calculus and analytic geometry, and having lived a serene life of hard work all the while to earn his living, he still was free to devote his remaining energy to his favorite amusement—pure mathematics, and to accomplish his greatest work, the foundation of the theory of numbers, on which his undisputed and undivided claim to immortality rests."

His hard-nosed views about philosophical grandstanding should not give the impression that he was a colorless and bloodless investigator who was only interested in dusty laws and mathematical equations. Indeed, Fermat was quite an accomplished man of letters, having mastered many of the languages and much of the literature of Europe. When he was not tackling mathematics, Fermat could often be found composing poems in any one of several languages, including French and Spanish. Although it would be too farfetched to suggest that Fermat was as skillful a poet as he was a mathematician, one can only wonder about the literary works that might have been produced by this versatile man had he devoted himself fully to poetry.

Fermat undoubtedly knew that he towered above most of the other mathematicians of his era. Indeed, many historians are inclined to give him the nod in the long-running dispute over whether he or Descartes discovered analytic geometry. Both men seem to have developed their ideas independently of each other; Fermat himself managed to best Descartes by about a year. But because Fermat was so hesitant to publish the results of any of his mathematical investigations, it would have been difficult for him to contest Descartes' own pivotal role in developing analytic geometry, even if he had been so inclined. As it was, Fermat and Descartes both made unique contributions to the subject, creating a rich intellectual heritage for future generations of mathematicians.

Fermat's discoveries of theorems in the theory of numbers seem to have come to him as sudden revelations whose truths he recognized intuitively. Thus he stated another important theorem in the theory of numbers as a marginal note, namely that a prime number of the form $4n + 1$ (e.g., the prime number $13 = 4 \times 3 + 1$) can be expressed in only one way as the sum of two squares (e.g., $13 = 2^2 + 3^2$). This theorem was proved later by Euler. Fermat's casual way of stating mathematical theorems is perhaps best illustrated by his theorem that the expression $a^{p-1} - 1$ is divisible by p when the numbers p and a are prime to each other (they have no common factor). He stated this theorem in a letter to a friend in 1640. Fermat was apparently not too greatly concerned about whether or not he received credit for a particular discovery.

Just as the development of certain branches of mathematics were stimulated by the needs of commercial and industrial growth, so did games of chance, such as gambling of all sorts, stimulate the development of the theory of probabilities, which has expanded far beyond its original designs and penetrated into every branch of science. In this area, the great French philosopher Blaise Pascal, together with Fermat, played an important role. They are, in fact, considered to be the founders of the mathematical theory of probabilities. Pascal began corresponding with Fermat on probability theory when a French nobleman approached him concerning certain games of chance and how to accumulate the highest scores in such games. Pascal's study of probability led him to the famous Pascal numerical triangle—in which the numbers give the probabilities of different, but basically related, events. This work can be illustrated by tossing a coin. Given n, where n is any integer, identical coins that are tossed simultaneously, each in exactly the same way, what is the probability for any particular combination of heads and tails when they land? If H is the probability of a single head and T is that of a single tail (H plus T must equal 1, because either a head or tail must occur), then the terms in the binomial expansion $(H + T)^n$ give the probabilities of the various head–tail combinations of the tossed coins. Thus, the probability of having $(n - 5)$ heads and 5 tails equals $n(n - 1)$ $(n - 2)(n - 3)(n - 4)/(2 \times 3 \times 4)(H^{(n-5)}T^5)$. Note that the sum of the exponents of H and T in each term is always n: the coefficients, such as $n(n - 1)(n - 2)(n - 3)(n - 4)/(2 \times 3 \times 4)$, are the numbers that appear in the Pascal triangle. Because H equals T for a perfect coin, each variable equals $\frac{1}{2}$. All the HT products in the binomial expansion, such as $H^{(n-5)}T^5$, equal $(\frac{1}{2})^n$; the probability for any particular combination of heads and tails, therefore, equals the binomial coefficient of that HT product divided by $2n$. As a specific example, we consider the tossing of five coins so that $n = 5$ and $(\frac{1}{2})^n = \frac{1}{32}$. The probability of obtaining five heads and no tails (or vice versa) is thus $\frac{1}{32}$; the probability of finding four heads and one tail is $\frac{5}{32}$; the probability of finding three heads and two tails is $5 \times 4/2^n = \frac{10}{32}$, and so on. These probabilities are called the *a priori* probabilities because they are given as the number of favorable events for a particular HT combination divided by the total number of possible events which is 2^n for n tossed coins. This corresponds exactly to our intuitive ideas about how probabilities should be calculated. These simple ideas about probabilities are fundamental to the theory of probability and to the vast modern theory of games that has evolved from it. The very important branch of physics called statistical mechanics is a sophisticated extension of probability theory to the analysis of physical phenomena from genetics to galactic structures, which we discuss in detail in later chapters.

Blaise Pascal was a mathematician whose life began with great promise but was cut short by illness and mental instability. He was born in Clermont, France on June 19, 1623. His mother, Antoinette, died while Blaise was still a young boy, prompting his father, Étienne—a local judge and a man who evidently enjoyed something of a reputation as a scientist—to move Blaise and his two sisters, Gilberte and Jacqueline, to Paris. Blaise's health was poor as a child; he seldom attended grammar school but instead received his lessons at home. Despite his recurrent bouts with sickness—which caused his father to fear, on more than one occasion, for his son's life—young Blaise was a precocious child who quickly mastered the classic texts customarily force-fed to the pupils of that day. Étienne refused to allow the boy to study mathematics because he believed that algebra and geometry would exhaust his son. But Blaise soon discovered geometry and joyfully set out to master the works of Euclid. Once his father discovered that Blaise's new affinity for mathematics would not harm his health, he strongly encouraged his son's interest and the boy became well acquainted with the works of many ancient and contemporary mathematicians. By age fourteen, Blaise was attending the weekly meetings of French mathematicians such as Roberval and Mersenne; they formed what eventually became known as the French Academy. He also began to devise his own elegant proofs. One of the earliest, a famous proof known as the "cat's cradle," was developed by Blaise before he reached his seventeenth birthday. He also composed his *Essay on Conics*, a work that was highly regarded by several leading scientists of that era, including Leibnitz, but was never published.

His eighteenth year was a turning point in his young life. His health began to deteriorate and he was tormented by acute dyspepsia and insomnia. He also invented the first calculating machine, the precursor of today's electronic computers. Although Pascal's device helped to boost his scientific reputation, it did not attract widespread attention. Pascal also conducted numerous experiments on atmospheric pressure, which later led to his discovery that the pressure of the atmosphere can be calculated in terms of weight. Finally, Pascal met Descartes, twenty-seven years his senior, and the two immediately formed a lasting, antagonistic relationship. Descartes himself did not care for Pascal because he believed that Pascal had filched some of his own ideas about atmospheric pressure. The two were also separated by religious differences; Descartes was a loyal Jesuit whereas Pascal was a devout follower of Cornelius Jansen, a fanatic who preached a message of unflinching intolerance for all other religious creeds. Descartes thought Pascal's religious leanings were idiotic and he did not hesitate to advise him of his opinion.

Pascal corresponded frequently with Fermat on the subjects of analytic geometry and physics. Not only did their letters act as a stimulus in the development of probability theory but they also helped to focus Pascal's attention on the scientific—as opposed to the metaphysical—aspects of his experiments with barometric pressures. But Pascal's personal life was less enjoyable. His sister, Jacqueline, had become a nun and had made it her ambition to win her sickly brother over fully to the Catholic Church. Whether it was Jacqueline or Pascal's own latent sexual hangups that contributed to his increasingly morbid religiosity is unclear. But between his dyspepsia and the rantings of his sister, Pascal himself began to lose interest in mathematics. Although his talent for the subject remained with him until the end of his brief life, he became increasingly preoccupied with contemplating the tragedy of the human condition. Pascal wrote his *Pensées*, a classic of French literature, which featured tortured rantings about mysticism, humanity, and sexuality. Whether Pascal lived as straitlaced a life as his writings seem to indicate is open to question because his own religious fanaticism waxed and waned. Indeed, it is likely that he indulged in occasional but spirited bouts of gastronomic and sexual orgies. But these flights of fancy were short-lived and the well-meaning Jacqueline was always there to point him back to the road of righteousness.

Following the death of his father in 1650, Pascal was appointed the executor of the estate. For the first time in many years, he was forced to deal with lawyers, bankers, and merchants. Whether the mundane affairs of the commercial world so revolted Pascal that he retreated back to the religious shroud that he had wrapped himself in is unclear. But his contempt for the secular world was then given a second jolt when Queen Christine of Sweden, whom Pascal had presented with a prototype calculating machine, declined to extend an invitation to him to come to Sweden and replace her tutor, Descartes, who had died earlier that year. The final straw occurred in 1654 when Pascal was driving a carriage and his horses bolted out of control, hurtling over the parapet of a bridge at Neuilly; the sharp jolt snapped the trace, leaving Pascal sitting alone atop his horseless carriage on the bridge.

Pascal took his miraculous escape to be a sign from heaven that he should give up mathematics and confine himself to religious contemplation. As he had already completed his great work on probability, it is doubtful that Pascal's move to Port Royal at the age of thirty-one, deprived the world of much in the way of new mathematical discoveries. Although he later completed an important paper on cycloids, his mathematical career was, for all purposes, over. Pascal instead threw himself into his famous *Provincial Letters*, a virulent attack on the Jesuits. Like *Pensées*, this new work came to be regarded as an important contribution to literature even though its impact was generally exaggerated by its supporters.

Pascal spent the last few years of his life in misery. His stomach flared up continuously and his teeth slowly rotted away, causing him to suffer great pain. He was seldom able to snatch more than a few moments of blissful sleep. Tortured by constant pain and periodic convulsions, Pascal became both physically and mentally exhausted. His plight was compounded by his own belief that his poor health was a form of divine punishment for his own real and imagined shortcomings as a human being. Indeed, Pascal seems to have regarded his brief flurry of activity when writing his paper on cycloids as a profane exercise that had imperiled his chances for salvation. Pascal's increasing mental instability was only fueled by Jacqueline, who had long been a resident at the Port Royal convent near his home. By the time he drew his last breath on August 19, 1662, Pascal had convinced himself that his own eternal damnation was assured and that he could hope for nothing better than eternal suffering.

Although Pascal made a number of important contributions to mathematics such as his theory of probability, he is remembered as much for what he did not accomplish, despite having been given the intellectual gifts that enabled him to be both a leading mathematician and essayist. How much more might Pascal have created had he not been imbued with his morbid spirituality which saw only dark, even when there should have been light? This question can never be answered. But historians seem to share a resentment toward Pascal based on the belief that he squandered his immense talents on worthless speculations about humanity instead of confining himself to the study of mathematics. Whether this judgment is too harsh cannot be determined because it is not clear that Pascal would have necessarily made any other mathematical discoveries even if he had managed to avoid becoming drawn into the quagmire of his theological speculations. But the scope and grandeur of the contributions he did make suggest that other mathematical discoveries might have been forthcoming.

A remarkable feature of this period in the story of mathematics is the important role that French mathematicians played in it. Newton, Leibnitz, and Euler, the dominant figures, were not French. But starting with Fermat and proceeding to Maupertuis, the father of the principle of least action and the calculus of variations, we have the outstanding mathematicians Lagrange, Laplace, d'Alembert, Pascal, Marquis de L'Hospital, Gérard Desargues, and a few minor figures to complete the French roster of mathematicians who left a deep imprint on the evolution of mathematics in the seventeenth and eighteenth centuries. But the brilliance of these dominant figures should not blind us to the contributions in that era of the lesser mathematicians, such as the Bernoullis, Huygens, and Wallis. The Bernoulli story itself began with the two brothers Jakob and Johann—born thirteen years apart—who were late contemporaries of Newton. Jakob began his intellectual life as a theologian and Johann studied medicine, but both were launched into mathematics by Leibnitz's published mathematical studies, and both became

Leibnitz's most famous students. Jakob later became the teacher of Euler's father and Johann became Euler's teacher.

Jakob was greatly influenced by Leibnitz and their correspondence over the years undoubtedly contributed to Jakob's important work in the early development of the calculus and its application to the solution of problems in physics and engineering. Thus, his mathematical theory of the catenary, the mathematical formula for the shape of a suspension bridge, is basic to the construction of suspension bridges. His use of polar coordinates in calculus—rather than the rectangular Cartesian coordinates—led him to the discovery of the equations of such curves as the logarithmic spiral, the Bernoulli lemniscate, and the isochrone—the curve in space along which a body under the influence of gravity falls with constant speed at each point of the curve. The solutions of such problems helped to persuade physicists of the power of mathematics to simplify and to lead to the solutions of complex physical problems. Jakob Bernoulli was thus one of the founders of mathematical physics. Together with Huygens and Pascal, Jakob laid the foundation of the mathematical theory of probability. Independently of Pascal, he discovered the importance of the binomial coefficients in the theory of probability, which he announced as the "theory of Bernoulli."

Jakob Bernoulli's younger brother, Johann, followed in the footsteps of his brother, working in trigonometry, analytic geometry, and calculus. Like Jakob, Johann was interested in physical problems and solved a very important problem which formed the basis for a branch of mathematics known as the calculus of variations, which we mentioned in connection with the work of Fermat and Maupertuis: the principles of least time and least action. Johann discovered the mathematical form of the curve of quickest descent connecting any two points in space of a particle moving under the action of gravity along such a curve. This curve, the brachystochrone, is such that the time of descent along an infinitesimally close curve is the same as along the brachystochrone; the variation in time between the two curves is zero. Hence, the name, the calculus of variations. This discovery later led the Bernoullis to the mathematics of the geodesic (shortest line) on curved surfaces, which plays a very important role in Einstein's theory of general relativity.

The Bernoullis' influence on mathematics extended beyond their own direct contributions because they were outstanding teachers. Johann was Euler's teacher. Johann's two sons, Nikolaus and Daniel, also taught Euler; they did most of their work in probability theory, astronomy, mechanics, and hydrodynamics. The older son, Nikolaus, died young, but his brother, Daniel, went on to become famous for his work in hydrodynamics, which led him to what is known as the "Bernoulli principle." This principle is crucial to airflow analysis and the design of airfoils such as airplane wings and propellers. The principle states that in a fluid, gas, or liquid streaming in a given direction, the lateral pressure decreases as the speed of the fluid increases. This principle, essentially a statement of the law of conservation of energy, tells us what the shape of an airplane wing must be to support the plane against the pull of gravity. The top surface of the wing must be convex (bulge upward) and the bottom surface must be flat (parallel to the ground). A wing shaped this way moving horizontally causes the stream of air flowing horizontally past the wing's upper surface to move faster than the air streaming past its lower surface. Thus the lateral pressure on the top of the wing is reduced with respect to that on the lower surface; an upward thrust is thus created which supports the plane. Daniel Bernoulli's study of hydrodynamics and his need to express his discoveries mathematically led him to the theory of partial differential equations to which he contributed the basic ideas. His work in this very important branch of mathematics enabled him to develop the mathematical theory of vibrating strings, the precursor of the wave theory.

Another person who was famous for his contributions to wave theory was the Dutch physicist, Christian Huygens, who proposed a wave theory of light in opposition to Newton's corpuscular theory. The basic idea he introduced, known as the Huygens principle, states that every point on the surface of an advancing wave is the source of new wavelets, which then form the new surface of the advancing wave. Thus each wave surface renews itself so that the wave advances. Huygens was never satisfied with a slipshod approach to the solution of problems and insisted on mathematical rigor in all his work, which probably set the proper tone for future mathematicians. By applying his mathematical skills to the study of light waves through small apertures he showed that his wave theory of light—unlike Newton's corpuscular theory—explained observed phenomena such as optical interference and optical diffraction.

Post-Newtonian mathematics was greatly influenced by the intimate relationship between mathematics and physics. Indeed, the need to solve physical problems, particularly in mechanics and in gravity, was probably the driving force behind the rapid development of post-Newtonian mathematics. The biographies of these remarkable mathematicians lists them as physicists and mathematicians, or as astronomers and mathematicians, as though they had studied mathematics to help them in their pursuit of science. Though primarily a physicist, Newton was the supreme mathematician of his era. If he had done nothing in physics, his mathematical discoveries would still have made him immortal. Leibnitz, on the other hand, was a full-time philosopher and mathematician, but only a part-time physicist. Fermat was primarily a mathematician, but still was enough of a physicist to discover the principle of minimum time, which was the first of a group of minimal principles, which are still of great importance and usefulness in modern physics. Philosophy and even a bit of theology played a role in the development of mathematics and mathematical physics; Maupertuis was led to his principle of least action by his belief that action is a kind of divine commodity that nature uses sparingly. Although Maupertuis introduced action as an ethereal concept, it ultimately acquired a physical reality which has enabled it to become one of the dominant concepts in our picture of nature and the universe.

Although the Bernoullis are listed as mathematicians in the history of mathematics, their fame as mathematicians stemmed primarily from their work in physics. This is also true of Huygens, who used mathematics primarily as a tool in his work in optics, particularly in his development of the wave theory of light. Indeed, his great mathematical skill enabled him to challenge Newton's corpuscular theory of light. The Bernoullis and Huygens were not, strictly speaking, post-Newtonian mathematicians, but their discoveries in physics and mathematics and their skills in combining physics and mathematics greatly influenced the post-Newtonians who were the first mathematicians to establish physics as an intellectual discipline, distinct from—but not independent of—mathematics. In spite of the recognition of this dichotomy, however, the dependence of physics on mathematics did not diminish but instead continued to increase. Indeed, as physics grew it became increasingly dependent on mathematics to serve as its language. This is evident in the works of Euler, Lagrange, Laplace, and their contemporaries, such as d'Alembert. Euler was primarily a mathematician but one of the most famous equations in dynamics is his partial differential equation that presents Newtonian dynamics in its most general and elegant form. This equation, as we have already noted, introduced the calculus of variations into physics. Lagrange, on the other hand, was as much a physicist as a mathematician. Modern mathematical or theoretical physics really began with Lagrange even though Newton was the first to formulate laws of physics mathematically as exemplified by his laws of motion and his law of gravity. Laplace inherited Lagrange's mantle and was the last of the great eighteenth-century figures who were both mathematicians and physicists.

One of the most important contributions the post-Newtonian mathematicians made to physics is the knowledge that mathematics can reveal new aspects of physical laws that actually lead to new discoveries and new theories which may not be immediately apparent in these laws as they were first formulated. The reason for this power of mathematics is that the mathematical manipulation—however complex and bold—of a simple law does not invalidate the law in any way, but instead leads to new features of it that greatly enrich our understanding of the law and our ability to apply it. In the following chapter, we illustrate this idea in the work of another group of remarkable mathematicians who constructed the mathematical foundation of modern physics and modern mathematics.

CHAPTER 8

The Golden Age of Mathematics

*In most sciences one generation tears down what another has built,
and what one has established another undoes. In mathematics alone
each generation builds a new story to the old structure.*

—HERMANN HANKEL

The eighteenth-century post-Newtonian era of mathematics was dominated by three great mathematicians, Euler, Lagrange, and Laplace, with the important activities of this era centered in France. The contributions by English, German, and Italian mathematicians were relatively meager during this period. It was at this time that an unbreakable bond was forged between mathematics, physics, and astronomy. Indeed, the title "mathematician" was applied to the pure mathematician, to the physicist, to the astronomer, and to the practitioner of all three disciplines indiscriminately. What we now call theoretical physics was then called mathematics or mathematical physics. But, in the nineteenth century mathematics acquired a complete life of its own and evolved quite independently of physics, proliferated explosively into so many different branches of mathematics that the mathematicians themselves were hardly able to follow the developments in one or two of these branches, let alone make significant contributions to physics or astronomy. Nevertheless, this golden age of mathematics was marked by the emergence of a few towering geniuses, such as Gauss, who contributed to both mathematics and physics.

This period of time has been called the "golden age of mathematics" because during this period the foundations of all branches of modern mathematics were laid. The eighteenth century had been characterized by the efforts of many mathematicians to consolidate and apply the calculus to a variety of problems. The nineteenth century, by contrast, was marked by the growth of a new mathematical freedom which broke all the constraints that had tied mathematics to physics and to Newtonian and pre-Newtonian mathematics. This does not mean that mathematical continuity was thus broken; nothing can be further from the truth. Instead, it means that mathematicians felt free to invent or develop new branches of mathematics that required the introduction of new arithmetic and algebraic laws or rules, leaving the basic principles of the calculus unchanged.

The nineteenth century also marked a shift of the center of mathematics from France eastward to Germany. Even though the French mathematicians were still leaders, German mathematicians, spearheaded by Gauss, began to challenge this French leadership. This shift was not without precedent. For example, even though the Newtonian era had been dominated by English mathematicians, they had allowed their leading role to slip away from them during the post-Newtonian period primarily because Newton's influence was so great that the English academic community considered it sacrilegious to try to alter or extend Newtonian mathematical concepts. Indeed, any movement to adopt the Leibnitz notation d for the derivative, as in dy/dx, rather than the Newtonian dot, as in $\dot{y}$, was severely criticized. Only when English mathematicians began to study continental mathematics independently did they begin to contribute significantly to nineteenth-century mathematics.

The post-Newtonian mathematics was spread eastward by Nikolaus Bernoulli, Johann's son, and by Euler. Nikolaus himself came to St. Petersburg, Russia, in 1725, and Euler followed him, remaining there until 1741. How important their presence was to Russian mathematicians can only be surmised, but it is interesting that the great Russian geometer Nikolai Lobatchevski was one of the first to develop a non-Euclidean geometry. Similar ideas were considered at about the same time by the Hungarian mathematician János Bolyai.

From the time of Galileo to this golden age, Italian mathematics lay dormant, but the geometrical concepts of the German geometers and the algebraic and function concepts of the French mathematicians finally swept into Italy in the nineteenth century, greatly stimulating Italian mathematics. This intellectual convergence marked the rebirth of Italian mathematics. At the same time, Scandinavian and American mathematicians began to contribute importantly to mathematics. In this chapter and the next, we shall attempt to delineate these developments, pointing out how they arose, their interrelationships, and their influence and applications to the physical sciences.

We begin with the mathematical discoveries of one of the most remarkable and tragic figures in the history of mathematical thought, the French mathematician Évariste Galois, who was born in 1811 and killed in a senseless duel in 1832. Considering the brilliance of his mathematical discoveries before he was twenty years old, we can only wonder in awe at what he might have accomplished had he lived a full life. Galois, essentially an algebraist, and greatly influenced by the algebraic discoveries of Lagrange, completed Lagrange's unfinished work on the solution of algebraic equations. In the process, he founded the theory of groups, one of the most profound branches of mathematics, with important applications to mathematics, physics, chemistry, and astronomy.

Galois' short life was one in which his potential for becoming perhaps the greatest mathematician of all time was snuffed out by both his own obstinacy and almost unbelievable bad luck. He was born in Bourg-la-Reine, a village outside of Paris, to Nicolas and Adélaide-Marie Galois. Nicolas was the town mayor and a vocal advocate of democracy and liberty. Adélaide-Marie was the daughter of a judge; she had received a rigorous classical education and was as independent in her thinking as her husband. Évariste himself was a happy child who liked to perform skits and sing songs for his family. Because no one on either side of the family had ever demonstrated any real talent for mathematics, Évariste had no real inkling that he could be a mathematician until he was well into adolescence.

This carefree childhood was shattered when Évariste, at the age of twelve, was sent to the lycée of Louis-le-Grand in Paris, which resembled a prison more than a school. Here, the students were abused verbally and beaten for even the most minor infractions. When the students finally rebelled against this heavy-handed regimen by refusing to participate in the chapel services, those persons believed to be the ringleaders of the makeshift revolt were expelled. Unfortunately for Galois, his conduct was apparently not serious enough to warrant dismissal from the school.

After the rebellion ended, the remaining students returned obediently to their classes. The revolt had little real effect on the way in which the school was operated, but it seemed to awaken a new spirit of independence in Galois. Though barely a teenager, Galois began to question everything, whether it was the justice of the existing social hierarchy in Paris or even the value of his classical education. More and more, the works of Aristotle and Virgil bored him. Mathematics, the only subject that really interested him, was little more than an appetizer in the school curriculum, especially when compared to the smorgasbord of Greek and Roman literature and philosophy offered by its instructors. But Galois was attracted to mathematics like a moth to light. He became smitten with a treatise by the French mathematician Legendre on geometry, which revealed mathematics to Galois as an art as opposed to a vocation for the very first time. Having seen the light, Galois decided to tackle the works of other great mathematicians such as Lagrange. Soon the mastery of university-level textbooks and scholarly papers became the focus of his waking hours; his school studies, which provided little intellectual challenge to him, occupied very little of his time. Not surprisingly, Galois' neglect of his studies did not endear him to his teachers.

That Galois was a special student was underscored by his ability to do the most complex mathematical calculations in his head. He seldom obeyed the command of a teacher to show his work because he found the use of pencil and paper to be too slow. Usually he offered the correct full-blown solutions without explanation to his dumbfounded teachers. These were the very same teachers who wondered how he could have such talents in mathematics and be such a poor student. But Galois was convinced by this time that he was a genius; he believed that he needed only to find the appropriate opportunity so that he might showcase his mathematical abilities and realize his dream of obtaining a university appointment. This movement seemed to arrive when Galois sat for the entrance examinations at the École Polytechnique, the training ground for the outstanding army engineers who would lead France's next generation of soldiers. Galois did not bother to prepare for the examinations, perhaps expecting that the examiners would recognize his mathematical brilliance. But the congratulatory handshakes that he expected to receive from them never came because he failed the examinations. Galois himself took the rejection very badly; his reaction was one of shock and bitterness. Whether the results were due to Galois' own failure to prepare or to examiners who were less intelligent than he has never been conclusively resolved although historians now favor the latter explanation.

In 1828, Galois met the mathematician Louis-Paul-Émile Richard, who was better known for his ability to teach advanced mathematics than for his own mathematical research. Richard was completely devoted to his students and would do whatever he could to help place them in universities or with the government. As his students included such towering figures as the astronomer Leverrier and the mathematician Hermite, Richard was well suited for the task of training brilliant mathematical minds. More important, he protected his students and provided them with personal and professional guidance. In Galois, Richard saw a mathematician who could become the

Gauss of France. Indeed, Galois might have fulfilled this promise had he lived long enough.

At Richard's urging, Galois published his first mathematical paper on continued fractions in 1829. Although Galois had spent the past year making very important discoveries in a number of areas such as the theory of equations, he had put off submitting a paper summarizing his original work until after he secured the promise of Augustin-Louis Cauchy, one of the most prolific mathematicians in history, to submit this paper on his behalf to the Academy of Sciences. Cauchy was well-intentioned but his memory did lapse on occasion and he forgot all about the promise he had made to Galois. As a result, Galois' paper gathered dust in Cauchy's office while Galois waited and waited for Cauchy to fulfill his promise. Needless to say, Cauchy never did submit the paper and Galois once again found himself the victim of forces beyond his control.

After taking and failing the entrance examinations at the École Poly-technique a second time, Galois resigned himself to pursuing a career as a schoolteacher. He was admitted as a university student but found little challenge in his studies, preferring to devote his time to solving advanced problems in algebraic equations. He composed a paper summarizing his results and submitted it to the Academy of Sciences in the hopes of winning the coveted Grand Prize in Mathematics. The receipt of such an award would catapult Galois to the forefront of Europe's ranks of mathematicians. Had his paper been reviewed, the Prize would have been his for the asking. But the black cloud over Galois appeared once again and the paper was lost by the Academy's secretary.

A person can bear only so many misfortunes and this incident with the Academy seems to have been the final straw that destroyed Galois' mathematical career. His entire life had been beset with a series of misfortunes that, when combined with his own lack of tact in dealing with authority figures, had created an unbelievable legacy of failure. This tragedy was all the more pronounced owing to Galois' staggering genius. But Galois no longer had even the slightest hope that his mathematical career could be salvaged; he apparently concluded that he was destined for greatness in some other area. Still shy of his twentieth birthday, Galois saw the scientific and educational system in France—which had deprived him of the honors and rewards he so richly deserved—as being representative of a pervasive social structure that oppressed the masses. This intense bitterness against persons of power and privilege emboldened Galois to enlist in the National Guard, a makeshift police force whose ranks consisted of many democrats and anarchists.

His own revolutionary sentiments fueled by the proud boasts of his fellow revolutionaries at a Guard banquet, Galois was goaded into offering a toast to King Louis Philippe, which was interpreted by loyalists as a threat against the ruler's life. Galois was thrown into prison and then forced to endure a trial in which he repeatedly attacked the court and the government itself. Fortunately for Galois, neither the court nor the jury believed that he—still a young man of twenty—deserved to be punished, especially because the testimony of the witnesses had not been in agreement as to the exact words that Galois was alleged to have said. The end result was a not-guilty verdict. Galois had won his freedom.

This freedom was to be short-lived, however, because Galois was soon arrested again as part of an ongoing effort by the monarchy to dismantle the republican organizations that continued to undermine popular support for the government. Now dubbed a "radical" by the authorities, Galois was thrown into prison but not charged with having committed any crime. The prosecutors eventually charged him with impersonating a soldier as he had been arrested while wearing a uniform after the National Guard had been disbanded. But Galois was apparently not so dangerous as to be consigned to life behind bars. A cholera epidemic led to Galois' transfer to a hospital and then, ultimately, his parole. But soon after his release, Galois was challenged to a duel by one of his political enemies. Sadly for the world of mathematics, Galois' own twisted sense of honor compelled him to accept the challenge. He met his opponent on May 30, 1832 and was shot in the stomach. His death was not instantaneous, instead a torturous day-long peritonitis-ridden purgatory. He closed his eyes for the last time on the morning of May 31 and was buried in a common grave. Only then did Galois finally find the peace of mind that had eluded him throughout his entire twenty-one years of life. But his mathematical legacy, though barely enough to fill a small notebook, was of such obvious quality and originality that it ensured Galois would always be remembered.

The most important problem in algebra that emerged in the eighteenth century for Galois and his fellow mathematicians was finding the roots of the general algebraic equation of the nth degree which we obtain by setting the nth degree polynomial $a_0 + a_1x + a_2x^2 + \cdots + a_nx^n$ equal to 0. The quantities a_i, the coefficients of the polynomial, are real integers. In 1799, the great German mathematician Gauss proved the fundamental theorem of algebra: that every algebraic equation of the nth degree has exactly n roots: the polynomial is zero for n and only n numbers $x_1, x_2, x_3, \ldots, x_n$ (not all necessarily different), which may be real, imaginary, complex, rational, or irrational. We are all acquainted with the general quadratic equation $ax^2 + bx + c = 0$ from our high school intermediate algebra courses, where we learned that the two basic roots x_1 and x_2 of this equation are

$$x_{1,2} = \frac{-b \pm \sqrt{b^2 - 4ac}}{2a}$$

These two roots are called algebraic roots because they are expressed algebraically, with radicals and in terms of ordinary algebraic operations. The roots x_1 and x_2 may be equal and real or complex. The quadratic equation and its roots were known to the Egyptian mathematicians some 4000 years ago but the cubic and higher-degree equations were not studied until the sixteenth century when H. Cardano of Milan wrote down the general algebraic expressions for the three roots of the cubic equation (now known as Cardano's formula) in 1545, even though historical evidence indicates that Professor S. Del Ferro of the University of Bologna had solved the general cubic equation earlier as had the Italian Tartaglia. Cardano and Ferro later found the roots of the quartic (fourth-degree) equation in terms of the coefficients but all attempts to find solutions (the roots) of the fifth- and higher-degree equations expressed in terms of a finite number of elementary algebraic operations (addition, multiplication, and radicals) failed. All attempts to find such solutions of algebraic equations of degree higher than four were finally abandoned in the nineteenth century when Niels Abel proved conclusively in 1824 that the general fifth-degree (or quintic) equation cannot be solved by means of radicals alone; the Italian mathematician Paoto Ruffini had given a partial proof of this theorem some years earlier.

Lagrange's work on this general problem serves as an introduction to Galois' great contribution to the theory of algebraic equations and his founding of group theory. Lagrange sought general criteria or relationships among the roots of an algebraic equation of any degree that would reveal whether or not such an equation can be solved by elementary algebraic operations. The general aspects of Lagrange's problem may be considered briefly, starting from the general nth degree equation expressed above as $a_0 + a_1x + a_2x^2 + \cdots + a_nx^n = 0$. If the n roots of this equation are x_1, $x_2, x_3, \ldots, x_n$, we can write the left-hand side of this equation as a product of n factors $(x - x_1)(x - x_2) \ldots (x - x_n)$, each factor containing just one of the roots. Note that if $x = x_1$, the first factor vanishes and the entire product disappears. This is true if x equals any one of the other roots so that this product is exactly equal to the left-hand side of our nth degree algebraic equation. If we multiply out the n factor of the product, we obtain relationships between the n roots $x_1, x_2, \ldots, x_n$ and the coefficients $a_0, a_1, \ldots, a_{n-1}$ of the equation. This process reveals the following relationships:

$$x_1 + x_2 + x_3 + \cdots + x_n = -a_{n-1},$$
$$x_1x_2 + x_1x_3 + \cdots + x_1x_n + x_2x_3 + \cdots + x_2x_n + x_3x_4 + \cdots = a_{n-2},$$
$$\cdots \cdots \cdots$$
$$x_1x_2 \ldots x_n = +a_0.$$

The sum of all the roots equals the negative of the coefficient of x^{n-1}, the sum of the products of the roots, taken two at a time, equals the coefficient of x^{n-2}, and so on, until we come to a_0, which equals the product of all of the roots. Suppose now that we interchange the roots in any way we please, interchanging, for example, x_1 and x_4 or x_2 and x_5. Such interchanges, called permutations, leave the relationships between the various sums of products of the roots and the coefficients $a_0, a_1, \ldots, a_{n-1}$ unchanged. We then say that the various sums of products of the roots listed above are invariant to permutations of the roots. Lagrange saw that these multiple sums of products of the roots do not give any hint or indication as to whether or not the equation can be solved algebraically. Lagrange therefore sought other functions of the roots that are invariant to permutations and do reveal something about the solutions of the equation and even lead to solutions of special equations. Lagrange showed that this approach leads to the solutions of the cubic equation but fails when applied to equations of higher degree. Lagrange ended his investigations into the solutions of the general equation of the nth degree with the discovery of certain basic theorems but with a confession of defeat.

At this point in the story of the algebraic solutions of nth-degree equations, Galois appeared on the scene and greatly extended and generalized Lagrange's results. He treated the permutations of the roots of an equation as a special case of a group operation. Because the group operation is a powerful mathematical tool that is used extensively by physicists and mathematicians, we discuss the group concept at this point, showing how Galois arrived at it from the permutation concept. The group concept may be illustrated in its simplest form as a set of permutations. We consider a set of elements—numbers, letters, etc.—to which we assign an arithmetic operation such as multiplication; we henceforth call this operation multiplication. This set is a group if:

1. the multiplication of any two of these elements gives another element of the set;
2. the set contains the identity—an element that leaves unchanged any element of the set it multiplies;
3. the set contains the inverse of every element in the set so that an element multiplied by its inverse gives the identity.

These three simple rules define the group concept completely. A few examples will illustrate the usefulness of this concept in mathematics and physics.

We first discuss the group of permutation because that is the group with which Galois first began his work. The group may be illustrated with just four elements, e.g., the four numbers 1, 2, 3, and 4 arranged in any order we please. By a permutation (the operation to which we subject the group) we mean any rearrangement of the four numbers, such as interchanging the positions of 1 and 2 or 3 and 4. The total number of such permutations is easily calculated: We may place any one of the four elements 1, 2, 3, 4 in position number 1, which gives us four choices. With any one of these four choices, we have three choices of the remaining three numbers for the second position. This gives us 4×3 or 12 ways of filling the first two positions. We are thus left with only two ways of filling the third place and one way of filling the fourth place. The total number of permutations of four elements is thus $4 \times 3 \times 2 \times 1$ or 24, written as 4! (4 factorial). If we apply a permutation P_1 to our four numbers and then apply the permutation P_2 to the new arrangement, we express this as the product $P_2 P_1$. This is equivalent to the single permutation P_3, which we could have applied directly to the initial arrangement 1234.

We may illustrate this idea further by considering just two simple permutations, each involving the interchange of only two of the four numbers in our arrangement. Let P_{12} be the interchange of the numbers 1 and 2 so that $P_{12}(1234) = 2134$. We now apply the permutation P_{24} (the interchange of 4 and 2 to this new arrangement: $P_{24}(2134) = 4132$, so that $P_{24}P_{12}(1234) = 4132$. This is equivalent to the permutation P_{124} applied to the arrangement 1234, which leaves 3 unchanged, but interchanges 1 and 2 and then 2 and 4. Thus $P_{24}P_{12}(1234) = P_{124}(1234)$. The product of the two permutations P_{12} and P_{24}, written in the order $P_{24}P_{12}$, equals the permutation P_{124}, which is one of the permutations in the group of the 24 permutations described above. This fulfills one of the conditions that the permutations constitute a group—that the product of any two permutations forms a permutation. We also include among our permutations the identity I, which leaves the arrangement of our numbers unaltered. Thus $IP = P$, where P is any permutation. Finally, for any permutation P in the group we also include the inverse permutation P^{-1}, which exactly reverses the permutation P so that $P^{-1}P = I$.

We have gone to some pains and into considerable detail in discussing the group properties of permutations because the group concept is of great importance to physics and mathematics. Before closing this discussion of the permutation group we consider the concept of the "product" of numbers of a group more carefully. This product is not to be interpreted as an arithmetic product such as $3 \times 4 = 12$; permutations are not numbers at all but operations so that the product of two operations means one operation followed by another of the same sort. Our specification of a group is therefore not complete unless we specify the nature of the mathematical operation. Thus we may consider all the positive and negative integers as constituting an infinite group if the operation is addition: the sum of any two integers is an integer. In this group, zero must be included. It is indeed the identity element of this "group" because adding 0 to any integer leaves the integer unchanged. The inverse of any element in this group is its negative. Galois did not treat the integers as a group but rather as a field; the concept of the "field," any manifold of mathematical elements of some sort, plays an extremely important role in modern algebra.

Because the operation in a group is called multiplication, we see that the order in which two elements in the group are multiplied (the product of the two elements) is important. We note, for example, that the order in which the two operations of firing and loading a gun are completed is important as far as the final result of these two operations is concerned. If P_1 and P_2 are elements of a group, the product P_1P_2 does not equal the product P_2P_1. As a general rule, groups are noncommutative. We illustrate this point with the two permutations P_{12} and P_{24} applied to the arrangement 1234 discussed above. We see that the product $P_{24}P_{12}$ applied to 1234 gives us an arrangement with 2 in the last place, whereas the product $P_{12}P_{24}$ applied to 1234 gives us an arrangement with 2 in the first place. Thus $P_{24}P_{12}$ does not equal $P_{12}P_{24}$. Groups whose elements do not commute when multiplied in pairs are called Abelian groups in honor of the brilliant Norwegian mathematician Abel who studied commutative groups in detail. His life paralleled that of Galois; they were born nine years apart, and died four years apart. They both contributed extensively to the theory of algebraic equations and to algebra in general.

Like Galois, Niels Henrick Abel (1802–1829) was a shooting star on the mathematical landscape who might have otherwise enjoyed a long and productive career had it not been for bad luck and the grinding poverty that dogged him throughout his life. He was not destined to enjoy the rewards of his mathematical legacy but he did make the most of his few short years on earth despite the indifference to his work displayed by most of the leading mathematicians of that time.

Niels was born in the remote village of Findö, Norway, where his father eked out a living as the town pastor. His mother, Marie Simonsen, was a handsome woman who was not very enthusiastic about playing the role of the dour pastor's wife. Unlike her pessimistic husband, Marie was extroverted and optimistic about life. Young Niels seems to have inherited both her good looks and her cheerful outlook; he needed this legacy in the coming years to overcome the many disappointments he would encounter in his brief career.

Niels and his six siblings seemed to thrive in the rugged winters that came each year to their mountainous land, but they also knew hardship and deprivation. Norway was an unforgiving land in the early nineteenth century; a person considered himself to be fortunate if he could find enough to eat on a regular basis. Despite the often bleak state of the family finances, Niels and his brothers and sisters went on with life as best they could, often studying their lessons or telling stories to each other around a blazing hearth in the family cabin.

His own mathematical gifts began to appear while Niels was still in his early teens. He was fortunate to be taught mathematics at the local school by a mediocre but enthusiastic mathematician named Bernt Holmboe. In much the same way that Isaac Barrow had encouraged Isaac Newton to focus on the subject of natural philosophy, Holmboe generated the jolt that Niels needed to waken his brain from its intellectual stupor. Under Holmboe's guidance, Niels was introduced to the works of Newton, Euler, and Gauss. Much to Holmboe's amazement and delight, Niels soaked up the works of the masters as easily as some of his classmates quaffed a pint of ale. Like Galois, Niels did not seem to find it very difficult to comprehend even the most complex

ideas of Lagrange or Archimedes. Within a very short time he had acquired
as extensive a knowledge of mathematics as almost any mathematician in
Europe.

The death of his father in 1820 forced Niels, the eldest brother, to
confront the prospect of providing for his mother and siblings, none of whom
had yet perished in the bitter Norwegian winters. Having developed something of a local reputation for his mathematical skills, Niels tutored pupils
of all ages and earned enough money to put food on the family table. Even
so, the meals were spartan and the hours demanding. Niels had to limit his
own mathematical researches to those few spare moments in the early
morning and late evening when he could be by himself. Through it all,
however, Niels remained confident of his own abilities and cheerfully
carried on with fulfilling his role as the family breadwinner as best he could.

Much of his researches during his later adolescent years focused on the
theory of infinite series and solving the general equation of the fifth
degree—a task that had confounded mathematicians for centuries. Although
Niels was not able to solve the equation, he did, after further review, prove
that such an equation was impossible to solve and thus saved many future
mathematicians fond of intellectual self-flagellation from squandering years
of their lives on a futile search for the solution. In deducing this proof, Niels
made a discovery using group theory that was almost as remarkable as any
solution even though his disappointment prevented him at first from recognizing the true value of his achievement.

Abel completed his undergraduate studies at the University of Kris-
tiania in 1822. His mathematical skills had already been recognized by
several of Norway's more prominent mathematicians, but Norway itself did
not rank very high in the nineteenth century pantheon of mathematics.
Abel's fame had not yet moved beyond the borders of his own country so that
he was totally unknown in France and Germany, then the leading centers of
mathematical studies. Abel himself dreamed of traveling to France so that he
could meet such titans as Legendre and Cauchy. To this end, Helmboe and
several of his colleagues spent the next two years begging the nearly
bankrupt Norwegian government to grant Abel enough money to cover the
costs of a year of study abroad. Finally, the government granted the request.

Before leaving Norway, Abel did manage to scrape enough money
together to finance a limited printing of his proof dealing with fifth-degree
solutions. Abel hoped that the memoir would serve as a letter of introduction
to Europe's leading mathematicians. Sadly, the impact of this memoir would

prove to be far beneath Abel's lofty expectations—at least with regard to his idol, Gauss, to whom he sent the first copy. Gauss, who was something of a prig when it came to acknowledging the greatness of other mathematicians, never bothered to read Abel's tract but instead tossed it in the trash with not so much as an acknowledgment to its hopeful author. Needless to say, Abel's opinion of Gauss changed greatly after he learned of the shabby way in which his brilliant paper had been received.

After deciding not to visit Gauss in Göttingen, Abel traveled to Berlin where he soon made the acquaintance of August Crelle, a prosperous civil engineer, who studied mathematics as a hobby. In Crelle, Abel found a friend as well as a colleague, one who recognized Abel's genius and his need for some assistance in establishing his reputation among mathematicians. As Crelle was the founder of the first journal devoted exclusively to mathematical research, Abel found a ready outlet for his own mathematical papers. Indeed, Crelle's journal might well have been named after Abel as nearly two dozen of Abel's papers appeared in the first three volumes. Although the journal was not widely known at the time, its circulation was enough to bring Abel's work to the attention of Europe's leading mathematicians. The quality of Abel's papers made it possible for him to repay his considerable debt to Crelle because he helped to establish the reputation of the journal itself as a showcase of original mathematical research.

Crelle introduced Abel to some of his mathematician friends and then sent Abel on to Paris to meet Legendre and Cauchy. Any thoughts Abel might have had about being warmly welcomed into the mathematical fraternity by these admittedly self-absorbed snobs were soon dashed. These shining lights of French mathematics either had not bothered to read Abel's papers thoroughly or did not fully understand the ramifications of his work. Abel thus spent much of his time in Paris listening patiently to Legendre's long-winded discourses on mathematical discoveries that Abel had, in many cases, long since surpassed.

After Abel's forgettable experiences in Paris, he was understandably eager to leave France. Having completed a brilliant paper which pruned the entire field of mathematical analysis of inconsistencies and errors, Abel left his paper in the care of Cauchy, who promised to submit it to the Paris Academy of Sciences. As with Galois, Cauchy promptly lost the paper—the one that Jacobi called the most important mathematical discovery of the century and which Hermite declared had given future mathematicians five hundred years of work—amidst the clutter of his own office. More than a year passed before Cauchy finally retrieved the paper at the urging of Jacobi. But it was not published for another decade, long after Abel had passed away and could no longer receive the well-earned accolades for this remarkable work.

While this tragic comedy of errors was being played out by Cauchy in Paris in 1828, Crelle tried to wrangle a professorship for Abel at the University of Berlin. Abel himself was trying to recover from what he believed was a stubborn cold but which was later diagnosed as tuberculosis. Abel ignored the doctor's warnings that he take better care of himself, finding that the yoke of his family's dependence gave him little choice but to continue teaching pupils while waiting for a professorship to materialize. But his health declined markedly and, after a long illness, he died on April 6, 1829, several months shy of his twenty-seventh birthday. The final irony of Abel's short but tragic life was that Crelle succeeded in securing the long-sought appointment for Abel at Berlin, several days before Abel's death. Abel and Galois are both recognized today as the founders of group theory.

Before leaving groups at this point, we note that mathematicians have divided groups into two general categories: the finite or discrete groups—those containing a finite number of elements—and the infinite or continuous groups. The group of any number n of permutations is an excellent example of a finite or discrete group, whereas the rotation of a body such as a cube is a good example of an infinite or continuous group. Continuous groups are called infinite because they contain an infinite number of elements or operations which are infinitesimally close to neighboring elements. Thus we can rotate a body by an infinitesimal amount so that the rotation group is continuous.

A simple example of a finite group in which the operation is arithmetic multiplication is given by the four units of the complex number system: 1, -1, i, and $-i$, where $i = \sqrt{-1}$. In this group 1 is the unit element and it is its own inverse and the product of any two elements is also an element of the group. Thus, $-1 \times -1 = 1$, $i \times i = -1$, $i \times i \times i = -i$, etc.

Galois' great contribution to group theory and the solutions of algebraic equations was his discovery that the application of the permutation group to certain relationships among the roots of the equation lead to criteria for the solution of the equation by algebraic procedures alone, in terms of square roots, cube roots, etc. We recall that Cardano in the sixteenth century had solved the general cubic equation, expressing the three roots in terms of algebraic functions of the coefficients of the equation. He had already shown that the general fourth-degree equation can be solved algebraically, but he was unsuccessful in stating or deducing anything about the solutions of fifth- or higher-degree equations. Here Galois applied his theory of groups to show that no algebraic equation of a higher degree than 5 can be solved alge-braically. As discussed above, Abel himself thought that he had found the general solution of the fifth degree in 1823, but he retracted his claim a year later in a famous paper in which he proved that the general fifth-degree equation cannot be solved in terms of algebraic radicals—square roots, cube roots, etc.

By this time, mathematicians had begun to study infinite series, which play very important roles in the theory of functions of complex variables. The important question which arises in connection with sums of infinite numbers of terms is whether they converge (are finite) or diverge (infinite). Two examples illustrate the properties of divergence and convergence—the harmonic series and the geometric series. The harmonic series is simply the sum of all the proper fractions formed by inverting the positive integers $1 + \frac{1}{2} + \frac{1}{3} + \ldots + 1/n + \ldots$. It is easy to show that this series diverges by comparing it term by term with a series that we know definitely diverges. We first rewrite the harmonic series, collecting the terms into groups: $(1 + \frac{1}{2})$ $+ (\frac{1}{3} + \frac{1}{4}) + (\frac{1}{5} + \frac{1}{6} + \frac{1}{7} + \frac{1}{8}) + \ldots$; and then compare each group in parentheses with the groups in parentheses in the infinite series $(\frac{1}{2} + \frac{1}{2}) +$ $(\frac{1}{4} + \frac{1}{4}) + (\frac{1}{8} + \frac{1}{8} + \frac{1}{8} + \frac{1}{8}) + \ldots$. Each group (in parentheses) sum in the harmonic series is larger than the corresponding group sum in the second series. Therefore, the harmonic series diverges (the total is infinite) because the second series diverges (each group in parentheses equals $\frac{1}{2}$ or more so that we have an infinite sum of such finite terms). We thus see that we can determine if a series diverges if, upon comparing it term by term with a series that does diverge, we find that each term of the given series is larger than the corresponding term of the comparison series. Because the harmonic series diverges we can use it as a comparison series to test for the divergence of other series. Thus, the series $1 + 1/\sqrt{2} + 1/\sqrt{3} + \ldots$ diverges because each of its terms is larger than the corresponding term of the harmonic series. We see that $1/\sqrt{n}$ is larger than $1/n$ because $\sqrt{n}$ is less than n if n is an integer.

The comparison test for divergence does not tell us anything about the convergence of a series. Even if the comparison test does not show that the series diverges, we may not conclude that the series converges. Finding the test for the convergence of a series therefore became one of the important goals in nineteenth-century mathematics. Abel, deeply concerned about the lack of rigor in the approach of the early nineteenth-century mathematicians to the mathematics of infinite series, expressed openly his contempt for the work being done on infinite series at that time. In a letter to a friend in 1826, he stated "Can you imagine anything more horrible than to claim that $1^n - 2^n + 3^n - 4^n + \ldots = 0$, n being a positive integer. There is in mathematics hardly a single infinite series of which the sum is determined in a rigorous way." He felt a great urgency about the need for rigor in this branch of mathematics and began the work that finally led to a rigorous theory of infinite series. Abel laid the foundation of a theoretical structure which was completed by Cauchy.

An important example of a convergent series is the geometric series

$$1 + \frac{1}{2} + \frac{1}{4} + \frac{1}{8} + \cdots + \frac{1}{2^n} + \cdots,$$

that is, the infinite series, each term of which is the reciprocal of a power of 2. One shows, with elementary algebra, that the sum of this series is finite; the series converges. If we call the sum of this series S, we have $S = 1 + \frac{1}{2} + \frac{1}{4} + \frac{1}{8} + \ldots = 1 + \frac{1}{2}(1 + \frac{1}{2} + \frac{1}{4} + \ldots)$, where we have taken the factor $\frac{1}{2}$ out of each term in the series after the first term 1. We now note that the sum of terms, $1 + \frac{1}{2} + \frac{1}{4} + \ldots$, in the parentheses on the right-hand side is exactly the original series and therefore equals S. We thus obtain the equation $S = 1 + \frac{1}{2}(S)$ or $S - \frac{1}{2}S = 1$ or $\frac{1}{2}S = 1$, so that $S = 2$. Because we now have a series that converges, we can use it as a comparison series for convergence. Thus, the series

$$1 + \frac{1}{2^2} + \frac{1}{4^2} + \frac{1}{8^2} + \cdots$$

converges because each term in this series is smaller than the corresponding term in the geometric series. We note that the series $1 - \frac{1}{2} + \frac{1}{3} - \frac{1}{4} + \ldots$, called an alternating series because the $+$ and $-$ signs alternate, converges. In fact, any alternating series in which the terms grow successively smaller, approaching 0 as the number of terms in the series increases, converges.

Abel did not limit himself to the study of numerical series but extended his investigations into the properties of infinite algebraic series or "power series," the general example of which is written as $a_0 + a_1x + a_2x^2 + \ldots + a_nx^n + \ldots$. Abel was interested in such series because certain mathematical functions that he was studying cannot be written in terms of simple algebraic expressions but can be expanded as power series. These power series are also very important in the solution of differential equations, most of which cannot be solved algebraically. The importance of this concept for science is that all natural phenomena can be described in the form of differential equations which, in general, can only be solved by infinite series. Abel did not discover the general criteria for the convergence of a power series, which involve the numerical values of the coefficients a_0, $a_1, a_2, \ldots, a_n, \ldots$ of the series. This discovery was made by Cauchy, who, perhaps more than any other mathematician, insisted on rigor in every branch of mathematics.

We have discussed above the main trends of early nineteenth century mathematical thought and emphasized the contributions of Galois and Abel to these two branches of mathematics. But other branches of mathematics such as the theory of numbers, probability theory, geometry, differential equations, differential geometry, the calculus of variations and mathematical physics also expanded rapidly during that era. The theory of numbers and probability theory have one thing in common: they did not evolve as parts of the mainstream of mathematics, rather they grew independently from geometry, algebra, mathematical analysis, and function theory. Nevertheless, the developments of number theory and probability are associated with mathematical investigations of all the great mathematicians from antiquity until today. The theory of numbers was and is especially esteemed by mathematicians, who consider it the "purest" branch of mathematics because it must be studied for its own sake rather than for its usefulness in science and technology. On the other hand, mathematicians dipped into the theory of probability owing to its usefulness in predicting the occurrence of natural events and in calculating the chance of winning at gambling. The theory of numbers did not grow as rapidly as did mainstream mathematics because the problems in number theory are exceedingly difficult. Probability theory, in contrast, evolved slowly because mathematicians, with few exceptions, did not consider it a serious branch of mathematics. Gauss was one who contributed significantly to the growth of both of these esoteric branches of mathematics.

The theory of numbers predates probability theory by many centuries because number theory is a direct outgrowth of arithmetic. It differs from pure arithmetic, however, in that it deals primarily with abstract numerical relationships rather than with numerical computations. In pointing out the very high abstract plane of the theory of numbers, which he called "higher arithmetic" Gauss stated: "The most beautiful theorems of higher mathematics have this peculiarity, that they are easily discovered by induction [intuition] while, on the other hand their demonstrations [proofs] lie in exceeding obscurity and can be ferreted out only by very searching investigations. It is precisely this which gives to higher arithmetic the magic charm which has made it the favorite science of leading mathematicians not to mention its inexhaustible richness, wherein it so far excels all other parts of mathematics." Simply stated, Gauss meant that theorems in number theory are easy to set up but exceedingly difficult to prove.

The following few examples of such theorems illustrate the truth of Gauss's statement:

1. The quantity $1 \times 2 \times 3 \times 4 \times \ldots \times (n - 1) + 1$ is always divisible by n if n is a prime but never when n has factors (is composite).
2. Every prime number of the form $4x + 1$ can be expressed as the sum of two integers in one and only one way.
3. Every even number can be expressed as the sum of two primes.
4. The expression $2^x - 1$ is a prime for a general formula for x.
5. The equation $x^n + y^n = z^n$ does not hold for integer values of x, y, z if the integer n is larger than 2 (Fermat's last theorem).

These theorems have stimulated some of the most intense and fruitless research in the history of mathematics. Even today, nearly every mathematician spends some time trying to prove theorems 3 and 5. Several of the more distinguished mathematicians have expressed their views concerning the status of the theory of numbers in the scheme of modern mathematics. The greatest of all twentieth-century mathematicians, David Hilbert, stated that, "In the theory of numbers, we have the simplicity of its foundations, the exactness of its conceptions and the purity of its truths; we extoll it as the pattern of the other sciences, as the deepest inexhaustible source of all mathematical knowledge, prodigal of incitements to investigation in other departments of mathematics. Moreover, the theory of numbers is independent of the change of fashion and in it we do not see, as is often the case in other departments of knowledge, a conception or method at one time given undue prominence, at another suffering undeserved neglect; in the theory of numbers the oldest problem is often today modern, like a genuine work of art from the past."

The mathematician L. E. Dickson in his *History of the Theory of Numbers* wrote that "the theory of numbers is especially entitled to a separate history on account of the great interest which has been taken in it continuously through the centuries from the time of Pythagoras, an interest shared on the one extreme by nearly every noted mathematician and on the other extreme by numerous amateurs attracted by no other part of mathematics." In a similar vein, the twentieth-century mathematician G. H. Hardy noted that "the elementary theory of numbers should be one of the very best subjects for early mathematical instruction; its subject matter is tangible and familiar; the processes of reasoning it employs are simple, general, and few; and it is unique among the mathematical sciences in its appeal to natural human curiosity. A month's intelligent instruction in the theory of numbers ought to be twice as instructive, twice as useful, and at least ten times as entertaining as the same amount of 'calculus for engineers.'"

The theory of numbers is difficult because it is a mathematics of integers so that one cannot use ordinary algebraic equations in number theory. Whereas the equation $y = 4x$ in algebra means that we can find a value of y for every value of x, the same equation in the theory of numbers means that given a definite integer for x, y must be an integer 4 times as large. Does this mean that number theory cannot deal with fractions? The answer is no! Gauss introduced a very elegant, compact, and efficient symbolism for working with relationships among numbers, whether they are integer multiples of each other or not. Gauss called these relationships congruences, which in a restricted sense, are a kind of equation. Consider the equation $\frac{33}{13} = 2 + \frac{7}{13}$, where 7 is the remainder of the division of 33 by 13. Gauss called this relationship a congruence among the integers 7, 13, and 33 which (following Gauss) is written as 33 = 7 modulo 13 or 33 = 7 mod13. This means that the numerical difference between the two integers 33 and 7 (33 − 7) is exactly divisible by 13. The statement of this relationship is as follows: "33 is congruent to 7 modulo 13." For each different divisor or modulers, we obtain a different congruence. Thus 33 = 5 mod14. A congruence cannot be treated as an operation but it is governed by certain arithmetic and algebraic rules that are similar to those that govern ordinary algebraic equations.

Thus, we may add or subtract the same integer from both sides of a congruence without altering the congruence, as in the example 33 + 5 = (7 + 5) mod13 = 12 mod13. We may also add or subtract two congruences with the same modules. Thus the congruence 33 = 7 mod13 and the congruence 35 = 9 mod13 may be summed to give the congruence 68 = 16 mod13. A congruence may also be multiplied by a common factor or divided by a common factor so that, on multiplying by 3, the original congruence becomes $99 = 21$ mod 13. Finally, we may raise both sides of a congruence to the same power without altering it. Thus $33^2 = 7^2$ mod13 or $1089 = 49$ mod13. As Gauss demonstrated, congruences are very powerful analytical tools that enable one to determine the factors of fairly large numbers and thus whether or not such numbers are primes. This is one of the most challenging problems in the theory of numbers.

As an example of the analytical power of congruences, we noted that they were used to show that the numbers given by $2^{2x} + 1$ for all integer values of x are not primes, as they were thought to be before Gauss introduced congruences. Using congruences one can show easily that for $x = 1, 2, 3, 4$, the numbers $2^2 + 1$, $2^4 + 1$, $2^8 + 1$, and $2^{16} + 1$ are primes but that the number $2^{32} + 1$ is a composite number, divisible by 641. This is but a single example of the beauty and richness of the theory of numbers.

Just as Gauss contributed important concepts and techniques to the theory of numbers, he also contributed significantly to probability theory. The story of probability theory, in its early years, paralleled that of number theory, primarily because both theories then dealt with numbers rather than with abstract mathematical concepts. That the same early mathematicians who were most instrumental in bringing number theory to the attention of the mathematical world also brought probability theory to a respectable mathematical status is also interesting. But probability theory always had very important practical uses in its application to the early theory of games of chance because gambling of all sorts played a very important role in the lives of the dominant people of society. It then spread from the calculations of the chances of winning points in games of chance to the calculations or estimations of the chances of the success of commercial ventures of all kinds and the calculations of statistical tables, particularly mortuary tables, which are so essential to the proper operation of insurance companies.

Until the last half of the nineteenth century, probability theory remained tied to games of chance, with some applications to the analysis of errors in scientific observations. Gauss then discovered the exponential law of error as expressed in his famous probability integral. But in the 1860s, the theoretical investigations of the superb British physicist James Clerk Maxwell into the behavior of gases brought probability theory into such prominence that it became one of the most powerful analytical tools in theoretical physics. Through the work of the great Austrian theoretical physicist Ludwig Boltzmann and the theoretical work of Einstein, probability theory acquired its own niche called statistical mechanics in theoretical physics. We will discuss this remarkable application of probability theory to physics later, but we note here briefly how Maxwell was led to the application of probability theory to the study of gases.

In the seventeenth century, Robert Boyle, a contemporary and friend of Isaac Newton, discovered that if the temperature of a gas is kept constant, the pressure in the gas varies inversely with its volume: the smaller the volume, the higher the pressure. This is known as Boyle's law of gases. This law was later generalized to include temperature changes as well as pressure and volume changes of the gas and, in its generalized form, this law states that, no matter how the temperature, pressure, and volume of a gas change together, the number obtained by dividing the product of the pressure and volume by the temperature remains the same; this is the famous gas constant.

When Maxwell began his studies of gases, the molecular theory of matter was a purely speculative theory, accepted only by a few physicists and chemists, including Maxwell. He saw that if a gas consists of huge numbers of particles (molecules) moving randomly, such gross gas features as pressure and temperature should be deducible from the behavior of groups of molecules by assigning to these molecules average dynamical properties obtained from probability theory. This analysis led him to what is now called the kinetic theory of gases, and statistical mechanics.

Returning now to the origins of probability theory, we see that Dante, in the sixth canto of his *Divine Comedy*, commented on the different probabilities of the various throws of three dice. The earliest attempt at a rigorous mathematical treatise on probability, however, seems to have been written in the sixteenth century by Cardan (Cardano), an inveterate gambler. This treatise is more in the nature of a gambler's manual than a mathematical treatise.

Kepler seems to have been the first to apply probability theory to the occurrence of natural events. In his book, *De Stella Nova*, published in 1606, he speculated about the causes of the supernova that had appeared in 1604 and tried to calculate the probability of the formation of a new star, which he assumed the 1604 nova to be, by the fortuitous contraction of atoms.

Galileo also dabbled in probability theory when he became interested in the number of ways a throw of three dice can produce the numbers 9 and 10. After a very careful and accurate analysis of all the cases that can occur, he showed that 27 are favorable to the appearance of the number 10 and 25 are favorable to the number 9 out of 216 possible cases—contrary to superficial expectations.

The work of Kepler and Galileo emphasizes the prevalence of probability analysis in the thinking of scientists as well as mathematicians. But the first serious mathematical treatment of probability began with Fermat and Pascal, whose general contributions to mathematics we have already discussed. Their contributions to probability theory stemmed from their analysis, done jointly through extensive correspondence, of the English Problem of Points which may be stated as follows: Each of two players in a game of chance needs a given number of points to win. If they separate without playing out the game, how should the stakes be divided between them? This question is equivalent to asking what the probability is—at any stage of the game—that each player will win.

The analysis and solution of the problem, carried out with great skill and rigor, marked the beginning of the modern theory of probability, when Pascal and Fermat were the most distinguished mathematicians of Europe. To place their work in the correct historical perspective we note that Descartes died in 1650, Newton and Leibnitz were still to become known and Huygens was born in 1629. Pascal died in 1662 at the age of 39 and Fermat died in 1665 at the age of 64.

An interesting and important aspect of the theory of probability in its relationship to mathematics is that it stimulated the mathematical interests of many countries and thus contributed greatly to the remarkable international brotherhood of mathematicians. Up until the end of the eighteenth century, the dominant mathematicians, except for Newton, were French, German, and Dutch. But the theory of probability was pursued widely by mathematicians in all countries because games of chance were prevalent throughout Europe. In Holland, Huygens wrote an extensive treatise on dice, analyzing various problems dealing with throws of pairs and triplets of dice involving two or three players. He also extended his analysis to cards and to the drawing of balls of various colors from bins containing many such balls.

In England, the algebraist John Wallis, who was to become professor of geometry at Oxford in 1649, published the first important book on algebra in England, in which he developed the algebra of combinations, which is basic to the calculation of the probabilities of competing events. The study of combinations was greatly extended by the Swiss mathematician Jakob Bernoulli after 1687. Bernoulli was one of the first to point out the importance of the binomial coefficients in calculating the probabilities of competing events—the famous coin-tossing problem. One of Bernoulli's most important contributions to probability theory—which stemmed from his study of combinations—is his formula for the number of permutations of n things taken p at a time. This formula is used extensively in the study of the physical properties of systems containing large numbers of particles such as the molecules of a gas. This formula and the solutions of many different problems in the theory of probability are contained in Jakob Bernoulli's book *Ars Conjectandi* which was published eight years after his death.

Of all the mathematicians who applied their skills to the analysis of probability theory, the French–English mathematician Abraham de Moivre did more than the others to bring about the acceptance of probability as a legitimate branch of mathematics. Born in France in 1667, he emigrated to England where he lived until his death in 1754. By the time he was elected a Fellow of the Royal Society of London in 1697, he was held in such esteem that Newton himself referred questions about mathematics to de Moivre, saying "Go to Mr. de Moivre, he knows these things better than I do." J. Coolidge, the author of *The Mathematical Theory of Probability*, praised de Moivre as follows: "In the long list of men ennobled by genius, virtue, and misfortune, who had found asylum in England, it would be difficult to name one who has conferred more honour on his adopted country than de Moivre." To students of trigonometry de Moivre is famous for "de Moivre's Theorem," expressed in the trigonometric equation $(\cos\theta + i\sin\theta)^n = \cos n\theta + i\sin n\theta$, where $i = \sqrt{-1}$ and n and θ are any two numbers. But de Moivre's first love was probability and the laws of chance, to which he devoted most of his mathematical skills. He is famous for his memoir "De Mensura Sortis" and his book *Doctrine of Chances*. In *Doctrine of Chances*, de Moivre analyzed and solved seventy-four different problems related to life annuities; his solutions of these problems became the foundation on which modern annuity and life insurance tables are based. With the vast growth of insurance companies of all kinds, the theory of probability grew explosively and spread out into every branch of human activity, so that today people hardly ever speak of the certainty of events but instead of the probabilities of their occurrence.

We have discussed here only the principal figures in the early history of probability theory but nearly all of the mathematicians from the time of Newton devoted some time to this subject. Thus, Leibnitz discussed the theory of combinations in a dissertation on combinatorial analysis and the theory of games of all kinds, which he found extremely stimulating and challenging to his inventive powers. He considered them to be of great educational value, stating "that men had nowhere shown more ingenuity than in their amusements and that even those of children might usefully engage the attention of the greatest mathematicians." Although he himself did not contribute directly to the theory of games, he suggested quite forcefully that a comprehensive and systematic treatise on games should be written and should consist of three parts:

1. those games that depend on numbers alone;
2. those games like chess that depend on position;
3. those games like billiards that depend on motion.

Today computers have been designed which do for games what Leibnitz proposed several hundred years ago.

Euler's principal contribution to probability theory dealt with the mortality and growth of mankind. Among the problems he solved, one involved a certain number of men aged n years. Euler asked himself how many of them would be alive after a specified number of years had passed. From this analysis and others like it, he deduced a formula for annuities on a life and thus laid the groundwork for the preparation of actuarial tables for life insurance purposes.

D'Alembert also dealt with the probabilities associated with life and death, emphasizing the importance of the different values we should assign to youth and old age, laying great stress on "the consideration that the additional years of life to be gained from a remote, and therefore less to be enjoyed benefit than the present years." He also distinguished between the physical life and the real life of an individual, defining the first as "the actual number of years lived" and the second as the years of good health.

Lagrange's contributions to probability theory dealt essentially with the error analysis of observations. He sought to find the probability that—in taking the average of n observations—the result shall be exact or the probability that the error of the mean shall not exceed a given value. Such problems are of extreme importance in scientific experiments of all sorts but also in sociology, economics, and historical research. That "facts" are burdened with errors and uncertainties has been known in science for many years. Scientists are thus careful to report their observations and experimental results with their estimated errors. Therefore they speak of the probable values of observations and measurements rather than the exact values, which can never be known. Thus, probability theory forces itself onto science and actually plays a very important role in the analysis of experimental results.

Laplace was the last of the great mathematicians of this post-Newtonian period who made important contributions to probability theory. In addition, he summed up in a precise mathematical theory what had been developed in a rather piecemeal fashion by both his contemporaries and earlier mathematicians. Most of his work on probability is embodied in his great treatise *Théorie Analytique des Probabilités*, published in 1812, which has been the standard of excellence in this area since that time. On the whole, the theory of probability is more indebted to him than to any other mathematician. Like the other mathematicians working in probability theory, Laplace began with the theory of games, but his approach differed from that of the others in that he formulated each problem in a precise mathematical form which enabled him to enunciate clearly, for the first time in the history of probability theory, the principle for estimating the probabilities of the causes of an observed event. He was also the first mathematician to apply probability theory to physical problems such as the frequency of the occurrence of comets moving in specified orbits—when under the influence of the sun's gravitational field—in accordance with Newton's law of gravity.

Laplace's application of probability theory to the observations of simple physical phenomena led him to the conclusion that natural events do not occur haphazardly, but are produced by constant natural causes. He indicated in this treatise that the correlation between natural events and constant causes of such events deduced from probability convinced him that the universe is governed by unchanging and eternal laws. This conclusion induced him to investigate astronomy mathematically, which led him to a number of deductions and explanations that are accepted today, such as the irregular lunar motion, the motions of the satellites of Jupiter, and the theory of tides. But Laplace did not stop with his application of probability theory to science; he applied it to the analysis of election results in politics, to the theory of errors and the method of least squares, to the probability of the occurrence of future events, to mortality tables, to psychology, to the duration of diseases and their influences on population growth, to epistemology and the origin of thought, and to the duration of marriages. That Laplace held the theory of probability in the very highest esteem and considered it to be the key to all mathematical sciences is clear from his characterization of it as a mathematics, which, though originating in the arithmetic of games, raised itself to the level of the most important realms of knowledge.

One of the most important questions that had to be answered before a self-consistent contradiction-free mathematical theory of probability could be developed dealt with a satisfactory mathematical definition of probability. Because no consensus about such a definition existed among mathematicians, many of them viewed the whole subject with distrust, and so it developed rather slowly as a purely mathematical discipline. But it received an enormous boost by its application to the solution of problems in physics, which ultimately led to the important branch of physics called statistical mechanics. The physicist is primarily concerned with calculating how many favorable events occur in an experiment as compared with the total number of events. The probability for the occurrence of the favorable event is the ratio of favorable events to the total number of events (favorable plus unfavorable). The physicist is not concerned with mathematical niceties in using his *a priori* definition of probability because it serves his purpose quite adequately. To the mathematician, however, this is too loose a definition because it does not properly define the essential conditions under which the physicist makes his observations (i.e., performs the experiment). This is, indeed, a fair criticism because many observations must be made under exactly the same conditions if the conclusion drawn from the observation is to be accepted. With the development of the quantum mechanics as the physics of the atom, the definition of the probability for the occurrence of atomic, subatomic, and microcosmic events changed. The quantum mechanics introduced its own definition of probability, which is independent of the concept of favorable versus unfavorable events. This quantum mechanical definition of probability, however, agrees with the *a priori* definition introduced above.

Although the theory of probability arose from the arithmetic analysis of games of chance, it soon spread into other branches of mathematics with one of its most important aspects expressed in algebra as Gauss's exponential law of errors. But it also spread into geometry with the development of geometrical probability, which became a favorite mathematical game of mathematicians during the last half of the nineteenth century. Among the problems that arose in this branch of probability is the famous "Buffon needle problem," in which the eighteenth-century French naturalist Comte Georges de Buffon proposed: Given a smooth table on whose surface a system of equidistant parallel lines is ruled, with d the separation between any two neighboring lines, what is the probability that a needle of length L less than d will intersect one of the lines if the needle is thrown randomly onto the table? This problem is particularly interesting because the probability equals $2L/\pi d$ so that this expression leads to an experimental way of determining the value of π by merely counting the frequency with which the needle cuts one of the lines. In 1901, an experiment of this sort was performed with 3,408 trials and the value of π was found to be 3.1415929 with an error of only 0.0000003.

CHAPTER 9

The End of the Golden Age
of Mathematics

*Mathematics is the queen of the sciences and number theory the queen
of mathematics.*

—KARL FRIEDRICH GAUSS

In our discussions of post-Newtonian and golden-age mathematics, we have
paid less attention to chronology than to the mathematical achievements of
the great mathematicians who lived in those periods so that we could present
the continuity of the subject matter from Newton's time to the early twentieth
century. This is not an easy task because mathematics did not grow linearly
like a corn stalk but branched out in many directions like a tree. Because
most leading mathematicians contributed to the growth of many of these
mathematical branches, it is almost impossible to delineate who did "this"
or "that" or who was dominant in any particular branch of mathematics.
Euler and Lagrange, for example, were clearly the dominant post-Newtonian
mathematicians because they constructed the mathematical bridge from
Newton to eighteenth-century mathematicians. We do not include Laplace
with them even though he was a leading eighteenth-century mathematician
and his years overlapped those of Lagrange, because he was not so much an
innovator in mathematics as he was an applier. Laplace demonstrated the
great usefulness of mathematics in the physical sciences. His *Traité de
Mécanique Céleste*, written in five volumes from 1799 to 1825, attests to his
great ability in this area and established mathematics as the technical
language of science. Indeed, these volumes unified his own investigations
and all previous work in mathematical physics and astronomy, culminating
in his famous nebular hypothesis of the origin of the solar system. Laplace's
belief in the infallibility of mathematics and its power to reveal the secrets of
nature led him to what is perhaps the ultimate expression of mechanical
materialism: "An intelligence which, for a given instant, knew all the forces
by which nature is animated and the respective positions of all the bodies

that compose it and which, besides was large enough to submit those data to analysis, would embrace in the same formula the motions of the largest bodies of the universe and those of the lightest atoms: nothing would be uncertain to it, and the future and the past would be present to its eyes. Human mind offers a feeble sketch of this intelligence in the perfection which it has been able to give to Astronomy." This was a fitting testament to leave to the great mathematicians whom we discuss in this chapter because they completed Laplace's work and extended mathematics to every branch of physics and astronomy, thus creating the vast intellectual enterprise which we now call "theoretical physics."

As we noted in the preceding chapter, Abel and Galois laid the foundation of what we called the "golden age of mathematics" by their insistence on mathematical rigor. Galois' application of groups to Lagrange's theory of equations projected algebra from a minor mathematical discipline to its present exalted state as the basic framework around which all other branches of mathematics are built. Hence, algebra is the basic analytical tool of all branches of mathematics. In introducing mathematicians to the algebra of groups, Galois extended enormously the role of mathematics as an analytical tool for physicists.

Abel, on the other hand, in his rigorous studies of the summations of infinite series and the criteria for their convergence, began what later became the theory of functions or, more generally, mathematical analysis. His insistence that no mathematical conclusions be drawn about any mathematical relationship and that no mathematical deductions be made without precise proofs at each step became the unwritten law for all mathematicians. Abel expressed his contempt for other procedures so forcefully that his call for rigor could not be ignored. He also contributed to the theory of groups in his study of what are now known as "Abelian groups," which constitute a subcategory of groups in general. The group theory work of Galois and Abel greatly influenced other mathematicians; the extent of its impact was expressed by the nineteenth-century mathematician Felix Klein in an address (now known as the "Erlangen program") in which he outlined a scheme for using groups to classify different branches of mathematics. We will return to Klein's work later, emphasizing here the great importance of the mathematical discoveries of Abel and Galois for the transition from the nonrigorous post-Newtonian mathematics to the rigorous mathematics that followed Abel and Galois.

This final phase of the "golden age of mathematics," like the initial phase, was dominated by French mathematicians. But English and German mathematicians were beginning to challenge this French dominance. Although British mathematicians were still bedazzled by Newton's legacy and partial to the Newtonian use of the dot notation for derivatives instead of Leibnitz's *d* notation, they began to develop a branch of algebra with important applications to geometry. This departure from pure Newtonian mathematics was led by three outstanding British mathematicians: Arthur Cayley, James Sylvester, and George Salmon. This esteemed group was later joined by the Irish royal astronomer, William Rowan Hamilton, whose presence further added to its luster. In time, the mathematicians William Kingdon Clifford and George Green, with their emphasis on mathematical physics, also became members of this select group.

With Karl Friedrich Gauss, perhaps the greatest mathematician of all time, at their head, the German mathematicians, particularly in the realms of geometry and analysis, made very important and impressive contributions to the rapid growth of mathematics during this period. Gauss attracted most of the young German mathematicians who later left their own indelible imprint on mathematics. Outstanding among these disciples of Gauss were Karl Gustav Jakob Jacobi, Peter Lejeune Dirichlet, and Georg Friedrich Bernhard Riemann. The mathematicians Weierstrass, Kronecker, Dedekind, Cantor, Steiner, Kummer, Frobenius, and Möbius operated on a somewhat lower mathematical level but their contributions were also essential to mathematical progress.

Most mathematicians rank Gauss with Newton and Archimedes at the very top of the mathematical pyramid even though they are usually unable to agree as to the precise ordering of the rankings. Gauss was not as prolific as Cauchy or Euler, but his work was consistently outstanding, characterized by penetrating insights and remarkable originality. His was a holistic view of mathematics in that he did not concern himself with distinctions between pure and applied mathematics. For Gauss, the two were necessary complements to each other.

His lineage was not noble. Johann Karl Friedrich Gauss was born in Brunswick, Germany on April 30, 1777, to Gerhard Diedrich and Dorothea Benz. Johann's father was a slow-witted, violent-tempered laborer who beat his children with a thick stick to keep them in their place. Johann's mother, by contrast, was a bright, personable woman who seemed to sense from the earliest years of young Johann's life that he was destined to achieve great things. But it was Dorothea's brother, Friedrich, a self-taught weaver, who was most crucial to Johann's early mental development. Friedrich peppered his young nephew with a steady barrage of questions about the world. These persistent queries not only stirred Johann's intellect but also caused him to wonder whether he was really meant to be a laborer like his father. Gerhard himself was a boorish man who seems to have viewed Johann as a threat to his mastery of the household; perhaps this envy explains Gerhard's repeated attempts to prevent Johann from obtaining an education. But Friedrich and Dorothea overcame the clumsy efforts of Gerhard to prevent Johann from rising above his station in life. Gauss would later repay his mother's kindness by personally caring for her in her old age—even though he had long since been recognized as the greatest living mathematician in the world—until her death in 1839 at the age of 97.

Johann's precociousness manifested itself at an early age. He taught himself how to read and do complex calculations in his head before he reached the age of four. His mathematical skills were aided by a photographic memory that enabled him to store vivid mental images of his earliest years and recall them, often decades later, as though they had just occurred. But he probably would have preferred to forget some of his childhood memories such as his early formal education.

Johann's first grammar school could easily have been mistaken for a prison as it was a dark, thick-walled building in which the students were mercilessly beaten by their instructors for the slightest violations of the rules. Indeed, Johann's father, Gerhard, probably would have heartily endorsed the institution's brutal "spare the rod, spoil the child" approach to teaching. But Johann's mathematical talents became apparent one day in an arithmetic class when he solved a problem requiring the addition of a hundred terms in his head. Even Johann's teachers, though inclined to thrash him for his impudence, realized that this student had certain intellectual talents. An assistant at the school, Johann Bartels, was assigned to work with the ten-year-old Johann on algebra. The two Johanns became good friends and their relationship quickly evolved from one of tutor–student to colleagues. Under Bartels's watchful eye, young Gauss became interested in infinite series and devised ways to bring a new level of rigor to the study of analysis. Over the next few years, Johann, though still barely a teenager, became consumed with uncovering flaws in the supposedly solid foundations of Euclidean geometry. He also began to wonder whether there might be other, equally plausible geometries.

Johann's aptitude for mathematics impressed his teacher Bartels, who sought financial assistance so that his student could continue his education beyond grammar school. To that end, Bartels arranged an interview for Johann with the Duke of Brunswick, a fair-minded patron of the arts and sciences. The Duke himself was taken with the young boy and agreed with Bartels that it would be nothing less than a tragedy if Johann could not complete his education. Accordingly, the Duke made one of the most profitable investments in the progress of human knowledge by agreeing to subsidize Gauss's education until it was completed.

Able to forget his financial worries for the first time in his life, Johann enrolled at Caroline College at the age of fifteen where he studied the classics. But like Galois, Gauss found that only mathematics truly inspired him. But he also had a gift for languages and it was only due to his greater love for mathematics that he did not become an outstanding scholar in linguistics or classical literature. Fortunately for mathematics, however, Gauss became enamored with the great works of Euler, Lagrange, Laplace, and Newton. At the same time, he conducted extensive investigations into higher arithmetic. After completing his studies at Caroline, Gauss matriculated at the University of Göttingen. By this time he had already invented the method of least squares—a process for approximating unknown quantities based on a large number of measurements. Gauss's studies of mathematics—like those of Newton—were prompted not by the lure of fame but instead by his own unquenchable desire to uncover the profound truths of that subject. Unlike Cauchy and Euler, Gauss was reluctant to publish his work. Were it not for a personal diary that he kept throughout his professional life, it is unlikely that his priority in a number of areas such as elliptic functions would have been established. Eric Temple Bell believes that Gauss's reluctance to publish his discoveries was based less on a fear of controversy than the desire to "leave after him only finished works of art, severely perfect, to which nothing could be added and from which nothing could be taken away without disfiguring the whole." This drive for perfection compelled Gauss to refrain from publishing accounts of his preliminary efforts in various branches of mathematics. The absence of unfinished or incomplete papers on particular mathematical discoveries made it that much more difficult for Gauss's contemporaries and successors to follow his reasoning because he presented each discovery as a single feat instead of the culmination of a series of more minor advances.

Gauss's tenure at Göttingen was exhausting; his demanding academic schedule left him little time for socializing with the other students. But

Göttingen marked the high watermark of his own career in mathematics, much in the way that Newton's two years at Woolsthorpe had provided him with the leisure time to make his fundamental theoretical discoveries in physics and mathematics. While at Göttingen, Gauss developed the non-Euclidean geometry that had tempted him for nearly a decade and put the finishing touches on his *Disquisitiones Arithmeticae*, his masterpiece on arithmetic. He fittingly dedicated the work to his longtime benefactor, the Duke of Brunswick, who had already granted a pension to the still-unemployed Gauss so that he could continue his work in mathematics. The Duke also financed the printing of Gauss's doctoral dissertation, which offered the first proof of what is known as the fundamental theorem of algebra. Although the *Disquisitiones* was only the first of Gauss's seminal works in mathematics, it immediately established him—according to no less an authority than Lagrange—as one of the greatest mathematicians in Europe.

This fame became even more widespread after Gauss undertook and completed the laborious calculations of the orbits of the newly discovered asteroids Ceres and Pallus. This venture into applied mathematics occupied Gauss for the better half of two decades and regrettably caused him to stray far afield from pure mathematics. But the calculations were extremely complex and had challenged the greatest mathematical minds of that era. They also offered Gauss an opportunity to exercise his arithmetic talents to their fullest and to complete one of the last remaining building blocks in the monumental edifice of Newtonian mechanics. More important, Gauss devised the method by which such computations could be completed in a matter of hours instead of days or weeks. These methods were discussed in a second treatise dealing with the motions of the planets and comets around the sun, which was published in 1809.

In 1805, Gauss married Johanne Osthof; she bore him three children, but the birth of the third child cost her her life. Devastated by the loss of his wife, Gauss was plunged into a depression that all but incapacitated him for a year. Life took a turn for the better in 1810 when Gauss married a longtime friend, Minna Waldeck, who later gave him two sons and a daughter. Little is known about Gauss's children but he is said to have been a tolerant father who was perhaps too reluctant to punish them for being naughty. After Gauss's own experiences at the hand of his father, however, this aversion toward corporeal punishment was to be expected.

Following the death of his longtime benefactor, the Duke of Brunswick, in 1806, Gauss was forced to look for employment. The Russian monarchy tendered an attractive offer to Gauss to come to St. Petersburg but the prospect of Gauss's leaving Germany for another rival nation such as Russia horrified Germany's mathematical establishment. Cognizant of the severe blow that Gauss's departure would deal to Germany's mathematical community, Alexander Humboldt and several of his colleagues arranged to have Gauss appointed as director of the Göttingen Observatory. His pay and quarters were not extravagant, but Gauss himself had never cared much for the "finer things" in life. According to Gauss's longtime friend, Sartorious von Walterhausen, Gauss was a simple man with simple needs: "A small study, a little work table with a green cover, a standing-desk painted white, a narrow sofa and, after his seventieth year, an arm chair, a shaded lamp, an unheated bedroom, plain food, a dressing gown and a velvet cap, these were so becomingly all his needs."

That Gauss did not pine for fine things was fortunate because Germany during these years was systematically pillaged by Napoleon's armies. Gauss himself was forced to pay a fine far beyond his means to the French occupiers to subsidize their presence in Germany. Although others such as Laplace came forward to pay the debt, Gauss was a proud man who would not accept charity from others, even though the debt itself amounted to little more than petty extortion.

For the next few years, Gauss buried himself in his work, pausing only long enough to calculate the orbit of a comet in 1811 that lit up the summer sky. In particular, Gauss examined the analytic functions of complex variables and formulated his theory of analytic functions of a complex variable, which was one of the most important mathematical advances of the nineteenth century. He also published important papers about the binomial theorem and quadratic binomial congruences. At the same time, he made important discoveries in mathematical physics and astronomy. Like Newton, Gauss seems to have been blessed with the ability to concentrate for prolonged periods of time on a single problem until it yielded to him. But Gauss was more preoccupied with the fundamental ideas of mathematics and the way in which such seemingly disparate topics as algebra, geometry, and the calculus are all related to each other. Newton, on the other hand, was eminently practical in his approach toward mathematics, looking at it as a means by which the physical universe could be modeled within a mechanistic framework.

Gauss's mathematical discoveries do not amply demonstrate his versatility as a scientist. He, like Newton, was mechanically inclined and invented a gravity-measuring device, a boon to the study of geodesy, and the bifilar magnetometer, an important tool for researching electromagnetic phenomena. More important for the world of communications was his invention, with Wilhelm Weber, in 1833, of the electric telegraph. There are very few mathematicians of Gauss's abilities who could also claim to have developed a device that would so revolutionize commerce and diplomacy. But Gauss was not interested in the practical uses of these inventions; he was more concerned with their value as models for his theoretical investigations.

Although Gauss was consumed by mathematics, he had a lifelong love affair with learning. He was a voracious reader, preferring the works of Sir Walter Scott and Edward Gibbon to what he viewed as the flighty babble of Byron and Goethe. Gauss also continued to acquire new languages throughout his years, mastering most of the languages of Europe, including Russian and Sanskrit.

Much has been said about Gauss's lack of interest in the careers of young mathematicians, particularly his failure to acknowledge the works of talented men such as Abel. Cauchy's work in the theory of functions of a complex variable and Hamilton's quaternions were but two of the more notable examples of his supposed failure to embrace his younger colleagues. To Gauss's credit, however, he did not try to undermine the efforts of these mathematicians to publish their results.

As Gauss passed through middle age, he became increasingly involved in public service, advising the Hanover government as to the best way to carry out its geodetic survey. Gauss's own facility for computations and his method of least squares proved to be invaluable to this endeavor and he was justly honored for his services to the state. This survey also caused Gauss to think about the properties of surfaces in small localities, thus leading him to what we now know as differential geometry. Gauss's own work greatly influenced that of Riemann, who in turn provided much of the mathematical scaffoldings for Einstein's general theory of relativity. Gauss also made important contributions to the field of conformal mapping, thus greatly aiding future generations of cartographers in their efforts to minimize the distortions in their maps. Indeed, Gauss's work pervades the entire field of mathematics; there are few branches of mathematics that were not affected and, in many cases, revolutionized by his unceasing efforts.

What of his later years? Gauss continued his work until the very end of his life. A year before his death on February 23, 1854, his declining health began to put a serious crimp in his working schedule. But no one lives forever and even Gauss was not nimble minded enough to avoid his own mortality. Persistent heart troubles and arthritis made it very difficult for him to write and the last few months of his life were a painful struggle to stay alive. But Gauss's death was peaceful and he died knowing that he had left the world an incredible—many would way, unsurpassed—mathematical legacy.

Despite the brilliant work of the British and German mathematicians, particularly that of Gauss, during this period (from about 1820 to 1870) we must still rank the French mathematicians of that era as preeminent even though none of them was of Gaussian stature. One may explain the burgeoning of French mathematics during this period by associating it with the French Revolution and the emergence of the "Age of Reason." Though a despot, Napoleon encouraged mathematics and science in general, so that mathematics flourished during his reign. A remarkable characteristic of the French mathematicians was their fondness for and deep attachment to mathematical physics. Indeed, Cauchy and Fourier began their careers as mathematical physicists, the former in analytical mechanics and optics, and the latter in the flow of heat.

Augustin Louis Cauchy was probably the most versatile French mathematician of that era. He was a child of the French Revolution, born on August 21, 1789, little more than a month after a Paris mob had stormed the Bastille. His father, Louis-Françoise, was a government lawyer, who had married Marie Madelein Dosestra, two years earlier. Augustin was the eldest of six children but his early years were anything but the idyllic childhood that one might have expected for the son of a man in Louis's position. France was then in the throes of Robespierre's bloody reign of terror and the nation's economy had all but collapsed. Often there was not enough food for the children. Moreover, the turbulence of the political landscape had led to the closing of the schools in Paris. Fears that he might be a candidate for the guillotine prompted the senior Cauchy to move his family to the town of Arceil. There in the country, Augustin spent the first decade of his life. But the bucolic country life did not prove to be any easier for the Cauchy family. Food was still difficult to get and Louis was reduced to scratching out a subsistence living as a farmer. What saved the family from starving was not Louis's clumsy gardening but the friendship he struck up with Pierre Simon de Laplace, who owned a nearby estate and periodically dropped in on the Cauchy family to offer his assistance. Laplace took an interest in Augustin, who always seemed to have his nose buried in one of the few textbooks lying around the family cottage. The famous mathematician's initial expectation that Augustin would talk about the usual subjects dear to most young boys was soon dispelled as Augustin had already acquired an extensive knowledge of the basic concepts of mathematics. As Augustin had never attended school, owing to the nation's political upheavals, his knowledge about subjects ranging from geometry to arithmetic was truly remarkable. Indeed, it was not long before Augustin was carrying on detailed conversations about advanced topics in mathematics that astonished even the normally unflappable Laplace.

After the Cauchy family moved back to Paris in 1800, following Louis's appointment as secretary of the Senate, Joseph-Louis Lagrange—who met with Louis from time to time—happened to strike up a conversation with the boy one day while he was sitting in his father's office. Like Laplace, Lagrange was amazed at Augustin's knowledge of mathematics and predicted that he would be one of the greatest mathematicians in Europe. But Lagrange, concerned that Louis might immerse the boy in nonstop mathematical study, cautioned the senior Cauchy not to let his son study mathematics until he was older so that he would not burn himself out and lose his enthusiasm—if not his aptitude—for mathematics. Accordingly, Louis encouraged Augustin's study of the classics, particularly after he enrolled at the Central School in Paris. Augustin's experience there was valuable in that he was exposed to a number of classical Greek and Latin writers for the first time but it was also difficult because he was not used to being around other boys. Augustin himself was somewhat awkward around his classmates, but he channeled any anxiety he might have felt into his academic work and collected most of the prizes the school had to offer for his work in Greek and Latin. After leaving the Central School, he studied mathematics for a year and then enrolled at the École Polytechnique, where he eventually completed his studies in civil engineering and was commissioned in Napoleon's army.

Cauchy's first assignment was Cherbourg, where he assisted in the design and construction of some of the rudimentary fortifications that were to be the prelude to Napoleon's planned invasion of England. Over the next three years, he divided his time between service to his country and his own personal mathematical researches. His interests were varied; his mind wandered across the mathematical terrain with abandon, searching for new propositions and ways to simplify existing proofs. He also composed papers on solid geometry and symmetric functions, laying the groundwork for his work in the theory of substitution groups.

After returning to Paris in 1813, worn out from the unending fortification activities along the western coast of France, Cauchy focused his energies on mathematics, producing important papers on definite integrals and proving one of Fermat's theorems. In 1816, he captured the attention of Europe's leading mathematicians by taking the grand prize offered by the Paris Academy of Sciences for a treatise on wave propagation. Soon afterward, he became a member of the Academy, taking the seat held by the mathematician Gaspard Monge, who had been expelled for his opposition to Napoleon.

Now recognized as one of Europe's leading mathematicians, Cauchy worked at a frantic pace that would have killed most ordinary people. According to Eric Temple Bell, Cauchy's "mathematical activity was incredible; sometimes two full length papers would be laid out before the Academy in the same week. In addition to his own research he drew up innumerable reports on the memoirs of others submitted to the Academy, and found time to emit an almost constant stream of short papers on practically all branches of mathematics, pure and applied." Cauchy married the former Aloise de Bure, who was to be his devoted companion for nearly forty years.

Perhaps because of the ease with which he churned out mathematical papers like a printing press, Cauchy founded a mathematical journal which served as a repository for those works that he felt did not merit submission to any of the leading journals. He also found an outlet in the *Comptes Rendus*, the Academy's weekly newsletter, which permitted him to disseminate his shorter papers to mathematicians and scientists throughout Europe in as brief a time as possible.

In 1830, King Charles was overthrown. Cauchy, who considered himself to be a loyalist, followed Charles into exile. After leaving the war-ravaged streets of Paris for the serenity of Switzerland, Cauchy then accepted a position as professor of mathematical physics at the University of Turin. Cauchy spent the next three years at Turin, continuing to generate a steady output of mathematical papers and treatises. The temperate Italian weather seems to have acted as a restorative and helped Cauchy, whose health was always fragile, to recover somewhat from his many years of incessant overwork. Cauchy might have continued his work uninterrupted until his death but for the decision by the deposed Charles in 1833 to reward

Augustin Louis Cauchy (1789–1857) (Courtesy AIP Emilio Segrè Visual Archives; E. Scott Barr Collection)

him with the dubious job of tutoring his young heir, the Duke of Bordeaux. The prospect of serving as a full-time tutor to Charles's ill-mannered child horrified Cauchy, but Cauchy's own sense of loyalty made it impossible for him to refuse the former king's request.

For the next five years, Cauchy lived in Prague, forced to wait on the young Duke like a clucking hen huddled over its prized chick. He had very little spare time as he was on call up to eighteen hours a day. Nevertheless, he did manage to snatch a few minutes here and there to write several important papers on subjects ranging from the dispersion of light to the elasticity of solids. Fortunately for his own sanity, his self-imposed purgatory ended in 1838, after five grueling and exhausting years of service as the royal tutor. Cauchy then returned to Paris where he accepted a vacancy at the Collège de France. Cauchy's refusal to take an oath of loyalty to Louis Philippe's government—owing to his belief that Charles had been wrongly removed from power—led to a political tug-of-war with the ruling authorities that lasted for several years. Despite the prospect of being branded a traitor by the new government, Cauchy stuck to his guns. The government, aware that it was being made to look increasingly foolish in the eyes of its citizens by its heavy-handed treatment of France's greatest mathematician, eventually relented and allowed Cauchy to assume his position at the Collége de France. Although the loyalty oath was abolished following the deposing of Louis Philippe in 1848, it was restored four years later in 1852 with the ascension of Napoleon III. But even the newly crowned emperor was not so foolish as to become involved in a public battle with Cauchy over something so pointless as a loyalty oath, so the matter was finally dropped for good.

Despite his occasional forays into politics, Cauchy remained focused on his mathematics. Like Euler, he seemed to have an almost machinelike ability to generate mathematical papers, many of which were of outstanding quality. By the time he died on May 23, 1857 at the age of sixty-eight, he had written nearly 800 papers, memoirs, and treatises, many of them hundreds of pages in length. This prodigious output prompted some of his critics, mathematicians and laypersons alike, to suggest that much of what Cauchy composed was mediocre at best. Because these criticisms were offered by persons who had their own axes to grind, Cauchy's reputation suffered somewhat during the first few decades following his death. The verdict of history, however, has been more charitable and Cauchy has since been recognized as a truly outstanding mathematician whose attention to rigor and intellectual versatility greatly influenced the development of modern mathematics, particularly analysis.

Cauchy's greatest contributions to mathematics were in the theory of functions of complex variables and to the criteria for the convergence of infinite series. His work on the theory of functions of complex variables was so extensive that he·is considered to be the father of this branch of mathematics, which has grown enormously and entrenched itself in every branch of physics, particularly in wave mechanics and in relativity theory. Because wave mechanics is of crucial importance both in optics (electromagnetic waves) and electronics (electron waves), the study of functions of complex variables is required of all physicists, astronomers, and chemists.

The study of complex function theory was greatly enhanced and stimulated by Gauss's introduction of the complex plane, which led to complex analytical geometry. Descartes initiated real analytic geometry by representing geometric curves on a plane by means of algebraic relations between the y and x coordinates of points on the curves. As noted earlier, the y values, measured along a vertical axis, are called the ordinates of the points and the x values are called their abscissas. The relationship between the y and x values for any one curve is written as $y = f(x)$ and y is then called a function of x. Because the y's and x's are real numbers, $y = f(x)$ is called a real function or a function of a real variable. Such functions dominated the study of the theory of functions because complex numbers were considered to be numerical curiosities which could have no application to the real world.

The theory of functions remained tied to the real Cartesian plane until the work of Gauss and Cauchy shifted the attention of mathematicians from real to complex functions. This change became an all-consuming occupation for many gifted mathematicians when Gauss introduced the concept of the complex plane, which made it possible to represent functions of complex variables as geometric curves in the complex plane. Like Descartes, Gauss introduced planes and set up coordinate systems on such planes. But the points along the vertical axis y of a coordinate system in a Gaussian complex plane represent imaginary numbers such as $4i$ or $3i$ (where $i = \sqrt{-1}$), whereas the points x along the horizontal x axis (the abscissa) represent the real numbers. Every point z on the complex plane then represents a complex number $z = x + iy$ and a curve drawn in this plane, which connects an array of points (an array of z values in this plane) may be described as a function of z. Because this function $f(z)$ is represented by an array of complex numbers, we may write it as $f(z) = u(x, y) + iv(x, y)$, where u and v are real functions of the real numbers (real variables) x and y. Any curve we draw in the complex plane can be represented as a real algebraic relationship between the real numbers x and y, even though the curve itself is an aggregate of complex numbers. A simple example illustrates this concept: A straight line in the complex plane x, y passing through the origin and making an angle of $45°$ with the x and y axes is given by the simple algebraic equation $y = x$. Because every point z on this line is a complex number $x + iy$, then every point on this straight line is given by $z = x + iy = x + ix = x(1 + i)$.

The theory of functions of complex variables involves two complex planes: the complex plane of the complex variable z (the x, y plane) and the complex plane of the complex function $f(z)$ (the u, v plane). Mathematicians then describe the functional relationship $f(z)$ as the mapping of the x, y plane onto the u, v plane; for every point z on the x, y plane, $f(z)$ defines a point on the u, v plane. The function $f(z)$ defines a specific mapping of a domain of the x, y plane on a domain of the u, v plane only if a specific function of z is given and z is restricted to a specific domain of points. We illustrate this concept using the function $F(z) = 1 + z + z^2$ and restricting z to the points on the 45 degree straight line in the x, y complex plane introduced above so that $z = x(1 + i)$. The $F(z) = 1 + x + ix + (x + ix)^2 = 1 + x + ix + x^2 + 2ix^2 - x^2 = (1 + x) + i(x + 2x^2)$. Thus $u = (1 + x)$ and $v = x + 2x^2$. This is equivalent to the functional relationship $v = 2u^2 - 3u + 1$, which is a parabola in the u, v plane with its own axis along the real axis v in the u, v complex plane. Thus, the points z along a straight line in the complex x, y plane are mapped on a parabola in the u, v complex plane. If the domain of z were defined by an x, y curve (a z-plane curve) other than a straight line, the mapped u, v curve would, of course, be different from that of a parabola.

Because the study of functions of a complex variable was pursued most intensively for functions of z that are sums of an infinite number of powers of z, Cauchy devoted a good deal of time to the study of such infinite series. He first proved the general theorem that any complex function of z can be expanded around any point z as a power series in z. He then investigated the very important question as to whether such sums (series) converge or diverge. A series is said to diverge (a divergent series) if it increases without limit (approaches infinity) becoming larger than any number as the number of terms in the series increases. If the sum always remains less than a definite value, the series is said to converge (a convergent series). Cauchy discovered a general criterion for the convergence of an infinite sum (series). If the infinite sum is $u_1 + u_2 + \cdots + u_n + u_{n+1} + \cdots$, then the sum converges if the ratio u_{n+1}/u_n for ever increasing n's remains less than 1. If this ratio is larger than 1, the series diverges.

This is but one of the important theorems in the theory of functions of a complex variable that Cauchy proved. Another important theorem deals with the integral calculus; Cauchy showed how to obtain the integral of complex functions along a closed path in the complex plane almost by inspection. This is known as the theory of residues. He also greatly extended the rigorous basis of the calculus, refining the concept of passing to the limit of an infinitesimal quantity (allowing the infinitesimal to become zero) on which all of differential calculus is based. In doing so, he gave a precise meaning to d'Alembert's concept of a limit. With this concept of a limit, calculus became a rigorous branch of mathematics; prior to that innovation, mathematicians had been puzzled by the haphazard process of allowing an infinitesimal to approach zero because a strict rule of algebra does not allow a denominator to become zero. With this precise delineation of the concept of a limit, Cauchy transformed the calculus from a somewhat amorphous mathematical formulation to a highly structured, rigorous discipline. He also discovered a beautiful theorem in the integral calculus of the functions of a complex variable. This discovery, called the Cauchy integral theorem, enables one to find the value of the integral of a function of a complex variable taken around a closed path in the complex plane in terms of the values of the function at certain discrete points lying inside the domain bounded by the path. This theorem is used extensively in various problems in physics.

Although Cauchy's incomparable contributions to almost every branch of mathematics are well known to current mathematicians, who properly rank him as one of the greatest mathematicians of all time, very few mathematicians or physicists know of his important contributions to mathematical physics. Cauchy flourished at a time when the rapid growth of experimental physics—electricity, optics, heat, acoustics, elasticity—demanded correct mathematical formulations of the physical phenomena involved and the solutions of the mathematical equations that describe such phenomena. Cauchy accepted this challenge, and, together with such other great French mathematicians as Poisson, Fourier, Legendre, and Hermite, made mathematics so indispensable a tool for physics that no physicist can consider himself as such without acquiring considerable mathematical skills. At the same time, the discoveries of these mid-nineteenth-century mathematicians placed France at the very peak of mathematical physics. Indeed, one may say that mathematical physics did not really begin until Cauchy and his colleagues had done their work. Among those mentioned above, Fourier ranks next to Cauchy in the importance of their contributions to mathematical physics. But Fourier's contributions were not as universal as Cauchy's except in the representation of a given function of a variable as an infinite sum of sines and cosines. With the exception of calculus, such series, called Fourier series, have had a greater impact on the development of mathematical physics than almost any other branch of mathematics. The story of Fourier series is somewhat complicated because the idea of solving certain types of equations, particularly partial differential equations, in terms of sines and cosines, goes back to d'Alembert, Euler, and Daniel Bernoulli.

Mathematicians as far back as the mid-eighteenth century thought of using sines and cosines because they were trying to describe the way a string vibrates (e.g., a violin string) when it is plucked. If we plot the distortion of the string when we first pull upward on it, we can express the shape of the string in an x, y Cartesian coordinate system as $y = f(x)$, where y is the height of any point x (the x coordinate of the point) of the string above the horizontal. If we let go of the string, every point on it moves, each with its own speed, toward the horizontal so that the string becomes distorted. The function $y = f(x)$ thus becomes distorted and this distortion changes with time. Hence, the function $f(x)$ changes from moment to moment as well as from point to point. We must therefore write this function as $y = f(x, t)$ so that y is a function of two changing variables x and t. The equation of motion that represents the motion of each point on the string is called a partial differential equation and its solution $f(x, t)$ shows how the shape changes from moment to moment. Before Fourier began his analysis of the mathematics of a vibrating string, Bernoulli and Euler noted that the mathematical forms of such vibrations are similar to the graphs of the sine and cosine of the variable x and so they tried to represent the function of the vibrating string by the sum of products of sines and cosines without much success. Lagrange went further than that and almost arrived at Fourier's great discovery.

Fourier did not begin his great synthesis purely as a mathematical exercise but as the solution of a physical problem—the conduction of heat in a solid. Heat flows in a solid because of the temperature difference from point to point and from moment to moment so that the equation describing the way the temperature changes is a partial differential equation for the temperature. Fourier saw that the solution of this equation can be expressed as a product of a function of time and a function of position x. He then showed that the function of x can be written as an infinite series of sines and cosines of x.

Although Fourier developed his trigonometric series to solve physical problems such as heat conduction, his most important contribution to the general mathematical theory of trigonometric series was to show that any part of a function can be expanded into a Fourier series. Thus if the function $y = f(x)$ is plotted in the x, y plane and a piece of it between two values of x, x_1, and x_2, is considered separately, this piece can be represented as the sum of an infinite number of sines and cosines within this x strip between x_1 and x_2.

The importance of Fourier series for physical phenomena can be understood from the simplification it introduces in analyzing sounds. No matter how complex a sound may be, it can be expressed as a sum of pure sounds (simple notes) which are just sine and cosine vibrations (single frequencies), each of which is called a harmonic. Thus, the Fourier series is really a harmonic analysis of any phenomena that can be represented as a functional relationship between two different entities.

The Fourier series led to another important concept—the Fourier integral, which is even more important in physics than the Fourier series. This integral is a simple and elegant mathematical procedure that enables us to transform the emphasis from one frame of reference to another. In considering the motion of a particle, for example, we may describe this motion in a spatial frame of reference—that is, in terms of the spatial path of the particle in a three-dimensional Cartesian coordinate system x, y, and z, or in what physicists call a momentum frame. A momentum frame is a three-dimensional frame, each point of which represents the three quantities mv_x, mv_y, and mv_z (called p_x, p_y, and p_z). The symbol m is the mass of the moving particle and v_x, v_y, and v_z are the particle's velocities in the x, y, and z directions, where the position of the particle in space is at the point x, y, z. The quantities p_x, p_y, and p_z are the components of the particle's momentum at the point x, y, z. The Fourier integral shifts our analysis from the coordinate space of the particle to its momentum space. This is very useful in studying the dynamics of a particle, particularly in quantum mechanics.

Most of the French mathematicians in that period directed their attention to the application of mathematics to mechanics and other branches of physics because they were connected with the École Polytechnique in its early years when that institution emphasized the study of mechanics. Poisson was perhaps the most influential and creative of these mathematicians in bridging the gap between pure mathematics and mathematical physics. He was so productive in mathematical physics and his reasoning was so profound in reshaping the mathematical forms of the laws of physics that his name appears more frequently in current theoretical physics than that of any other of his contemporary mathematicians. Among his most important discoveries was a reformulation of Newton's laws of motion into a form now called the Poisson brackets, whose importance for theoretical physics is indicated by the way the great twentieth-century theoretical physicist Paul Adrian Maurice Dirac used them to formulate quantum mechanics into a very productive form. Poisson also discovered the "Poisson equation" (a partial differential equation) of the gravitational field, which Einstein used, almost a century later, as the basis for his formulation of his general relativistic field equations that are used extensively in modern cosmology.

Here we have emphasized the work of the outstanding mid-nineteenth-century French mathematicians and their decisive contributions to the rapid growth of theoretical physics. This occurred at a very propitious time in the rapid burgeoning of physics which stemmed from such new branches of physics as electricity and magnetism, optics, the structure of matter, the theory of gases, and thermodynamics. Up until that time, mathematical physics had been dominated by Newtonian mechanics and gravitational theory. It was not that new kinds of mathematics had to be introduced but rather that the classical mathematics of Newton and Leibnitz had to be enlarged and molded to fit the needs of the theoretical physicist. Thus, optics and the wave theory of light required new kinds of equations—wave equations—as did the study of acoustics. The development of electricity and magnetism greatly strengthened the concept of the force field so that mathematics had to be expanded to represent the properties of fields, particularly electromagnetic fields. In a similar vein, the introduction of molecules as the basic constituents of gases led to the kinetic theory and the statistical mechanics of gases, which necessitated the expansion of probability theory into the kind of mathematics that theoretical physicists could use to study systems, such as gases consisting of vast numbers of elementary particles.

In this growing intimacy between mathematics and theoretical physics, the motivating force that stimulated this affinity was the needs or demands of the physicist, who often took the first step in molding the known mathematics to the need of the physics. But the physicists rarely developed any new mathematics, contenting themselves with adapting the existing mathematics to their needs. But the pure mathematicians, with rare exceptions, again, did little to develop pure physics. It is therefore all the more remarkable that the collaboration between physicists and mathematicians advanced both disciplines enormously. We note here that such outstanding physicists as Ampere (the great pioneer in electromagnetic research), Malus (the discoverer of the polarization of light), Fresnel (who demonstrated the truth of Huygens's wave theory of light), and Sadi Carnot (who discovered the basic principles of the heat engine and the second law of thermodynamics), used mathematics extensively. In this interdependence of physics and mathematics, the mathematician could continue playing his beautiful and exciting games without the physicist, but the physicist could not proceed without the aid of the mathematician. But physics has always been more exciting than mathematics because physics is more than a game and therefore much more challenging than mathematics. The reason for this intellectual superiority of physics to mathematics is that the constraints that nature has put on physicists forces them to hew to a very narrow but infinitely long path. Mathematicians, by contrast, can meander in a vast boundless maze governed by laws that the mathematician lays down according to his fancy.

We have been discussing the great French mathematicians, who not only gave France its mathematical preeminence but also set the course of mathematics for the next century. Their research began to spread to other European countries owing to the growth of science and technology in these nations. Germany was in the forefront of this group of countries that began to challenge France's intellectual leadership. This challenge was due primarily to the emergence of Gauss. Like Archimedes and Newton, Gauss was almost as great a physicist as he was a mathematician. Like Newton, Gauss made his great mathematical discoveries while still very young—between the ages of 14 and 17—and then spent the rest of his life enlarging on these discoveries and extending them to all branches of mathematics. We have already discussed Gauss's life and his contributions to the theory of numbers, his contributions to algebra, and his discoveries in complex numbers, but we now turn to what is perhaps his most memorable work, and, as far as physics is concerned, his most important. This is his work in geometry, which led to the discovery of the two different kinds of non-Euclidean geometry: hyperbolic geometry, which is associated with the Russian mathematician Lobachevski and the Hungarian mathematician Bolyai, and elliptical geometry, which was discovered by Riemann, a student of Gauss. Although Gauss himself did not develop the complete mathematics of non-Euclidean geometry, he constructed the mathematical framework around which the main body of non-Euclidean geometry had to be built.

To understand the need for non-Euclidean geometry, we point out briefly the weakest link in the chain of axioms on which Euclidean geometry is built—the famous parallel axiom (the fifth axiom), which states the following: Given a straight line and a point outside this line, one and only one straight line (in the plane determined by the given line and point) can be drawn through the point parallel to the given line. As far back as the fifteenth century, mathematicians were uncomfortable with this axiom and so they tried to prove it using the other axioms without success. Gauss was thus attracted to it as a challenge, but he soon became convinced that the parallel axiom is, indeed, an axiom and that it, therefore, cannot be proved; it must be accepted as a truth if one works with Euclidean geometry. But by that time, the work he was doing on the geometry of surfaces indicated that a geometry without the parallel axiom but with a substitute axiom can be constructed and used in the geometry of surfaces. (The following brief discussion of Gauss's work on surface geometry shows us the mathematical way to the general non-Euclidean geometries.)

Gauss began this work by analyzing very carefully the role of the coordinate system in geometry and by insisting that one must introduce the most general coordinate system possible and express the infinitesimal distance between two infinitesimally close points in terms of these general coordinates, called Gaussian coordinates, of the points. The mathematical formula for this infinitesimal distance, called the line element of the surface, depends on the geometry of the surface. If the surface is flat (a plane), for example, the square of the infinitesimal distance ds^2 between the two infinitely close points in Cartesian coordinates x and y is $ds^2 = dx^2 + dy^2$, where dx and dy are the infinitesimal differences in the x and y coordinates of the two points. This is just the infinitesimal or differential form of the theorem of Pythagoras, which means that the same formula applies to the square of the distance between any two points on the surface regardless of how far apart they may be. As Gauss pointed out, the formula for ds^2 stated above applies to two nearby points on a flat surface (a plane, for example) regardless of their location on the surface so long as the coordinate system is

Cartesian. By using different non-Cartesian coordinate systems, one can make the formula for ds^2 as complicated as one pleases. In general, however, mathematicians use coordinate systems in which the algebraic formulas for various geometric entities are as simple as possible.

Gauss made his great contributions by insisting that geometers use the most general coordinate system to describe any surface. In such a Gaussian coordinate system, the formula for ds^2 (the square of the line element) contains certain parameters that can be chosen to describe any kind of surface. Gauss thus laid down the basis for all different possible geometries of a surface. This geometry is now called Gaussian differential geometry because the formula for its line elements ds^2 is expressed in terms of the differentials du and dv of its Gaussian coordinate u, v, where the axes u and v are not straight lines like x and y in the Cartesian coordinate system, but sets of intersecting curves of any kind one pleases. Gauss then wrote the formula for the square of the line element as $ds^2 = g_{11}du^2 + g_{12}dudv + g_{21}dvdu + g_{22}dv^2$, where the g's are not constant numerical coefficients but functions of the positions of the two infinitely close points on the surface. As the two points are shifted on the surface, the g's change, indicating that either the geometry of the surface changes from infinitesimal region to infinitesimal region or the changes are due entirely to the type of coordinate system used, without any change in the geometry.

Gauss then developed a mathematical procedure for distinguishing between the changes in the g's resulting from the non-Euclidean character of the surface and changes stemming from the type of coordinate introduced on the surface. He reasoned that if the changes in the g's stem from the departures of the surface from the flatness (point to point) of the surface rather than from the quirkiness of the coordinate system, one should be able to construct from the g's a mathematical expression that permits one to distinguish between purely coordinate effects and real, intrinsic, non-Euclidean geometrical characteristics of the surface, which, in general, may change, from point to point. He discovered the mathematical (algebraic) formula of the g's he was seeking and showed that this formula gives the curvature of the surface at any point. If this curvature (called the Gaussian curvature) is zero at every point on the surface, the surface is Euclidean (flat). The expression for ds^2 is called the metric of the surface so that the g's determine the metric and, therefore, the geometry.

As long as one limits oneself to two-dimensional surfaces (two coordinates) the geometry of the surface is described by just the four g's: g_{11},

g_{12}, g_{21}, and g_{22} or since $g_{12} = g_{21}$, the three g's: g_{11}, g_{12}, and g_{22}, but mathematicians soon built up Gauss's ideas and concepts to three-dimensional space and then to four-dimensional space, which Einstein incorporated into his theory of relativity (the physics of a four-dimensional space-time manifold). But Riemann went beyond this point and developed n-dimensional geometry which we will discuss later. In three-dimensional geometry, the number of g's is 9 and in four-dimensional geometry the number is 16.

Gauss went beyond pure mathematics and did important work in astronomy and physics. Thus, he became interested in theoretical astronomy, following Piazzi's discovery on January 1, 1801 of the asteroid Ceres (the first asteroid to be observed), even though it soon disappeared from sight. Gauss then calculated the complete orbit of Ceres from three distinct observations of it that Piazzi had made, thus making it possible for Ceres to be located again very easily. Gauss continued with his work in astronomy and laid the mathematical basis for the theory of perturbations, which is used extensively in calculating the planetary orbits around the sun and the orbits of any objects such as comets, satellites, and meteorites moving in gravitational fields.

Gauss's analysis of the intrinsic geometry of a surface led him to the study of geodesy (the geometry of the earth's surface and surveying), and to the development of triangulation as the general method for measuring distances on the surface of the earth. Since the surface of the earth is spherical, Gaussian coordinates must be used to study its geometry. Because a small piece of the earth's surface may be taken as approximately flat, we may introduce a Cartesian coordinate system for that small area, but as we go from small area to small area, we must constantly change from one Cartesian coordinate system to another. Gauss's intrinsic geometry of a surface shows how to relate these constantly changing local coordinate systems to the overall correct Gaussian coordinate system. Because the geometry within each local Cartesian coordinate system is an infinitesimal Euclidean geometry, Gauss's intrinsic geometry of a surface is essentially an integration of or a sum of an infinite number of infinitesimal Euclidean geometries.

In his later years, Gauss was drawn ever more strongly to physics and he did very significant theoretical work in electricity and magnetism; physicists honored him, in time, by naming the unit of the strength of the magnetic field the gauss (G). He also did extensive theoretical work in electrostatics and is famous for what is now called Gauss's theorem, which enables one to calculate the quantity of electric charge enclosed within any kind of surface. Gauss's theorem states that the magnitude of that electric charge equals the sum of all the electric field intensities taken over the entire surface. As noted above, he was also deeply interested in the practical aspects of electricity and magnetism, performing many experiments in terrestrial magnetism, and on the propagation of electrical signals along wires. These experiments, as pointed out above, led him and Wilhelm Weber, at the University of Göttingen, to the invention of the electric telegraph in 1833–1834. This discovery predated Samuel Morse's invention of what he called the magnetic telegraph by three years, which is essentially the same as the electric telegraph.

We saw that Gauss, like Newton, abhorred controversy and refused to publicize mathematical discoveries that might lead to a conflict of ideas between him and other mathematicians. His letters and his diaries show that he had made many mathematical discoveries that he did not publish. In 1800, for example, he discovered certain important functions known as "elliptic functions" and by 1816 he had extended his early geometry to encompass most of the features of the non-Euclidean geometries discovered later by Lobachevski, Bolyai, and Riemann. Gauss's diaries show that he did not accept the Kantian thesis that the Euclidean geometry of space must be accepted as an absolute, *a priori* truth. He insisted that the nature of the correct geometry of space must be determined by observation as an experimental fact. In line with this approach, Gauss proposed an experiment which could not be performed with the crude technology of that era. He suggested that the three angles of a triangle, formed by the three lines connecting three different stars, be measured to see if their sum is 180 degrees as demanded by Euclidean geometry. This experiment was never performed, but it underscores Gauss's rejection of any *a priori* concept about the correct geometry of space.

This discussion leads us to the other great mid-nineteenth-century geometers Lobachevski, Bolyai, and Riemann, who continued Gauss's work on non-Euclidean geometry and brought it to its modern form. The works of Lobachevski and Bolyai did not stem directly from Gauss's work but rather from the rejection of Euclid's fifth postulate (the parallel axiom), which had troubled mathematicians from the time of its pronouncement by Euclid to the discovery of non-Euclidean geometry in the nineteenth century. For almost two thousand years, outstanding mathematicians had tried to prove it as a theorem rather than accepting it as an independent axiom so that the notions of a non-Euclidean geometry were implicit in the failure of these proofs. In his *A Concise History of Mathematics*, Dirk Struick notes that attempts to derive Euclid's fifth axiom go back to Ptolemy; such mathematicians as Nasir-al-din in the Middle Ages and Lambert and Legendre in the eighteenth and nineteenth centuries, respectively, devoted much time to it. These attempts were not abandoned until after Gauss had developed his surface geometry which, as we have already stated, convinced Gauss that non-Euclidean geometries are valid but can be constructed only if the fifth axiom of Euclid is replaced by another parallel axiom. Why Gauss stopped at this point, going no further than introducing the label "non-Euclidean geometry" is not clear; he certainly had the mathematical ability to do so, but was probably so involved in mathematical physics that he had no time to devote to non-Euclidean geometry.

Here we come to a remarkable change in the attitudes of the mathematicians about their own discipline. We recall that during this period the demands of science for new and improved mathematical techniques and discoveries that could be used to solve problems in physics and astronomy were so great that mathematicians were drawn to mathematical physics and astronomy rather than to pure mathematics. Of all the branches of mathematics, however, non-Euclidean geometry appeared to be the least promising as a theoretical tool for science and the least rewarding for advancement in a career of mathematical physics. As an intellectual adventure, developing a non-Euclidean geometry was, indeed, very attractive, but the idea that it could be useful to science in any way was dismissed out of hand. How, it was asked, could "real flat Euclidean space" be distorted into non-Euclidean curved space since three-dimensional space has no higher-dimensional space into which to be distorted from flatness? This response assumes that the only reality is Euclidean space and that non-Euclidean space is, at best, a mathematical idealization. In this sense, a two-dimensional surface, as a geometrical manifold, differs from three-dimensional space because a two-dimensional surface can be twisted or distorted in three-dimensional space in any way that one may want. From the point of view of a mathematician like Gauss living on a surface without knowledge of a third dimension, the geometry of the surface may be taken to be non-Euclidean. But Gauss knew that a mathematician in a three-dimensional space can describe all the properties of the surface using three-dimensional Euclidean geometry. But Gauss must have seriously questioned the reality of three-dimensional Euclidean geometry when, as we previously mentioned, he proposed measuring the three angles of an astronomical triangle to determine whether or not their sum is 180 degrees.

If Gauss was deterred from developing a complete theory of non-Euclidean geometry because he doubted its reality or its practical usefulness, Bolyai, Lobachevski, and Riemann were quite the opposite. Being unhappy with Euclid's parallel axiom, each, in his own way, decided to see how far he could go without Euclid's fifth axiom. Simple geometrical considerations convinced each one that leaving the axiom out altogether would not do because it would be impossible to prove certain basic geometrical theorems. The parallel axiom makes it possible for one to prove, for example, that the sum of the three angles of a triangle is 180 degrees and that the circumference c of a circle equals 2π times its radius r. But the restrictions imposed on Euclidean geometry by its basic elements, the line and the point, make it an ideal intellectual concept rather than a real one because neither a point nor a line has any reality. Thus Euclidean geometry ultimately would have to give way to a real geometry which could only be produced by replacing Euclid's fifth axiom by another "parallel" axiom. Bolyai and Lobachevski, independently, carried this out in a way which led to what we now call hyperbolic geometry. Riemann, by contrast, replaced Euclid's fifth axiom in quite a different way which led to what we now call elliptical geometry.

Although Lobachevski and Bolyai share the credit for constructing the first kind of non-Euclidean geometry, they worked independently of each other, without any collaboration. Because Bolyai was Hungarian and could not read Russian, he did not know, when he did his work, that Lobachevski, who was Russian, had already written a book in Russian, in 1829, presenting his mathematical model of a non-Euclidean geometry. Although Lobachevski's book was translated into German in 1840, it failed to arouse the interest of most mathematicians; Gauss himself did find it interesting, noting that it paralleled some of his own earlier ideas. To Bolyai, however, the revelation was devastating because it robbed him of his feeling of triumph at having achieved a very important mathematical breakthrough.

Janos Bolyai's story in this interplay of ideas is of particular interest because it hints at how precarious the standing and reputation of any but the very top practitioners are in a field as intellectually demanding as mathematics. Bolyai's father, a mathematics teacher in Hungary, and Gauss had studied mathematics together at Göttingen. The father, Farkas Bolyai, felt free to consult Gauss about Janos's "strange" ideas about a new geometry, which had appeared in the appendix of a mathematics book Farkas had written. Gauss praised Janos's work highly, indicating that he, having explored these same ideas himself earlier, and written them down in his diaries, could not honestly credit Janos with prior discovery. This revelation and Lobachevski's book left Janos so destitute of self-esteem and self-confidence that he never published anything else in mathematics. That three different and geographically widely separated mathematicians should arrive at the same revolutionary mathematical discovery almost simultaneously is, in itself, quite remarkable. It is as though great theories in mathematics and science are biding their time and waiting to be discovered and, then, the paths to such discoveries are revealed to a few different minds at the same time.

Janos had been warned by his father not to be ensnared in the fruitless quest for a proof of Euclid's fifth axiom. As a result, Janos turned to constructing a geometry with a different kind of parallel axiom. This approach was also considered by Lobachevski but we do not know what led him to these ideas. In any case, both Bolyai and Lobachevski replaced Euclid's parallel axiom by the following statement: Given a straight line and a point outside this line, an infinite number of straight lines can pass through the given point parallel to the given straight line. This leads to a self-consistent non-Euclidean geometry called Lobachevskian or hyperbolic geometry.

This parallel axiom cannot apply to a flat surface nor to what we ordinarily call straight lines on a plane. The concept of a straight line must be replaced by the concept of the shortest distance between two points (called a geodesic). In Lobachevskian geometry, straight lines are curves. Triangles consist of three such curves with their concave sides out. In a triangle of this sort, the angles are smaller than they would be if the sides were straight in the Euclidean sense. Thus, in Lobachevskian geometry, the sum of the three angles of a triangle is smaller than 180 degrees and the circumference of a circle is larger than 2π times its radius. A saddle surface, which has two concave sides (no convex or flat side), is an excellent example of a hyperbolic surface.

We turn now to Riemann's contributions to the development of non-Euclidean geometries, which, in part, stemmed from Gauss's tutelage at the University of Göttingen, where Riemann received his Ph.D. degree in mathematics in 1851. Riemann, like Lobachevski, abandoned Euclid's parallel axiom, replacing it by the axiom that no parallel lines exist. Indeed, all straight lines intersect in Riemannian (elliptical) geometry. The geometry on the surface of a sphere is an example of elliptical geometry; "straight lines" (shortest distances) on a sphere are arcs of great circles (circles whose radii equal the radius of the sphere) and they all intersect each other. Because the three sides of a triangle on a sphere are the arcs of three great circles, the sum of the three angles of such a triangle is greater than 180 degrees. One can also prove that the circumference of a circle on a sphere is always smaller than 2π times its radius. The circles of longitude (the meridian circles) on the earth's surface are the equivalent of straight lines and all intersect each other at the north and south poles.

Whereas Euclidean geometry imposes exact arithmetic relations among the geometrical entities that define figures such as triangles (the sum of the angles must equal exactly 180 degrees) and circles (circumferences must equal exactly 2π times their radii), Lobachevskian and Riemannian geometries allow a range of values for the sum of angles of a triangle and for the relationship between a circle's circumference and its radius. Thus, Euclidean geometry is like a line on a plane that divides the points on the upper half from those on the lower half. Euclidean geometry can be pictured as the limiting case of both Lobachevskian and Riemannian geometries; as we pass from one Riemannian geometry to the next, in infinitesimal steps, in a direction such that in each succeeding example of Riemannian geometry the sum of the three angles of a typical triangle decreases, we instantaneously pass through Euclidean geometry when the sum equals 180 degrees. We then enter Lobachevskian geometry for which the three angle sum is less than 180 degrees and decreases as we move along.

Perhaps the most important contribution made by Riemann to non-Euclidean geometry was his development of n-dimensional non-Euclidean geometries. He introduced the concept of the Riemannian curvature of n-dimensional geometry and wrote down the mathematics that one has to use to calculate this curvature at any point. This innovation ultimately led to tensor calculus, which is an extension of what is called vector analysis. Riemann's development of the geometry of n-dimensional space became the basis of Einstein's general theory of relativity and tensor analysis became its mathematics. The general theory of relativity, as we shall see, is a four-dimensional non-Euclidean theory of gravity.

Although Riemann is best remembered for his great contributions to non-Euclidean geometry, he did outstanding work in every branch of mathematics. His prodigious output was all the more remarkable because his life was short and his productive professional career lasted only fifteen years following the granting of his Ph.D. But he made the most of his brief time on earth and revolutionized every area of mathematics in which he worked.

That Georg Friedrich Bernhard Riemann would become a mathematician, let alone one of the outstanding original theoreticians of his age, was not foreshadowed by his humble origins. The son of a pastor, he was born in Breselenz, Hanover, Germany, on September 17, 1826. His mother, Charlotte Ebell, also gave birth to five other children, thus giving Georg's father even more reason to ask for heavenly blessings. Like Cauchy, Georg's youth was one of deprivation and he and his siblings often went hungry owing to the poverty that gripped the German countryside. Georg's parents did all they could, often foregoing meals so that their children could at least have some food on their plates, but George and his brothers and sisters all suffered from malnourishment and poor health. The Riemann family, like the Abels in Norway, still managed to enjoy each other's company. Despite the boisterous household, Georg was a quiet, retiring child who had an almost pathological fear of public speaking. He was also a very bright boy who received his early instruction in arithmetic and history from his father. Games and puzzles were among his favorite pastimes.

Young Georg received his earliest public education at the Hanover Gymnasium. He was a loner who did not socialize easily with the other children. But Georg excelled in his studies, particularly mathematics. Two years later, he entered the Gymnasium at Lüneburg, where he struck up a friendship with the director of the school. The director recognized Georg's mathematical abilities early on and granted him certain special privileges such as access to his private library. One of Georg's earliest interests, the theory of numbers, was whetted by his reading of Legendre's dense *Theory of Numbers*. He also benefited greatly from the classic works of Gauss, Abel, and Cauchy. Upon leaving the Gymnasium, Riemann spent a year at Göttingen before moving on to Berlin, where the brightest stars in the constellation of German mathematics could be found, including such luminaries as Jacobi, Steiner, and Eisenstein. Like a child in a candy shop, Riemann moved from geometry to the theory of numbers to analysis and back again, immersing himself in the newest discoveries in mathematics. Riemann considered his two years in Berlin to be a grand adventure in thought, which concluded with his return to Göttingen in 1849 to finish his doctoral work.

Before finishing his doctoral research, Riemann became acquainted with some of Göttingen's leading experimental physicists, particularly Wilhelm Weber. The two became good friends and Riemann learned firsthand about the differing perspectives that mathematicians and physicists bring to their respective professions. This exposure to the world of laboratories and experimental techniques instilled in Riemann an appreciation for the real-world applications of scientific theories and a keener sense of the need to be cognizant of the real-world counterparts of pristine mathematical models.

Riemann submitted his doctoral dissertation, which dealt with the theory of functions of a complex variable, to the legendary Gauss for his review. Not known for demonstrating his enthusiasm for the works of others, Gauss nevertheless praised Riemann's dissertation as "evidence of the author's thorough and penetrating investigations in those parts of the subject treated in the dissertation, of a creative, active, truly mathematical mind, and of a gloriously fertile originality." Although Gauss voiced a minor complaint about the clarity of Riemann's work, he was clearly taken by the dissertation which he viewed as a "substantial, valuable work, which not only satisfies the standards demanded for doctoral dissertations, but far exceeds them."

To secure a position as a lecturer at the University of Göttingen, Riemann was required to submit a qualifying essay. The examiners, headed by the venerable Gauss, asked Riemann to write about the foundations of geometry, a subject which Riemann was not prepared to take on with his usual thoroughness. Because he was also unable to tear himself away from his investigations of mathematical physics, Riemann took on the Herculean task of trying to complete both tasks at the same time. Before he could complete the essay, however, the rigors of constant overwork coupled with his own frail health resulted in his physical collapse. The illness dogged Riemann for much of the spring of 1854, but he slowly shook it off and finished the preparation of his lecture. Finally, the day that he had both anticipated and dreaded arrived, and Riemann presented his lecture on the foundations of geometry to Gauss and other members of the faculty. Their response was warm, far better than the timid Riemann had dared hoped. Gauss himself was visibly impressed.

Riemann had hoped to parlay his success into a university position but a paid academic appointment continued to elude him. He was despondent but continued his mathematical investigations, focusing on aspects of

Georg Riemann (1826–1866) (Courtesy AIP Emilio Segrè Visual Archives; T. J. J. See Collection)

mathematical physics such as differential equations and Abelian functions that were of particular interest to him. Regardless of the topic, however, Riemann's approach was always intuitive and global; he did not have the patience or the need for the densely methodical approach of a Legendre.

Fortune finally smiled on the impoverished Riemann in 1857 when he was appointed assistant professor at Göttingen and awarded the princely sum of three hundred dollars per year. This salary might have enabled Riemann to live in some comfort for the first time, but three of his sisters came to live with him soon thereafter, thus necessitating that the pot be divided four ways. But Riemann did not rail against fate for having quadrupled the size of his household. He was grateful for their company and their constant encouragement of his researches.

In 1859, his work in mathematics and in electrodynamics was officially recognized by the German government when it appointed Riemann to be the director of the Göttingen Observatory, the same position once held by Gauss. Riemann was also elected to several foreign scientific academies. Having taken up residence at the Observatory, Riemann was able to take a wife, the former Elise Koch. Although the two were devoted to each other, Riemann soon fell ill with pleurisy. Although he eventually recovered, the illness left him weak and listless. For the next two years, he divided his time between Göttingen and Italy, hoping that the milder Italian winters would help him to recover his health. Despite his recuperative efforts, however, his strength slowly slipped away. Death came quietly on July 20, 1866, in the Italian village of Salasca while Riemann sat quietly in a meadow, holding his wife and daughter in his arms.

Riemann's second love, after non-Euclidean geometry, was the theory of functions of a complex variable in which he completed certain important theorems that Cauchy had left only partially proved. In particular, he developed the concept of Riemann surfaces for complex functions of the real variables x and y. This work shows that if the complex function $f(x, y)$ equals $u + iv$, where u and v are real functions of x and y, then more than one u, v surface may exist on which the x, y plane is projected and which are attached to each other at their common origin $u = v = 0$. This was the beginning of modern topology. Riemann also studied and clarified the mathematical properties of certain important mathematical functions which appear in the mathematical literature as extremely important discoveries including the study of elliptical modular functions, the study of the convergence of trigonometric series, the study of the foundations of mathematical analysis, the precise definition of the integral concept, and so on. But in all of this his crowning glory was and will always remain Riemannian non-Euclidean geometry.

Although Gauss and Riemann were the best of the mid-nineteenth-century German mathematicians, a "second tier" group of very competent mathematicians must be mentioned, including Peter Dirichlet, who succeeded Gauss at Göttingen in 1855, Karl Jacobi, famous for his contributions to Newtonian dynamics; Leopold Kronecker; Richard Dedekind; Hermann Grossmann, to whom we return later in our discussion of vector and tensor analysis; August Möbius, famous for the Möbius strip; Karl Weierstrass; and, finally, Georg Cantor, the founder of the mathematics of transfinite numbers. Most of the mathematicians we have just mentioned turned their mathematical skills to the study of the functions of complex variables. They were so dominant in this field that Germany became the center of research in complex function theory. Weierstrass, in particular, became famous for his superb mathematical reasoning and for his beautiful lectures as a professor at the University of Berlin. He set the pattern for rigor and precision in mathematical reasoning that became the standard for all mathematicians who followed him.

We now turn to the non-French and non-German mathematicians who began to make their mathematical presence felt, not only in pure mathematics, but also in mathematical physics and in the general field of applied mathematics. Outstanding among these noncontinental mathematicians was the famous Irish mathematical physicist William Rowan Hamilton, who spent his entire life in Dublin, where, in 1827, at the age of 21, he became Royal Astronomer of Ireland, having already become professor of astronomy at Trinity College. In spite of all his administrative duties at the Irish Royal Astronomical Observatory and later as president of the Irish Academy and his teaching obligations, he did superlative research in pure mathematics and theoretical physics, particularly optics. At the age of thirteen, Hamilton had begun thinking deeply about optics, to which he returned in his early twenties. He developed completely his initial discovery of the relationship between optics and dynamics of particles moving in a force field such as gravity. His careful analysis of optics revealed to him its surprising similarity to mechanics. This work led him, in conjunction with Karl Jacobi—of the famous Hamilton–Jacobi equation—to the discovery of one of the most important equations in theoretical physics and the basis of modern quantum mechanics.

Hamilton was the greatest scientist ever produced by Ireland and also one of its most tragic figures. He was born in Dublin on August 3, 1805 into a well-to-do Irish family. William's father was a successful businessman but something of a pompous windbag. His gracious mother, Sarah Hutton, was the most likely source of her young son's genius, having descended from a family noted for its professional achievements. Although both his mother and father undoubtedly had some influence on William's early years, their legacy is uncertain as William was sent at the age of three to live with his uncle, the Reverend James Hamilton, who presided over the village of Trim.

William's uncle was no intellectual slouch, but took a great—some would argue—obsessive interest in Greek and Latin. He was also an accomplished linguist, having mastered a dozen languages ranging from Hebrew and Sanskrit to French and German. Not surprisingly, he believed that William should pursue a similar course of study and, consequently, steered the boy through a rigorous instruction program which enabled him to master Latin, Greek, Arabic, Sanskrit, and Hebrew by the time he was ten years old. Young William's facility with foreign tongues became a topic of conversation among the locals; he was asked at the age of fourteen to compose a statement welcoming the Persian ambassador on a visit to Dublin. Although the translation was a fine example of sickeningly sweet adulation, it called attention to the boy's genius.

An encounter with the American Zerah Colburn, a boy known for his ability to solve problems involving very large numbers in his head, was of pivotal importance in William's young life. He was intrigued with Zerah's ability and, after several conversations, uncovered most of his tricks. Thus attracted by the magic of numbers, Hamilton put aside his study of languages and turned his ravenous mind to mathematics, devouring the works of Newton, Lagrange, and Descartes. While still a teenager, Hamilton taught himself the calculus and familiarized himself with the important concepts of mathematical astronomy.

Hamilton's first formal education outside the home of his uncle began when he enrolled at Trinity College, Dublin, after having breezed through the rigorous entrance examinations and served notice to the other students and the faculty that he was not a country bumpkin but a force to be reckoned with. While most of his fellow students were preparing for their first week of classes, Hamilton was putting the finishing touches on his ideas about optics, which later formed the basis for one of his greatest mathematical works. Hamilton's treatise on rays confirmed the initial suspicions of the teachers and students that Hamilton was an extraordinarily talented young man. Indeed, his undergraduate education at Trinity was something of an afterthought because learning came so easily to him. Hamilton won every academic prize offered by the school and, in recognition of his brilliance, was offered the chair vacated by the resignation of Dr. Brinkley, a professor of astronomy. The governing board of the college had passed over several prominent professional astronomers to select Hamilton, who was then only twenty-two years of age. The Board's confidence was soon justified by Hamilton's publication of the first part of *A Theory of Systems of Rays*, which, according to Eric Temple Bell, "does for optics what Lagrange's *Mécanique Analytique* does for mechanics and which, in Hamilton's own hands, was to be extended to dynamics, putting that fundamental science in what is perhaps its ultimate, perfect form." In composing this treatise, Hamilton introduced a number of mathematical techniques that later became cornerstones of mathematical physics. Hamilton's mathematics would prove to be especially helpful to physicists such as Erwin Schrödinger in formulating a wave theory of the quantum mechanics in the 1920s. Although Hamilton's work was densely mathematical, its worth was recognized by Europe's leading scientists and mathematicians. Hamilton's study of optics thus caused him to be regarded as one of the world's leading mathematicians before he had reached the age of thirty.

Hamilton's study of optics and dynamics may have marked the pinnacle of his career. Certainly it was one of the most satisfying times of his professional life. He was young, brilliant, and wined and dined by the leading scientists, politicians, and businessmen of his era. But there was a dark side to Hamilton which soon began to manifest itself after he married Helen Bayley in 1833. Because Hamilton's proposal of marriage had been motivated more by the desire to settle down than by any real affection for his bride, theirs was largely a marriage of convenience. But Helen was soon afflicted by a debilitating disease which left her bedridden for the rest of her life. Hamilton, perhaps feeling guilty about his lack of interest in Helen, waited on her dutifully, but soon found greater solace in the bottle. What began as a dinner habit soon became an unending bout with alcoholism in which Hamilton himself was often unable to control what he said or did at even the most august public affairs.

Despite his alcoholism, Hamilton continued to pursue his mathematics and bring honor to Ireland. At the age of thirty, he was knighted and was thenceforth known as Sir William. Eight years later, the British government

William Rowan Hamilton (1805–1865) (Courtesy AIP Niels Bohr Library)

decreed that he should be awarded an annual pension that continued for the rest of his life. These public honors pleased Hamilton and were received as he finished his work on his quaternions, a subject that Hamilton believed was his greatest contribution to mathematics. This work was prompted by Hamilton's desire to develop an algebra that can describe three-dimensional rotations in space.

But Hamilton soon became mired in a swamp of Kantian philosophical babble about space being a form of "sensuous intuition," and foundered in trying to prove that algebra was somehow fundamentally connected with our notion of time. But Hamilton did invent a noncommutative algebra which was a true testament to his genius. His methods also showed other mathematicians how to devise an infinite variety of algebras by altering or otherwise manipulating specific postulates. Hamilton's death on September 2, 1865 at the age of sixty-one cut short a brilliant career; he left behind literally hundreds of unpublished papers, most of which were strewn about his office and home with reckless abandon. It was only several decades after his death that a systematic effort was undertaken to compile a complete edition of his collected works.

The mathematical formulation of the laws of optics evolved along two different routes, one called geometrical optics and the other physical optics. Geometrical optics stems from Fermat's principle of least time which deals with rays of light and states that a ray, in passing through different physical media (air, glass, water, etc.), moves along the path which requires the shortest time. This principle is used in tracing rays of light through lenses to design a good system of lenses for a camera, telescope, or any other kind of optical instrument. Physical optics, on the other hand, treats light as a wave and describes its propagation by means of a wave equation. This treatment is essential for describing the interference between two beams of light that start from a single source and then on passing through mirrors, is split into two rays which recombine again. Hamilton reformulated the mathematics of geometrical optics to show that a ray of light, moving in a straight line, behaves like a free particle, which also moves in a straight line, in accordance with Newton's laws of motion. He then showed that the path of a light ray can be computed from a single mathematical quantity that he called the characteristic function of the ray. Its mathematical properties are quite similar to those that physicists call the action in the dynamics of a particle. This concept was introduced in the eighteenth century by Maupertuis, whom we discussed in an earlier chapter.

At this point Hamilton took a very important speculative step to unite mechanics and optics. The similarity between the ray trajectories of geometrical optics and the Newtonian paths of particles suggested to him that it should be possible to express the laws of mechanics in a mathematical form closely resembling that of geometrical optics. He was further encouraged in this very productive flight of fancy by his knowledge that just as the laws of geometrical optics can be deduced from Fermat's principle of least time, the laws of Newtonian mechanics can be deduced from Maupertuis's principle of least action. Hamilton then showed that his characteristic function for geometrical optics is formally similar to the action function in mechanics introduced by Maupertuis. According to Hamilton, the motion of a particle in classical Newtonian mechanics can be described as though it were a ray in geometrical optics. This was the first step in the discovery of the remarkable unification of the laws of nature; that particles of matter and rays of light share a kind of corpuscularity that was not then understood. The full significance of this discovery was not fully grasped by physicists until the beginning of the twentieth century when Planck and Einstein, the first experimentally, and the second theoretically, discovered that light has a real corpuscular nature—the photon. But this advance left an asymmetry between light and matter: Whereas Hamilton's work shows that light has particle properties in addition to its wave properties, no one had shown, at that time, that particles themselves have wave properties. The wave properties of particles were finally discovered in the 1920s so that the circle of ideas that Hamilton had begun to draw in the 1820s was completed a century later.

In developing his characteristic function for optics and then relating it to the action function in mechanics, Hamilton had to reformulate Maupertuis's principle of least action to bring it as close as possible mathematically to Fermat's principle of least time. Hamilton thus had to enlarge the Maupertuis action concept by adding another term to it. To understand better this idea, we picture a particle of mass m moving for an infinitesimal moment dt along an infinitesimal stretch of its path represented by infinitesimal amounts, dx, dy, and dz of displacement along the x, y, and z axes of a Cartesian coordinate system. If $p_x = mv_x$, $p_y = mv_y$, $p_z = mv_z$ are the components of the particle's momentum p along its path, and E is its energy during the time dt, then, according to Hamilton, its infinitesimal increase ds in its action s is $ds = p_x dx + p_y dy + p_z dz - E dt$. Hamilton's great contribution to the action concept was the inclusion of the term $-E dt$, which never occurred to Maupertuis. This is one of the most far-reaching steps in the story of physics, though it was not recognized as such at the time. Its great importance lies in its combining three components of space, dx, dy, and dz, with a component of time dt to express the principle of least action. A noteworthy feature of this achievement is that each of the three components of space has the corresponding component of the momentum of the particle as its coefficient and the element of time has the negative of the energy multiplying it.

During this very productive period of his mathematical career, Hamilton had discovered that he could reformulate Newton's laws of motion in terms of the momentum and energy of a particle by introducing the energy of the particle in terms of its momentum p and its position q (here p stands for the three components of the particle's momentum and q for the three coordinates of its position). To elucidate this concept further we consider the total energy E of a single particle moving in some kind of force field (e.g., gravity). E is then a sum of its kinetic energy (energy of motion) and its potential energy (energy which depends on its position q in the field). Hamilton replaced the mathematical expression $(\frac{1}{2})mv^2$ (where m is the mass of the particle and v is its velocity) for the kinetic energy of the particle by the expression $p^2/2m$, where p is the momentum, mv, of the particle so that the total energy E becomes $p^2/2m + V(q)$. When expressed in this form E is called the "Hamiltonian" H of the particle so that the velocity v of the particle does not appear explicitly in the Hamiltonian H. Offhand, this change may appear to be a mere mathematical formality, empty of any physical significance. This is not so; the change is of deep significance. Indeed, the Hamiltonian became, and is now, the basis of quantum mechanics, which must replace Newtonian mechanics in the dynamical equations of an electron in an atom.

Another important advantage of Hamiltonian dynamics over Newtonian dynamics is that the Hamiltonian formulation can be applied quite easily to aggregates of particles and to force fields themselves as though they were particles. Using the entity H, Hamilton replaced Newton's laws of motion by two sets of partial differential equations, which may be expressed symbolically as $\dot{q} = dH/dp$, $\dot{p} = -dH/dq$, where the dot means rate of change with respect to time and the symbol d means the partial rate of change. These equations exhibit a very remarkable feature about the dynamics of a particle which is not at all apparent in the Newtonian formulation: namely, the symmetry between the momentum p of a particle and its coordinate position q. Owing to this symmetry, p and q are called conjugate dynamical variables. In 1927, the German physicist Werner Heisenberg discovered that such variables obey his famous uncertainty principle and are governed by a noncommutative algebra: pq is not equal to qp.

As we noted above, Hamilton demonstrated that the laws of optics and dynamics can be deduced from a single general minimal principle, of which the principle of least time (optics) and the principle of least action (dynamics) are special cases. This brought the calculus of variations into theoretical physics as a very powerful mathematical tool. All one has to do to analyze the dynamics of a system is find the action S that applies to it, then find its minimum by applying the calculus of variations to it and thus discover its dynamics. This is essentially the goal or purpose of theoretical physics. Finding the correct mathematical expression for the action is, however, often quite difficult. For particles moving with energy E in a known force field, however, Hamilton and Jacobi wrote down the famous Hamilton–Jacobi partial differential equation for the action S which is obtained from the following properties of the action: $dS/dt = -E$, $dS/dq = p$, and $E = H$. If we now replace E by $-dS/dt$ and p by dS/dq wherever it appears in H, we obtain the Hamilton–Jacobi partial differential equation for S: $dS/dt + H(dS/dq, q) = 0$, where $H(dS/dq, q)$ is merely mathematical shorthand for the statement that H is some kind of function of dS/dq and q.

We have devoted these pages to Hamilton's work because his mathematical discoveries, perhaps more than those of any of the mathematicians of that period, have played, and still play, such an enormous role in modern physics, astronomy, and cosmology. However, other mathematicians throughout Europe such as Arthur Caley, James Sylvester, George Green, Eugenio Beltrami, Volterra, Gregorio Ricci-Curbastro, and Tullio Levi-Civita began to contribute to the global dissemination of the new developments in mathematics.

In this chapter we have built a bridge which will take the reader from the classical mathematics of Newton, Lagrange, Euler, Cauchy, and Gauss to modern mathematics—the focus of the remainder of this book.

CHAPTER 10

The Beginning of Modern Mathematics

It is no paradox to say that in our most theoretical moods we may be nearest to our most practical applications.

—ALFRED NORTH WHITEHEAD

In the previous chapter we noted that all of the basic principles of mathematics had been well formulated by 1850 and mathematicians had focused their attention on consolidating their gains by imposing rigor and strict rules of mathematical procedures on their mathematical ventures. The ragged edges of proofs of basic theorems were being smoothed away and skepticism, if not distrust, of doubtful results was being strongly encouraged. At the same time, many mathematicians were turning to mathematical physics where mathematical rigor was not as necessary nor as exacting as in pure mathematics. But the practical, experimental physicists were not always happy about this intrusion into their domain because they were fearful that the formalism of mathematics, which was becoming increasingly complex and abstract, would obscure the physics. To most physicists, mathematicians were busily trying to learn more and more about less and less. But as physics advanced and new experimental discoveries required new explanatory theories that could be properly formulated only mathematically, advanced mathematics became indispensable to the growth of physics. This handmaiden role that physics demanded of mathematics became increasingly more urgent with the new discoveries in such diverse fields as electromagnetism, the kinetic theory of gases, and so on.

The mathematical models of most of the theoretical, mathematical physics stemming from these discoveries could be, and were, adequately constructed with the classical mathematics that had been developed up to that time. But a drastic change occurred in the nature of the mathematics that theoretical physicists had to apply to the quantum theory and the theory of relativity—two great physical theories discovered in the first decade of the twentieth century. The remarkable mathematical aspect of these theories, without which probing the universe, from its subatomic to its cosmic dimensions, would be impossible, is that they have features that can be understood only in a mathematical sense because they seem to have no real physical counterpart. Put more bluntly, some of the deductions about the universe to which the mathematics leads us make no sense from the physical point of view. Indeed, if we did not have the proper mathematics to guide us, we would know nothing about these strange, unphysical features which are of both a subnuclear and a cosmological character. Here the only reality seems to be the mathematics, as some outstanding physicists have argued; they insist that looking for any deeper reality behind the mathematics is fruitless. Mathematicians themselves do not take this point of view, but instead accept mathematics as a magnificent intellectual enterprise, without having to justify its existence as the embodiment of ultimate truth. That the phenomena in the universe can be described mathematically is, in itself, something of a mystery but also quite wonderful.

We have pointed to the year 1900 as a watershed year in the development of theoretical physics, marking the transition from the classical theoretical physics of Newton, Lagrange, and Gauss to the modern theoretical physics and relativity theory of Planck and Einstein, respectively. But the mathematical transition was not as revolutionary as that which had occurred in physics. The pre-twentieth-century mathematics kept step with the growth of the pre-twentieth-century physics, adjusting itself or expanding as required by the physics. Before describing these mathematical expansions in some detail, however, we note them first and then relate them to the relevant physics.

With the growth of the field concept in physics as well as the conservation laws of momentum and energy, vector analysis arose as the pertinent mathematics for dynamics and fields. Vector analysis is really an extension or generalization of the Euclidean geometry of three-dimensional space, which had originated some years before Riemann's geometrical theories but its importance for physics as well as for geometry was not recognized or fully understood until Hermann Grassmann, a mathematics teacher at the Gymnasium in Stettin, wrote his book *Theory of Extensions* in 1844. Some forty years later, the American physical chemist Josiah Willard Gibbs, then a professor of theoretical physics at Yale, extended Grassmann's work, with special emphasis on the application of vector analysis to the solution of physical problems. Much of the modern vector analysis nomenclature was introduced by Gibbs.

The rapid growth of vector analysis led to a greater concentration by mathematicians on the role of the coordinate system in mathematics and the development of the theory of transformations of coordinate systems. The British mathematicians Arthur Cayley and James Joseph Sylvester were the primary mathematicians in this field, which later became extremely important in the theory of relativity. This work finally grew into what mathematicians now call transformation theory—a very important branch of algebra. The promulgation of James Clerk Maxwell's electromagnetic theory of light required the development of the differential and integral calculus of vectors and a deeper understanding and mastery of partial differential equations. During the last quarter of the nineteenth century, thermodynamics arrived as a full-blown, self-consistent theory on the wings of a special or restricted branch of the differential calculus. This led to the very important concept of entropy. Finally, the molecular theory of gases—the kinetic theory—required the growth of classical probability theory into the statistical mechanics of modern physics.

Having indicated the connection between each of the various branches of physics as they evolved, and a particular branch of mathematics, we now consider each of these branches in somewhat more detail. The invention or, rather, construction of vector analysis as a mathematical tool or technique for handling three-dimensional problems in dynamics was the natural outgrowth of using a coordinate system to describe the motion of a particle. A line drawn from the origin 0 of the coordinate system to the particle is called the vector position $\mathbf{r}$ of the particle. This is a mathematical shorthand for specifying the positions of the particle by a single directed physical entity rather than by its three coordinates. This entity is the vector $\mathbf{r}$ which has both magnitude (its length represents the distance of the particle from the origin of the coordinate system) and direction (given by the angles the vector makes with any two of the three Cartesian axes x, y, or z). The laws of addition and multiplication of two or more vectors are not the same as those in the arithmetic of numbers and in standard classical algebra. Thus the vectorial sum of two vectors of magnitudes 3 and 4 is not 7; the sum of these two vectors can range from 1 to 7. The product of two vectors is also quite different from that of ordinary algebraic quantities. Two different kinds of products are defined in the multiplication of two vectors. In one of these products, the result is an ordinary number, not a vector; if the product of the two vectors in this kind of multiplication is zero, we may not conclude that one of the vectors must be zero as in arithmetic or in ordinary algebra. Another type of product, called the vector product, is also introduced in vector algebra. In this type of multiplication of two vectors, the product itself is a vector. But the product is noncommutative in that $\mathbf{A} \times \mathbf{B}$ does not equal $\mathbf{B} \times \mathbf{A}$ if $\mathbf{A}$ and $\mathbf{B}$ are two different vectors.

The beauty and utility of vector analysis is that a vector in a given space (e.g., three-dimensional space) can be expressed as an ensemble or sum of entities of the same kind which are independent of each other. Thus the position vector $\mathbf{r}$ represents three independent distances—one along each coordinate axis in a rectangular Cartesian coordinate system. This enables us to represent three different independent distances by a single directed distance by choosing the correct magnitude and the right direction for this single distance.

The vector concept was extended by physicists to represent any physical entities that have both magnitude and direction such as velocity, acceleration, forces, fields, and so on. If we now consider any vector, whatever physical entity it may represent, in any Cartesian coordinate system, we say that the vector is the sum of three mutually perpendicular, independent components of the given vector, each a vector in its own right. The importance of the vector concept in physics and mathematics is that if we alter the coordinate system by transporting it to another origin or rotating it in any way, the components of the vector change but the vector itself remains unaltered. Mathematicians and physicists therefore say that the vector is an invariant and hence represents something intrinsic—that is, something physically important. Physicists therefore express the laws of nature in vector form because laws must be intrinsic truths and, hence, invariant.

The great British physicist James Clerk Maxwell made the most important and elegant use of vectors in the history of mathematics and physics when he wrote down his famous partial differential equations of the electromagnetic field in terms of the electric and magnetic field vectors. This is probably also the first important example of vector calculus. We note that every term in any algebraic equation involving vectors must be a vector.

The introduction of vectors opened up the whole field of transformation theory, which became one of the most powerful mathematical tools in the theoretical physicist's workshop because it enables the theoretician to choose, from all possible statements about the universe, those that are intrinsically true and therefore candidates for laws. One need only subject the statement to a transformation of coordinates; a necessary condition that it be true is that it remain unaltered no matter how the coordinate system is altered. We may illustrate this concept by considering a vector $\mathbf{A}$ in a Cartesian coordinate system with axes x, y, and z and suppose that the x, y, and z components of $\mathbf{A}$ are A_x, A_y, and A_z so that we have the vector sum $\mathbf{A} = A_x + A_y + A_z$. Suppose now that we introduce a new Cartesian coordinate system x', y', and z' by rotating our initial coordinate system. In this new coordinate system, the vector $\mathbf{A}$ remains unaltered but it has components $A_{x'}'$, $A_{y'}'$, $A_{z'}'$, which are, in general, different from the components in the initial coordinate system. Each of the new components of $\mathbf{A}$ can be expressed as algebraic sums of the components. As an example, we may have $A_x = a_{11}A_{x'}' + a_{12}A_{y'}' + a_{13}A_{z'}'$ with similar equations but with different coefficients such as a_{21}, a_{22}, a_{23}, for $A_{y'}$ and $A_{z'}$. We thus have three simultaneous algebraic vector equations with nine coefficients a_{11}, a_{12}, $\ldots$, a_{32}, a_{33}, which define the transformation. Mathematicians write this expression symbolically as $T\mathbf{A} = \mathbf{A}'$, where $\mathbf{A}$ and $\mathbf{A}'$ are the same vector, but with different components.

The quantity T, which stands for the transformation, is a square array of the nine coefficients a_{11}, a_{12}, $\ldots$, a_{32}, a_{33}; it is called a matrix. If we apply two different transformations, T_1 and T_2, one after the other on our coordinate system, we find that such a product T_2T_1 is equivalent to applying a single transformation T_3 to our initial coordinate system. We may therefore write $T_2T_1 = T_3$ so that these transformations form a group because the product of any two equals another one. This is also true of matrices because the product of any two matrices containing the same number of elements is a matrix of the same kind. Thus the study of matrices, led by Cayley and Sylvester in England, became very important to physicists.

The algebra of matrices is noncommutative; the product T_2T_1 is different from the product T_1T_2. Matrices are very important in the quantum mechanics of atomic structure; they represent physical entities such as the position and the momentum of a particle in Newtonian dynamics. In the example given above of the transformation of a coordinate system, the coefficients a_{11}, a_{12}, $\ldots$, a_{32}, a_{33} are written as a matrix having three rows and three columns of coefficients.

To illustrate that very different branches of mathematics are related to each other, even though they may appear quite unrelated, we show that vectors and the rotations of coordinate systems can be represented by complex numbers in a very elegant way. We consider the complex plane x, y where the points on the x axis represent real numbers 1, 2, 3, . . . and the points on the y axis represent imaginary numbers i, $2i$, $3i$, where $i = \sqrt{-1}$ is the imaginary unit. Any point on the complex plane now represents the complex number $x + iy$, where x and y are any real numbers we may choose. If we now draw a line from the origin 0 to this point, this line can be treated as a vector whose direction is given by the angle it makes with the x axis and whose magnitude (length) is simply

$$\sqrt{x^2 + y^2}$$

All the rules of vector algebra and vector arithmetic can be derived from the algebra and arithmetic of complex numbers.

But before we can go further with complex numbers and use them to represent rotations, we must simplify the algebra of rotations considerably. If we multiply a real number x on the real x axis by i, it becomes the pure imaginary number xi and therefore must lie on the imaginary y axis. If we then picture a vector along the x axis, with its tip at the point x on the real axis, and multiply it by i, it must then be along the y axis with its tip at the point xi. This means that multiplying by i is equivalent to a rotation of 90 degrees or $\pi/2$ radians. But by de Moivre's theorem (discussed in Chapter 2) $i = e^{\pi i/2}$ where e is the base of natural logarithms. If we now rotate our vector counterclockwise on the complex plane from the x axis through the angle θ, the rotation is equivalent to multiplying our vector by $e^{i\theta}$, where θ is expressed in radians. A rotation in a plane is thus given by multiplication by e raised to an imaginary exponent. This operation simplifies the theory of rotations and rotational transformations of coordinate systems considerably because exponentials are easy to handle algebraically. Thus vectors, rotations, and complex numbers are related to each other in a very simple way, thus illustrating the beauty of mathematics.

To show the impact of all of these mathematical relationships on theoretical mathematical physics, we return to James Clerk Maxwell's towering discovery of the electromagnetic theory of light and the partial differential wave equation of the electromagnetic field. This story began in the 1820s when the Swedish physicist Oersted discovered that a wire carrying an electric current is surrounded by a magnetic field. This means that electric charge plus motion produces magnetism. In the 1830s Michael

Faraday, the great British experimental physicist, discovered that a moving magnet produces an electric current in a wire past which the magnet is moving so that magnetism plus motion produces electricity.

Maxwell then took the next step by combining Oersted's and Faraday's discoveries into the single concept of the electromagnetic field. He pictured this field as consisting of two oscillating or rotating vectors: the electric vector E and the magnetic vector H, which are perpendicular to each other and oscillate 90 degrees out of step but at the same rate. Using the complex plane model of a vector one can then represent the time variation of both E and H by multiplying each of them by e^{iwt}, where w is the number of oscillations per second and t stands for time. If one now pictures the complex plane in which E and H are oscillating as moving perpendicular to itself at the speed of light c, the mathematical model that represents this phenomenon is the formula for a wave. As Maxwell's deductions were of a purely mathematical nature, it is astounding that it leads to a correct physical description of one of the most remarkable phenomena in nature, namely, the physical nature of light and, indeed, of all radiation from gamma rays to radio waves.

We have described above the intimate relationship between pure mathematics and Maxwell's electromagnetic theory of light and how the mathematical manipulation of Maxwell's field equations leads to wave equations of the electric and magnetic field vectors. This illustrates an important feature of mathematical physics—the mathematics reveals physical features of the phenomena that are not apparent without the mathematics. Any mathematical deductions from mathematically expressed physical laws must also be laws. From Maxwell's electromagnetic field equations alone, we cannot see that the electromagnetic field vectors E and H are propagated as a wave with a definite speed; we must perform some additional mathematical operations on these vectors to obtain this astounding result. If Maxwell's field equations are correct, then the wave equation of the electromagnetic field so deduced must also be correct—it must also be a law. This technique shows how mathematics may be used to generate new physical laws. The experimental physicist, of course, does not accept the mathematical deduction as a law; he does so only if the deduction can be experimentally verified. The German physicist Heinrich Hertz demonstrated the truth of Maxwell's theory experimentally by using a spark coil to generate electromagnetic waves and then picking them up with another coil. He also showed experimentally that his spark-coil-generated waves behave exactly like light waves; this was the beginning of wireless telegraphy and all facets of modern telecommunications.

Mathematicians performed another important service for electromagnetic field theory by introducing the electromagnetic field potentials to represent the electromagnetic field in place of the field vectors. The French mathematician Poisson had introduced the concept of the gravitational field potential in place of the gravitational field at a point, and had shown that the potential obeys a second-order partial differential equation; the solution of this equation for any given distribution of masses gives the gravitational potential at all points of space. From these spatial values of the potential, one can deduce the gravitational field strengths at any point in space. The mathematics of the potential concept introduced for gravity was carried over unaltered to the electromagnetic field, simplifying the mathematics of the electromagnetic field enormously. The Poisson equation holds for the electromagnetic field with the masses replaced by electric charges. The advantage of working with the potentials of fields is that the potentials are scalars (quantities without directions) so that directions are not involved, whereas fields are vectors and therefore more difficult to handle mathematically.

The contribution of mathematics to the development of thermodynamics, another important branch of physics but much simpler than electromagnetism, parallels its contributions to electromagnetism. Thermodynamics is perhaps the most universal and self-sufficient branch of physics, based, as it is, on two simple laws, called the first and second laws of thermodynamics. The first law is the restatement of the most basic law in physics—the principle of the conservation of energy. But the first law states the conservation principle in such a mathematical form to make it most useful in thermodynamics. The second law introduces the somewhat mysterious concept of "entropy," from which the mathematics strips away the mystery. Entropy thus becomes a mathematical entity which can be described by its mathematical properties even though it cannot be measured directly like a length or like temperature. Here again we see the usefulness of mathematics to the physicist; it enables him to incorporate nonphysical concepts into his description of nature.

To describe in the simplest way the important role of mathematics in the formulation of the laws of thermodynamics, we consider a perfect gas confined to a cylinder topped by a piston which is free to move along the cylinder in either direction in response to physical changes in the gas. We know that the gas consists of freely moving, random motion molecules, which produce the macroscopic properties of the gas. Pursuing an investigation of that sort mathematically to deduce the observed macroscopic gas properties from its molecular properties is called the gas kinetic theory. Pushed to its ultimate conclusion the kinetic theory states that the macroscopic properties of the gas stem from Newton's laws of motion applied to the molecules.

We mention the kinetic theory of gases here merely in passing, devoting ourselves entirely to the thermodynamics of gases, which limits itself to the macroscopic gas parameters: temperature T, volume V, and pressure P. The relationship between the pressure of a gas and its volume, with the temperature of the gas kept constant, was first studied and discovered by Robert Boyle, a friend of Newton's. This discovery, called Boyle's law, initiated a whole series of experiments on gases which led, some years later, to the discovery of the general law of gases by Charles and Gay-Lussac which applies to a gas, where all three parameters—P, T, and V—are changing together. The general gas law states that the three parameters P, V, and T are related and must change together in such a way that if two of them are given, the third can be immediately determined. This law is true regardless of whether the gas is heated, cooled, compressed, or expanded. The mathematician states this law by saying that the three parameters are functionally related, expressing it as an algebraic equation $f(P, V, T) = 0$; this is called the "equation of state" of the gas. This remarkable equation reveals that all gases, regardless of their chemical nature, obey exactly the same equation of state, namely, $PV/T = $ constant. This equation states that the product of a gas's pressure and volume, PV, divided by its temperature T never changes; this is the famous gas law and it holds true regardless of what we do to the gas in our cylinder. Of course this relationship may be manipulated mathematically: P may be expressed as a function of T and V, or T as a function of P and V, or V as a function of P and T. These various forms of the gas law are used by chemical engineers; mathematics thus extends the usefulness of the gas law enormously.

With these basic ideas about gases understood, we can now describe one of the most remarkable technological discoveries ever made for which mathematics deserves the lion's share of the credit. This discovery was first announced to a skeptical society of physicists in 1842 by the medical doctor Julius Mayer, who argued that heat must be considered as a form of energy.

Presented entirely from the physical point of view, the concept of heat as a form of energy was rejected. Once this statement was put in the proper mathematical form, however, it finally gained universal acceptance, becoming the "first law of thermodynamics." This is an excellent example of the way mathematics sharpens and clarifies our understanding of physical concepts which may appear fuzzy without the mathematics. This also accounts for the long gap of almost 200 years between the discovery of Newtonian physics and the birth of thermodynamics. After all, heat and energy were known entities even before Newton's time but no one understood how to relate heat to energy mathematically. Once this relationship was uncovered, a whole new branch of physics was revealed.

To see how the mathematics came together to produce thermodynamics we return to our gas-laden cylinder with a fixed weight on the piston to keep the piston in place. We soon discover, however, that keeping the piston in place with the single weight is impossible. Careful observation of the weight reveals that it keeps bobbing up and down, and still more careful observation of the weight and our environment shows that the weight bobs up when the room gets warmer and bobs down when the room gets cooler. Because the gas warms up and cools off with the air in the room, we conclude that the weight bobs up when the gas gets warmer and bobs down when the gas gets cooler. Expressed differently, the weight rises when heat is supplied to the gas and bobs down when heat flows out of the gas.

We now place a thermometer in contact with the gas to monitor its temperature from moment to moment and inject a small heating element in the gas with which we can supply a tiny amount of heat (as tiny as we may wish) to the gas. This amount may be expressed as dQ—Q stands for heat and d is the symbol for infinitesimal. We then discover that the temperature of the gas, as shown by the thermometer, rises by an infinitesimal amount dT and the weight rises by an infinitesimal amount dh, where h is the height of the weight above the ground. How are we to interpret these very striking phenomena? Clearly, the flow of heat produces two effects: part of the heat stays inside the gas, raising its temperature, and another part of it raises the weight. Since these two phenomena are produced by heat, their physical essence must be the same—which must also be the same as that of heat. Now we know exactly the physical character of the upward motion of the weight—it is energy. The gas is pushing the piston upward, raising the weight, thus doing work on the weight and increasing its energy. We thus conclude that the increase in the temperature of the gas means that the energy of the gas itself, called its internal energy V, also increases. Hence, heat is a form of energy.

We can now incorporate these ideas into a single elementary algebraic equation which relates work W, heat Q, and internal gas energy U to each other: $dQ = dU + dW$. This is the differential algebraic statement of one of the most celebrated laws in physics—the first law of thermodynamics. This law is a bookkeeping statement about energy; it states that energy may appear in different forms but whatever its form and however it may be interchanged among bodies in any phenomena, the total amount must remain the same. This is just the principle of the conservation of energy but is rephrased to include heat as a form of energy.

Why was this simple equation so momentous in the development of modern technology? It states that work can be obtained from heat, an idea which had not been known or, at least, properly understood. This equation performed two great services for mankind because it initiated the industrial revolution by replacing the work of humans and animals with machines driven by heat, and it destroyed the institution of slavery because it made it unprofitable to use expensive human labor to perform work that machines could do more efficiently, more quickly, and more accurately. It appeared that the simple algebraic equation that expresses the first law of thermodynamics promised mankind untold wealth because heat is all around us, but this optimism was soon tempered by the discovery of the second law of thermodynamics by Rudolf Clausius in Germany and William Thomson (Lord Kelvin) in England. This second law places severe restrictions on the way heat can be changed into work. Although the second law is a physical principle, its full significance can be grasped only through its mathematical formulation because it introduces a new physical entity into physics—entropy—which can neither be perceived by our senses nor measured directly. Entropy is a purely mathematical entity.

We can best introduce the second law by using a thermodynamic two-dimensional coordinate system, which, of course, has nothing to do with a spatial coordinate system. We recall that a thermodynamical system such as a gas can be completely described by just two of its three parameters P, V, and T (pressure, volume, and temperature). We may therefore picture a two-dimensional coordinate system on a P, V plane with P as the ordinate and V as the abscissa. Each point on the plane represents a state of our gas, if the thermodynamical system we are dealing with is the gas in our cylinder. The P and V values at a point are the coordinates of our gas, from which, using the equation of state, we can calculate the temperature T of the gas. As the gas changes, these changes can be represented by a path on the P, V plane, at each point of which, the pressure, volume, and temperature of the gas have very definite values; as discovered experimentally, the algebraic combination PV/T must, of course, have the same value at every point on the plane. This is the experimental law of Charles and Gay-Lussac discussed previously.

Suppose now that the condition in the cylinder changes from moment to moment with heat flowing into the gas and out of the gas as the gas does work on the weight. We then discover a very important difference between the way the internal energy U of the system changes and the way the heat entering and leaving the gas and the work done by or on the gas change. Whereas the change of U of the gas between any two fixed points on the PV plane, regardless of the path that connects them, is always the same, this is not true of either the heat exchanged or the work done. The heat exchanged between the gas and its environment and the work done on the gas or by it depends on the path in the PV plane connecting the two points. The internal energy U of the gas is a state function whereas the heat and the work done are not state functions. Because a state function is an intrinsic property of the gas, physicists were interested in finding other such functions and they discovered that the entropy S of the gas is one such function. A "state function" describes a physical state of the gas which does not depend on how the gas is brought to that state.

In our discussion of the first law of thermodynamics we noted that it is just a bookkeeping constraint on the energy in a system: energy must be conserved but nothing in this principle constrains the direction in which a process involving the flow of heat may proceed. In other words, the first law does not show any favoritism between the two processes. Heat $\rightarrow$ Work and Work $\rightarrow$ Heat. But nature is not so evenhanded; it favors the second process by a large margin over the first. Work is generally changed completely into heat but heat can never be changed completely into work. The second law of thermodynamics is the statement of this constraint on the direction of processes in nature. It states, essentially, that processes in nature are irreversible. We come now to the role of entropy in this story and we show how elementary mathematics leads us to it. To this end we return to the first law of thermodynamics which we wrote algebraically as $dQ = dU + dW$, which is an equation of differentials and the first step in writing down a differential equation.

But we cannot express this as a true differential equation which can be solved to obtain expressions for Q, U, and W in terms of the parameters P, T, and V of the gas because neither Q nor W can be written as a function of P, T, or V. But this problem did not disturb the mathematicians who studied differential equations because they had discovered a simple mathematical procedure for changing an equation of differentials into a soluble differential equation. This procedure involves finding a certain factor, called an "integrating factor," which, on multiplying the recalcitrant equation of differentials, changes it into a soluble differential equation. It is like waving a wand and changing the ugly duckling into a beautiful swan. The integrating factor that changes the equation stating the first law of thermodynamics into a simple differential equation is $1/T$. If we multiply the equation $dQ = dU + dW$ by $1/T$, we obtain $dQ/T = dU/T + dW/T$, where dQ/T equals the differential dS of the quantity S, the entropy. Mathematicians thus write the differential equation for S as $dS = df(P, V, T) + dg(P, V, T)$, where S, f, and g are now definite functions of P, V, and T so that the differential equation for S can be solved. This is the mathematical statement of the second law of thermodynamics.

The mathematics now reveals some very remarkable properties of S and of the second law. The entropy can never decrease regardless of what happens to the system being studied. In general, it increases but, under very restricted conditions, it may remain constant. The total disorganization in a system increases as S increases, but organization and disorganization must go hand in hand so that the entropy in some parts of a system must decrease while it increases in other parts, with the increase exceeding the decrease. As the entropy increases, the total information we have about a system decreases. Because the net entropy of any system must increase as the system evolves, this outcome must be true of the entire universe so that its entropy is always increasing. In any kind of nuclear reaction, atomic reaction, or molecular reaction, the reaction must proceed in such a direction as to produce a net increase in entropy. Because the system evolves along states of increasing probability, the larger the entropy of a state, the larger is its probability, which is expressed by the simple equation S (entropy) $= k \log Y$, where Y stands for probability and k is a basic universal constant. As entropy can go in only one direction, processes in nature can go in only one direction so that all natural phenomena are irreversible.

This discussion of the laws of thermodynamics and the properties of entropy illustrate the revelatory power of even simple mathematics and reveal its beauty. Much of what we described above could not have been deduced without the mathematics, although it may have been surmised from the physical principles alone. But here physicists can be as adamant in demanding rigor as are mathematicians and so thermodynamics, as a mathematical system, is probably the most rigorous of all branches of physics. Nowhere else in the world of physics has so much been deduced from so little.

Before returning to the story of pure mathematics, we consider another branch of mathematics—the theory of probability—which is ideally suited for application to physics. We noted in our chapter on probability that the theory grew out of games of chance in the eighteenth century. In the last quarter of the nineteenth century, theoretical physicists began to use probability theory to analyze the physical behavior of and the processes in physical systems containing large numbers of members such as the molecules in a gas. The branch of physics that evolved from this theory, called statistical mechanics, uses much of the mathematics of probability theory with the limitations imposed on it by the laws of physics. The first important application of the basic mathematical features of probability theory to physics problems goes back to Daniel Bernoulli, who applied it to the theory of errors. But this theory rapidly grew beyond its usefulness in the hard sciences and spread into every area in which the statistical analysis of data is important. Bernoulli started from the common practice of taking the arithmetic average of a set of observations of a given phenomenon as the most probable value for the actual occurrence of the phenomenon. This is indeed so if each observation is of equal weight, but Bernoulli did not accept this assumption. Instead he argued that observations with small errors are more probable than those with large errors and so he altered the arithmetic mean by introducing a weighted mean, assigning larger statistical weights to small-error observations than to large-error observations. This change led to a much more complicated averaging formula than the simple arithmetic average; this Bernoulli formula contains an infinite series, which the English mathematician James Stirling had published earlier and is similar to a formula that de Moivre had already published.

Both Euler and Lagrange contributed to the theory of errors; Euler in proposing a formula for the maximum value of a series of errors and Lagrange in his analysis of ten different problems involving errors in measurements and observations of a general kind. In analyzing these problems he was interested in obtaining general formulas for calculating the probabilities that the errors in the mean values of different sets of observations would lie between any two designated values. He obtained the remarkably simple formula $(h - z)(dz)/h^2$ for the probability that the error in the mean lies between z and $z + dz$ if negative and positive errors are equally likely and that the error for every trial or observation lies between $-h$ and $+h$. This is remarkable because it is an exact formula for the probability that the error in the mean lies between assigned limits for a definite but very reasonable assumption about the occurrence of single errors.

From what we have already said about the theory of errors, we may conclude that the problems it presented in its early days were irresistible to the greatest mathematicians and that it would attract their attention. It is no wonder then that Laplace contributed to this theory. But he began by redefining or, more precisely, by defining the mean of observations as such a value that the true value is equally likely to lie above or below this mean. Laplace then suggested another equivalent definition: It is the minimum of the sum of the errors, each multiplied by its probability of occurrence. Being greatly interested in astronomy, Laplace proposed that astronomers adopt this definition of the mean in publishing their observations. Assuming that positive and negative errors are equally probable and assuming the formula $(n/2)e^{-nx} dx$ for the probability that an error lies between x and $x + dx$ for n observations, Laplace produced a formula for the error in the mean.

In our discussion of the theory of errors, we have assumed that an exact numerical value of the particular physical parameter that describes the phenomenon we are studying or observing exists but we must not conclude that we can actually measure this exact value of the parameter; a perfect physical measurement is a fiction because every measurement is burdened with errors. But we can always calculate the mean value for all our separate measurements and accept this value as the "best value," which we hope is as close as we can possibly get to the "true physical value." The question that naturally arises here is whether we have any mathematical procedure that can use the calculated mean value to tell us the error of any particular measurement. Gauss found the solution to this problem and presented it in a form called Gauss's exponential law of error: The probability that an observed measurement of a quantity has an error in the infinitesimal region $x - dx$ to $x + dx$ is $(a/\sqrt{\pi})e^{-a^2x^2} dx$ (e is the exponential) where the average error of a single measurement is $1/a\sqrt{2}$. This is one of the most famous expressions in mathematical statistics. We call this formula the law of error but it is not a law in the sense that Newton's law of gravity is a law of nature or that the theorem of Pythagoras is a law of geometry. A law of error does not exist in nature; there is no natural principle which can tell us how errors in measurements are distributed around a mean value. Gauss's formula is then a very useful numerical device to tell us about the trustworthiness of distinct measurements.

Recognizing the great importance of probability theory in analyzing processes in physical systems containing large numbers of particles (molecules in a gas), physicists were quick to apply it to gases. The first among these scientists was Maxwell, who deduced the formula for the distribution of the molecular velocities in a gas at a given temperature T: $dn = n(m/2kT)^{3/2}e^{-mv^2/2kT}dv$, where n is the total number of molecules in the gas and dn is the infinitesimal number of such molecules with velocities in the velocity interval from $v + dv$ to $v - dv$. Here m is the mass of a single molecule, T is the temperature of the gas, v is the velocity of any given molecule, and k is the universal physical constant called the Boltzmann constant. We recognize Maxwell's distribution formula as a special form of Gauss's error formula, and from it, we see that the average velocity of the molecules in the gas is $\sqrt{3kT/m}$. This is an amazing result which we could never have deduced without the mathematics. This formula permits us to relate the macroscopic parameters P, V, and T of a gas to its molecular properties. Indeed, Maxwell's distribution formula is one of the great mathematical props of the molecular theory of matter. This was further strengthened by Einstein's use of probability theory and statistical methods to explain Brownian motion, so that probability theory became an indispensable tool of the theoretical physicist. But all of this did not happen overnight; its fulfillment required the individual efforts of Gibbs, Maxwell, and Boltzmann. Their labors culminated in the development of statistical mechanics which we now describe briefly.

With Maxwell's discovery of the formula for the distribution of molecular velocities in a gas, Gibbs and Boltzmann, independently, saw that Gauss's error function and Maxwell's distribution law can be generalized to obtain statistical values for the various macroscopic physical parameters that describe a system. To this end they introduced the concept of the phase space of a system, which in its simplest form for an ensemble of particles is a six-dimensional space which is obtained from the ordinary three-dimensional configuration of space by adding three other fictitious dimensions called the momentum space. The physical state of each particle in the ensemble is then described by assigning the three space coordinates x, y, and z to the particle at any moment (its position) and the three components $p_x = mv_x$, $p_y = mv_y$, and $p_z = mv_z$ of its momentum at that moment. As the particles move about, their positions and momenta change so that the state of the entire ensemble (e.g., a gas) changes. Using the mathematics of probability theory, as previously described, one can then calculate the most probable state of the system and, thus, mathematically, determine its macroscopic behavior. The introduction of six-dimensional phase space to analyze the statistical behavior of a system containing a large number of particles (an ensemble) reduced statistical mechanics to the geometry of phase spaces, which can differ from ensemble to ensemble. In an appropriate phase space, the behavior of an ensemble is then represented by points (one point for each member of the ensemble) in this space. In addition, the way these points move about, forming a kind of fluid, describes the changes in the physical state of the system. The probability that the ensemble is in a particular state at any moment is then given by the distribution of the phase points in the phase space at that moment.

We have spent considerable time discussing the mathematics of thermodynamics and statistical mechanics because they lend themselves most readily to an analytical discussion that reveals, in a very simple way, the interdependence of physics and mathematics. We see here that whereas the mathematician can get along very well without the theoretical physicist, the reverse is not true. The theoretical physicist desperately needs the mathematician; this need becomes ever more pressing as the physics itself becomes more obscure. We face the serious danger here, however, of allowing the formalism of the mathematics to obscure the physics and even to lead us astray into fruitless mathematical speculations.

During the remarkable transition period of the last half of the nineteenth century—when many of the outstanding mathematicians were turning to physical problems and mathematics itself was losing its pristine character—a resolute minority in the mathematical community continued to reject the taint of practicality that they felt might ultimately destroy mathematics as a purely intellectual enterprise. The leaders of this group were the mathematicians who were working in number theory, in complex variables, in pure algebra, in symbolic logic, and in the foundations of mathematics. Among these mathematicians were Leopold Kronecker, Karl Weierstrass, Ernst Kummer, Georg Frobenius, Richard Dedekind, and Georg Cantor. Kronecker, the leader in this group, had founded what became known as the

Berlin school of mathematics, which was devoted to the arithmetization of mathematics. The aim of the Berlin school, as announced by Weierstrass, though not always explicitly stated, was to reduce the principles of analysis to the simplest arithmetic concepts. In short, the members of the Berlin school believed that all mathematics was arithmetic. Kronecker expressed his love for arithmetic in his statement that "The dear God created the integers, all else is man-made." He replaced Plato's statement that "All is geometry" by the statement "All is arithmetic," and asserted that God always "arithmetizes." But one man, Georg Cantor, stood out among these mathematicians because he was a revolutionary arithmetician who left an indelible imprint on number theory by inventing or introducing the truly original concept of transfinite numbers. Cantor's ideas were much too bizarre for the arithmeticians and for Kronecker in particular, who considered Cantor to be a heretic.

Georg Cantor was one of the most misunderstood and revolutionary thinkers in the history of mathematics. The son of a prosperous merchant of Danish origin whose travels had taken him across Europe, Georg was born in St. Petersburg, Russia on March 3, 1845. Georg's mother, Marie Bohm, was a talented artist. Marie's artistic legacy to her son manifested itself not in musical or artistic ability but instead in his passion for mathematics. As a child, Georg was attracted to medieval theology, particularly philosophical discussions about continuity and the infinite. But he also enjoyed music and art.

Like most great mathematicians, Georg's talent became evident while he was still a young boy. In the beginning, Georg was tutored privately at home. After the family moved to Frankfurt, Germany, in 1856, Georg attended private school before enrolling at the Wiesbaden Gymnasium in 1860. Despite his father's insistence that Georg prepare for a "practical" career as an engineer, the boy was seduced by pure mathematics. Georg's realization that he was a mathematician and not an engineer distressed him greatly because he wanted to please his stubborn father. But Georg's father gradually accepted his son's plans to pursue mathematics, displaying a willingness to compromise that was not very common in the German households of that time.

After spending a year at the University of Zürich, Georg then transferred to Berlin, where he undertook a rigorous program of study in mathematics and physics. Georg did not find physics very appealing so it was not long before he was studying mathematics almost exclusively. His teachers at Berlin included such noted mathematicians as Weierstrasss and Kronecker. Georg was a brilliant but conventional student. His work to that point offered little indication that he would soon startle the mathematical world with his shockingly original ideas. Indeed, his doctoral dissertation dealt with a problem raised by Gauss in his *Disquisitiones Arithmeticae* regarding the solution of an indeterminate equation. Although this work fit in nicely with the prevailing orthodoxy held dear by German mathematicians, it did encourage Georg to consider more thoroughly the foundations of mathematical analysis, particularly infinite series. Bringing both the rigor of a logician and the skepticism of a philosopher to the table, Cantor began to construct his theory of infinite sets. He published his first paper on set theory before his thirtieth birthday. He also married Cally Guttmann, who later gave him six children.

Cantor's radical paper on infinite sets was not warmly received by Germany's leading mathematicians, most of whom were intellectually conservative and still laboring in Gauss's long shadow. Because Cantor's ideas seemed to call into question the validity of much of modern mathematics, he was not welcomed by the mathematicians who had invested so much of their careers and their reputations in shoring up conventional mathematics. It was this animosity that kept Cantor from obtaining the position he coveted at the University of Berlin and prompted such bitter attacks by his one-time mentor Kronecker. This controversy seems to have been instrumental in causing Cantor's career to languish at the University of Halle—an institution that ranked low on the academic pecking order—where he passed his entire professional life as a professor of mathematics.

Despite the initial hostile reaction to his paper, Cantor did find some distinguished supporters, including Dedekind, who was himself considered to be on the fringe of mathematics by his more hidebound colleagues. Cantor took solace in Dedekind's kind words and continued his work, resolving to let the chips fall where they might. Like Cantor, Dedekind was an original thinker whose career had suffered due to his failure to tow the party line. The tragedy for both men was that they were victims of a sort of mathematical bigotry, no less insidious than bigotry based on racial or sexual features. Had they lived at a different time or in a different nation in which the academic mentality was not so reactionary, their careers might have been much more rewarding.

As the subversive nature of Cantor's work became clear, the mathematical establishment rolled out the heavy artillery in the form of Leopold

Kronecker, the former financial genius turned mathematician, who saw Cantor's work in set theory as nonsensical because it offered no way to construct the transcendental numbers that were at the very heart of his ideas about infinite sets. Kronecker believed that Cantor was offering a collection of mathematical wares that would undermine, if not destroy, the logical foundations of mathematics itself. Accordingly, Kronecker waged an ongoing battle against Cantor and his ideas that was extraordinarily bitter. Because Cantor was already on the sidelines of German mathematics, he was at an inherent disadvantage in this war of words and reputations. Furthermore, Cantor's gentle temperament did not lend itself to a prolonged struggle for intellectual supremacy. As a result, the outcome of this battle was never truly in doubt. Kronecker used his considerable resources and reputation to prevent Cantor from securing a position at Berlin and even criticized Cantor personally while lecturing his students.

Even though Cantor eventually had the last word and mathematicians everywhere came to accept his work in set theory as one of the most important contributions to mathematics in the modern era, Cantor gained little satisfaction from this change in his fortunes. A series of nervous breakdowns had effectively ended his mathematical career before he celebrated his fiftieth birthday even though he had stopped doing important mathematical work nearly a decade before. But Cantor did live to see mathematicians take up the gauntlet and approach their subject with the rigor and thoroughness that he had brought to the study of set theory. He died in a Halle asylum on January 6, 1918 at the age of seventy-three.

Cantor began his study of aggregates (sets) at Halle which later became the modern theory of sets and led to the theory of the functions of real variables. His ideas about set theory are contained in his "Foundations of a General Theory of Domains." He also presented his theory of transfinite numbers, which stemmed from his attempt to understand the concept of infinity. Up until his investigations, all infinities were considered to be the same because, it was argued, an infinite number of things of one sort is no different in number from an infinite number of things of another sort. But Cantor showed that infinities do differ from each other. This becomes apparent, however, only if we introduce a scheme for counting the number of things in each group and comparing the two counts. Cantor did just that and showed than an infinite number (an infinite hierarchy) of infinities exists. He began with the basic concept of an aggregate or set of objects to which he assigned a number, called its cardinal number, which is, in a sense, a designation of the quantity of objects in the set. He then developed an arithmetic of such sets, laying down the rules for their addition, subtraction, multiplication, etc., which, with very important exceptions, follow the rules of ordinary arithmetic. This was the beginning of modern set theory, called, in German, "Mengenlehre," or theory of aggregates. Set theory is accepted today as the logical basis of all mathematics.

To arrive at his concept of transfinite numbers, Cantor analyzed the concept of counting, which he defined as an operation establishing a correspondence between the elements of one set and those of another. Two sets are said to be equivalent if for each element in set A we can associate an element in set B so that a one-to-one correspondence exists between the two sets of elements. Cantor assigned the same cardinal number to each of these sets. The counting operation is then defined by Cantor in terms of a very special set of entities: the integers of our number system. To count the elements in a set, we then pull out any member of the set and attach to it the integer 1, without, in any way, differentiating this element from any other element. The number 1 is then the cardinal number of this element. Proceeding in this way we can, following Cantor, assign increasingly larger cardinal numbers to these elements. If no largest cardinal number exists for the set, we call the set an infinite set. But, again following Cantor, we must be careful about how we define this concept of infinity which, as Cantor demonstrated, has many facets. Cantor called a set denumerable if each element in the set can be counted—that is, placed in a one-to-one correspondence with a definite integer. Cantor assigned the cardinal number $\aleph_0$ (aleph zero or aleph null) to such a set. He called aleph zero the smallest transfinite number.

In carrying out such a counting operation, we are also arranging an order. Indeed, we can imagine all the elements in a denumerable set arranged in a line, so that each element is assigned to a definite point on the line. Because this is equivalent to arranging the elements into an ordered sequence, we follow Cantor, and say that the positional numbers assigned to the elements are the ordinal numbers of the set, which emphasizes their places in the sequence rather than their quantitative aspects. We can count the elements in a set only if we can specify its ordinal number, regardless of its numerical or cardinal aspect. Cantor then went on to show that the transfinite numbers themselves can be arranged in an ordered sequence starting with $\aleph_0$ and then going on to $\aleph_1, \aleph_2, \aleph_3, \ldots$, and so on. Thus all the integers and fractions, taken together, are denumerable, so that their set is assigned the transfinite cardinal number $\aleph_0$ (aleph zero). The irrational numbers, however, cannot be counted and so the set of all irrational numbers has the transfinite cardinal number $\aleph_1$, which is larger than $\aleph_0$. Cantor went on to show how to construct sets of ever larger transfinite cardinal numbers. This was the origin of transfinite arithmetic.

This discussion of transfinite numbers may be concluded by illustrating, in an elementary way, the difference between the arithmetic of ordinary numbers and that of transfinite numbers. If we remove the infinite set of all even integers (transfinite cardinal number $\aleph_0$) from the infinite set of all integers (transfinite cardinal number $\aleph_0$), we are still left with an infinite set of transfinite number $\aleph_0$. Thus $\aleph_0 + \aleph_0 = \aleph_0$ and $\aleph_0 - \aleph_0 = \aleph_0$; the product of $\aleph_0$ and $\aleph_0$ is still $\aleph_0$. The same is true if we raise aleph zero to any power; we still get aleph zero.

The close of the nineteenth century was marked by two distinct and somewhat divergent trends in mathematics. One important group of mathematicians was turning ever more inward, deeply concerned with establishing a unifying philosophy. Another influential group was expanding its work into the field of applied or practical mathematics, which became increasingly more attractive because the rapidly advancing technology which stemmed from new scientific discoveries required for their growth very skillful mathematicians. Thus, at the end of the nineteenth century theoretical physicists, pure mathematicians, and even engineers were called "mathematicians." By that time, the center of mathematical studies had begun to drift toward Germany and pure mathematicians were becoming increasingly concerned that mathematics might become so contaminated by its practical aspects that it would lose its pristine quality and no longer be considered as the only domain of pure thought. To lay down a program that would lead mathematicians along the proper path, a group of mathematicians, led by Felix Klein, met at the University of Erlangen, Germany, in 1872, to announce what has since been called the "Erlangen Program," to keep mathematicians from straying and nonmathematicians from preempting the role that rightfully belongs to mathematicians. This program did nothing of the sort but it did orient mathematics, to some extent, in the direction Klein wanted it to go, and actually benefited mathematical physics greatly.

Felix Klein, who had studied mathematics at the University of Bonn, where geometry was emphasized, was primarily a geometer. Concerned about becoming too narrowly focused, he decided to broaden his mathematical interests and traveled to Paris in 1870 at the age of twenty-two to learn more about analysis from the reigning French mathematicians of that period. There he met the Norwegian mathematician Sophus Lie, who, shortly before that time, had turned to mathematics as a second career, influenced by Galois' theory of groups and his application of group theory to the solution of algebraic equations. Lie believed that the study of groups represented a universal approach to all branches of mathematics. The study of groups consists of two branches: continuous (infinite) and discontinuous (finite) groups. Lie was devoted to the study of continuous groups and Klein to the study of discrete groups. It was clear to both Klein and Lie that groups would play an important, indeed dominant, role in the development of twentieth century mathematics. This perception of the influence of groups on mathematics led Klein to his formulation of the "Erlangen Program."

To describe this program as briefly and as simply as possible, we recall our previous discussion of groups, defined as collections of elements: matrices, mathematical operators, permutations, etc., which obey definite multiplication rules. Among the most important groups are those which change one kind of description of a given mathematical or physical system to another kind of description. Such a change was originally called a substitution by the French mathematicians but is now universally called a "transformation." Obviously, the system itself being described mathematically does not change in such a "substitution" or "transformation"; only its mathematical description changes. This means that the intrinsic features of the system must not change; these unchanging intrinsic features are called the "group invariants of the system." The branch of mathematics now called the "theory of invariants" emerged from this phase of group theory. We can most easily describe the mathematical and physical properties and functions of a group by considering its application to a geometric system. This is most apparent in the special branch of geometry called projective geometry, which is important in art (perspective), architecture (blueprints), and mapping (geographical maps, geodesy, etc.).

Projective geometry was applied unknowingly by the ancient Greek and Roman architects as well as the early Egyptians and the Mayans. But the first formal presentation of projective geometry as a branch of mathematics was proposed in 1636 by Gérard Desargues, a French architect, in a book titled *An Attempt to Deal with the Intersection of a Cone with a Plane*. In this book Desargues presents the famous "Desargues theorem," which became the basis of perspective, a form of descriptive geometry, in which a three-dimensional object is correctly represented by a two-dimensional drawing; it consists of the lines that connect all the points on the plane produced by rays that emanate from a fixed point (the eye) outside the plane, then pass through points of the objects, and pierce the plane. The great German artist Albrecht Dürer used perspective extensively for his remarkable engravings and drawings.

Very little attention was paid to Desargues' discovery until a century later when Gaspard Monge, director of the École Polytechnique, published his own lectures on descriptive geometry, which contain Desargues' basic concepts. We may thus consider Monge's lectures as the formal beginning of projective geometry because they contain the basic concepts on which modern projective geometry rests. Monge also greatly stimulated the development of analytical geometry, applying calculus to the study of surfaces and space curves. These ideas later evolved into differential geometry, which owes so much to Gauss, who did so much to expand Monge's ideas. The students of analytic geometry today learn about it from textbooks that greatly resemble Monge's lectures.

Although Monge presented the basic concepts of projective geometry, it became a recognized branch of geometry only after Monge's most talented student, Jean Victor Poncelet, devoted all of his efforts to the development of projective geometry along the lines of Desargues' original work. Poncelet's volume, *A Treatment of the Projection of Figures*, is the first recognized textbook of projective geometry which laid down the basic laws of perspective and projection.

We have discussed projective geometry in some detail because it is an excellent example of the representation of a special transformation group: the group of mathematical operators which transform three-dimensional objects into two-dimensional, flat configurations. Our eyes do the same thing when we look at any object, producing what we call perspective. In perspective, the projection, as we have already stated, is produced by nonparallel lines that diverge from the eye. But parallel lines also produce projections. Thus, the rays of light from the sun produce projections which we call shadows, whose dimensions and shape depend on the slant of the sun's rays and the surface on which the shadow is cast. We know that the shadow of a tennis ball on the surface of the court is a perfect ellipse. Thus, this kind of projection is a mathematical transformation of one kind of shape into another. All such transformations constitute a group.

As other specific examples of groups which, in addition to being mathematically interesting, have practical applications, we mention the architect's blueprints, which are the projections of a spatial structure onto three mutually perpendicular planes. Similarly, mapping is the projection of the figures on the surface of a sphere, the continents of the earth, onto a flat surface, the map; this technique goes back to the famous sixteenth century Flemish geographer, cartographer, and mathematician Gerard Mercator. The observational astronomer tries to reconstruct the three-dimensional reality of what he sees as a projection of objects (e.g., stars, planets, galaxies) on a fictitious surface he calls the sky. But of all the transformation groups mathematicians have introduced and studied, the groups of transformations that change coordinate systems are, by far, the most important and interesting because, as we shall see, they play a fundamental role in all branches of physics and mathematics, particularly in the theory of relativity.

Returning now to Felix Klein and his famous "Erlangen Program," we must admire his genius in recognizing "group theory" as the mathematician's guide to a complete understanding of the nature of geometry and that all geometries must be related to each other by definite transformation groups. The group of coordinate transformations which leave invariant the theorem of Pythagoras, for example, represents Euclidean geometry. To justify his group theory approach to geometry and its power, Klein demonstrated that non-Euclidean geometries are projective geometries with a special kind of "Pythagorean theorem." Klein completed his investigations into the nature of groups by showing that the theory of groups leads to a unification and synthesis of the various branches of mathematics into a single self-consistent logical system based on an impeccable philosophy. Through group theory, Klein thus brought mathematics to the peak of rigor that Galois and Abel had sought and Cauchy and his colleagues had continued.

Klein left Erlangen in the 1880s to replace Gauss at Göttingen as the outstanding German mathematician of his day. Under his leadership, Göttingen became the world center of mathematics, attracting the most talented students; it remained the preeminent institution in that discipline until it was destroyed by Hitler. By the end of the nineteenth century, with the discovery of the quantum theory and the theory of relativity, all the important branches of mathematics, as pure intellectual disciplines, had reached their maturity and could be used as tools by the physicists who needed them for their search for an understanding of nature and its laws. More specifically, mathematics could now be used by physicists for making models of physical structures—both on the microscopic and macroscopic levels. At the same time, a group of pure mathematicians, following the lead of George Boole and Augustus de Morgan in England and G. Frege in Germany, continued their search for an algebra of logic (Boolean algebra, symbolic logic). This search led, in the first decade of the twentieth century, to the definitive work in this area by Bertrand Russell and Alfred North Whitehead as presented in their monumental three-volume work, *Principia Mathematica*. The beginning of the twentieth century was also marked by the rise of the two great mathematicians Henri Poincaré of France and David Hilbert of Germany; they set the mathematical tone for the first half of the twentieth century, a subject to which we now turn.

CHAPTER 11

Modern Mathematics and the New Physics

The golden age of mathematics—that was not the age of Euclid, it is ours.

—C. J. KEYSER

Mathematics did not experience a fundamental revolution in its most basic concepts as physics did during the first three decades of the twentieth century. The most fundamental change in mathematics occurred in the mid-nineteenth century with the introduction of non-Euclidean geometry, the full implications of which stemmed more from what physicists did with it than what mathematicians did. The revolution in physics was shattering in the sense that it altered our basic concepts of space, time, motion, energy, and the continuity of the physical processes in nature. Moreover, it erased the sharp division between particle and wave, on which classical physics is based, replacing it with the wave–particle dualism concept. The revolution in physics also affected the kind of mathematics that has to be used to describe the physical processes in microscopic physical systems such as molecules, atoms, and nuclei. But the mathematics of macroscopic systems such as the universe itself was also affected by the revolution in physics.

As specific examples of the changes that the new physics demanded of mathematics, we note that Newton's basic law of motion, $F = ma$, that expresses the force F acting on a body of mass m required to produce an acceleration a of the body, does not work when applied to electrons in an atom. Newton's simple algebraic law of motion had to be changed to the kind of mathematical law that incorporates the dual wave and particle properties of the electron. In addition, Maxwell's electromagnetic equations, which describe the relationship between electric charge and the way a charge radiates electromagnetic waves, had to be amended to account for the non-radiating behavior of an electron when in an atom, where it behaves as though it were stationary. Two, apparently disparate, mathematical proce-

273

dures or techniques were introduced to replace Newton's equation of motion $F = ma$ with a new kind of equation for the electron to account for its wavelike properties. Werner Heisenberg, the discoverer of the uncertainty principle, one of the most important principles in nature, replaced Newton's law by what is now called the "matrix mechanics"; it describes the position and momentum of an electron using two sets of different matrices. Erwin Schrödinger, very unhappy with the model of the atom offered by Niels Bohr, introduced a "wave equation" for the electron to account for its motion in an atom.

A casual consideration of these two different mathematical descriptions of the same physical phenomena is very disconcerting because the "matrix mechanics" is essentially a mathematically discontinuous theory whereas the "wave mechanics" is a mathematically continuous theory. Moreover, the matrix mechanics requires a noncommutative algebra; the Heisenberg uncertainty principle then follows, as a necessary mathematical consequence of the noncommutative algebra. But the Schrödinger wave mechanics does not contain any specific reference to a noncommutative algebra. How, then, can these two different kinds of mathematics be reconciled? Can the mathematics be misleading and say two different things about the same phenomena?

Both mathematicians and physicists were in a state of consternation about this apparent incongruity because it appeared to undermine the deep trust that physicists had placed in mathematics. If this trust had been misplaced, then the future of theoretical physics was in deep jeopardy because mathematics had been the trustworthy tool of physicists from the time of Newton. But this was a baseless fear because the British theoretical physicist Paul Dirac, as well as Schrödinger himself, showed, shortly after Schrödinger's wave equation appeared, that the matrix mechanics and the wave mechanics are completely equivalent. Dirac went further and showed that both matrix mechanics and wave mechanics are subsumed under a more general mathematical discipline called quantum mechanics in which the physical entities are represented by noncommutative algebraic operators, which may appear as matrices in one frame of reference (the Heisenberg frame) or as differential operators in another frame (the Schrödinger frame).

As propounded and emphasized by Dirac, the mathematics of the quantum mechanics, as an algebra of operators, leads to conclusions about the behavior of particles that completely contradict some of the basic features of particles that formed the foundations of classical Newtonian physics. Thus the quantum mathematics forces us to accept the conclusion that a physically indivisible particle of matter, like an electron, or a physically indivisible particle of energy, like a photon, can simultaneously be moving along two different spatially separate paths and then, at the will of the observer, reappear instantaneously as a single particle again. Here, again, the mathematics forces us to accept what appears to be a nonphysical reality, which, under ordinary conditions, must be dismissed as an oxymoron, but in the quantum mechanics is a profound, but mysterious, truth.

Dirac discovered a universal mathematics of quantum mechanics which can be applied to the electromagnetic field as well as to the electron; he thus altered Maxwell's electromagnetic theory, changing it into a quantum mechanical field theory that can be applied to the emission and absorption of photons by electrons. He did this by introducing special noncommutative mathematical operators to describe the interaction of a charged particle with the electromagnetic field. The algebra of these operators then led to a complete mathematical quantum mechanics, called the quantum electrodynamics, of electromagnetism, which had a profound influence on field theory.

These new developments in theoretical physics required the very close cooperation of diverse twentieth-century mathematicians. David Hilbert, a German, and John von Neumann, a Hungarian Jew who had come to the United States to escape the Nazis in 1935, were among the most outstanding of these scholars. Hilbert introduced and developed a special kind of vector geometry of functions (infinite dimensional space) called Hilbert space, which geometrized the quantum mechanics. Von Neumann, on the other hand, carried out a complete theoretical analysis of the foundations and the basic tenets of quantum mechanics which he published as *The Mathematical Foundations of Quantum Mechanics*, one of the most influential books in physics. Von Neumann's later work in pure mathematics marks him as one of the greatest mathematicians of all time.

As the second specific example of the influence of mathematics on the development of physics, we briefly discuss the impact of non-Euclidean geometry on physics, in particular, on our understanding of the nature of gravity as described by Einstein in his general theory of relativity, which is essentially a theory of gravity. Einstein based his extension of the special theory of relativity to the general theory on his hypothesis that the laws of nature should be the same in all frames of reference regardless of how they are moving. Starting from this basic principle of invariance, he argued that any inertial reaction of an observer to a change in the motion of his frame of reference (acceleration of the frame) is equivalent to the action on him of a gravitational field.

This remarkable principle, now known as the "principle of equivalence," is supported by the behavior of bodies falling freely in a gravitational field, as first observed by Galileo: all bodies in a vacuum at the same point in a gravitational field (e.g., at the same height above the earth's surface) experience exactly the same acceleration regardless of their masses. This is contrary to what we might naively expect because the weight of a body is the pull of gravity on it. According to Newton's laws of motion, the acceleration of a body is proportional to the force acting on it. We might therefore expect heavier bodies to experience a larger acceleration than lighter bodies because the pull of gravity is greater on the heavier than on the lighter bodies. That all the bodies, regardless of their weights, fall with the same acceleration indicated to Einstein that gravity is not really a force but a geometric distortion of four-dimensional space-time produced by the matter in the universe. Bodies falling freely then move the way they do because they all follow the distortion of space-time in exactly the same way. This thinking led Einstein to non-Euclidean geometry because only such a geometry can describe the kind of space-time distortion we can identify or, alternatively, relate to the force of gravity.

By 1915, non-Euclidean geometry had expanded considerably beyond the seminal work of Riemann. In the early 1870s, for example, the British mathematician, William Kingdon Clifford, who died in 1879 at the age of 33, had become deeply interested in Riemann's geometric discoveries and had wondered about their application to the motions of bodies in a non-Euclidean space. In fact, he had begun to develop such a theory of motion, using some of Hamilton's quaternion concepts, but he died before he had progressed beyond a schematic outline of his ideas. Riemann's theory of non-Euclidean spaces was a natural challenge to the group theorists, particularly to Klein and Lie, because they saw that the geometric properties of such spaces can be understood through or revealed by the invariants of such spaces under various groups of transformations of such spaces. The search for a mathematical representation of the invariants of an n-dimensional space led to the extension of the vector concept to the tensor concept, which was very important in Einstein's theory of gravity. Just as in ordinary three-dimensional space the invariant quantities are vectors, so in n-dimensional non-Euclidean space the invariants are tensors. Thus tensors, or in mathematical jargon, tensor analysis, became the mathematics of general relativity. All one had to do, then, to be sure that one had discovered a basic law of nature, was to express it in its most general tensor form. If one did so, the statement or principle was a law; otherwise, it was not a law.

By the time Einstein announced his general theory of relativity in 1915, tensor analysis had been fully developed. Indeed, all that was left for him to do was to "pluck it off the shelf" and use it to formulate the general theory of relativity; a monumental intellectual edifice that is considered by many to be the greatest single creation of the human mind and certainly the most beautiful of all scientific theories. Einstein's promulgation of the general theory and its presentation of gravity as space-time curvature was not readily accepted by the physics community, but mathematicians embraced it very warmly, perhaps because general relativity cannot be understood except through its mathematical formulation. Physicists were thus forced to immerse themselves in advanced mathematics more than ever before. The curvature of space-time cannot be perceived from a physical point of view because the physical perception of curving is that of distorting a structure into a higher dimension. Thus we have no problem accepting the idea of curving a line in a plane or a plane in a three-dimensional space but curving four-dimensional space-time makes no physical sense to us. We cannot conceive of a fifth- or sixth- or tenth-dimensional space in which to curve our four-dimensional space-time. The mathematician is not bothered by such physical objections or niceties because all that matters to him is that the geometrical properties of the figures or structures he can perceive in four-dimensional space-time be deducible from the axioms he has assumed for four-dimensional space. That the model of the four-dimensional geometrical manifold which can be deduced from the geometer's axioms fits or describes the geometry of real, physical space-time is all to the good and makes the geometer all the happier, but his main concern is that a mathematically self-consistent four-dimensional geometry can be constructed. The physicist, however, cannot accept the geometer's model unless the physical deductions from the model agree with his experiments or observations. All such experiments and observations during the last fifty years of the twentieth century have confirmed, with incredible accuracy, the general theory of relativity and, therefore, that real four-dimensional space-time is governed by Riemannian, not Euclidean, geometry.

As the general theory of relativity became more acceptable, mathematicians from many countries began to contribute to the growth of non-

Euclidean geometry and to the development of the mathematics of relativity theory. These efforts began with the mathematics of the special theory in the first decade of the twentieth century. From the point of view of the pure mathematician, the problem of formulating the special theory of relativity into a self-consistent mathematical structure was primarily a group theory problem: to find the group of transformations that describe the physical invariants of the special theory as proposed by Einstein.

We can best understand the relationship between mathematical transformation theory and the theory of relativity by first considering, briefly, the pre-Einstein concepts of space and time, and why and how a very important discovery about the speed of light forced Einstein to change these concepts and, owing to this change, introduce his special theory of relativity. We consider two events (e.g., two explosions) occurring at two distinct points, separated by a distance s and by a time interval t. All of Newtonian physics is based on the assumption that every observer moving with uniform motion in a straight line, regardless of his speed or direction of motion, will measure the same distance s between the two events and exactly the same time interval t. The motion invariants in Newtonian mechanics are thus space by itself and time by itself—absolute space and absolute time.

To see how this is expressed in terms of group theory, we consider the transformation from one coordinate system to another if we picture one coordinate system as fixed in space and the other as moving with constant speed v along the common x axis of the two systems. If x is the position along the x axis of an event as measured from its origin 0 by the observer in the fixed coordinate system at the moment t and x' is its position as measured from its origin $0'$ by the moving observer at the same time t, then the transformation from the fixed to the moving coordinate system is given by the equation $x' = x - vt$, where t is the time that has elapsed between the moment of coincidence of the origins 0 and $0'$ and the moment of the measurements of x and x'. If we now consider all possible positions x, and all possible speeds v, and time t, we have an infinite number of transformations which constitute a continuous group, called the Galilean group of translations. The distances and the time intervals between events are the invariants of this group. These are also the invariants of all possible rotations of the coordinate system; such transformations constitute the group of all possible rotations. Certain dynamical quantities of a moving particle, however, are not invariants of the Galilean group of transformations; quantities which depend on the velocity of the body, including the velocity itself are not invariants of the Galilean group. We all know that the speed at which an automobile is passing our automobile depends on the speed v of our automobile. Our experience with automobiles then tells us that the speed at which a beam of light passes us should depend on how fast (our speed v) we are moving in the direction of the moving beam.

But here our experience misleads us because measurements made on the speed of light in 1887 by the American scientists Albert Michelson and Edward Morley proved conclusively that the speed of light c is the same in all coordinate systems, regardless of how they are moving. To Einstein, this invariance proved that the Galilean group of transformations do not apply to the real world and that a new group of transformations would have to be introduced, the invariants of which are not distances and time intervals separately but special mathematical combinations of distance s, time t, and the speed c of light. Einstein pointed out that the constancy of the speed of light for all observers requires that the algebraic combination $s^2 - c^2t^2$ be an invariant of the correct group of transformations. This means that the speed of light must be present in the algebraic expressions of these transformations, and that the transformations must apply to time as well as to space so that space and time are thus woven into a single four-dimensional space-time fabric or manifold. The transformations that perform this miracle of merging space and time are called the Lorentz–Einstein transformations—probably the most famous and far-reaching algebraic expressions in the history of science. All of the magical results stemming from the special theory of relativity can be deduced—using nothing more than elementary algebra—from the Einstein–Lorentz transformations.

In the mathematical formulation of the special theory of relativity, we see perhaps the most beautiful and instructive example of the symbiosis of two distinct intellectual disciplines: mathematics and theoretical physics. Without the relevant mathematics, the profound physical consequences of the special theory of relativity could never have been deduced, and its application to the solution of the most important and perplexing problems in the universe would have been impossible. But the theory of relativity also greatly stimulated the development of new areas of mathematics and attracted the outstanding mathematicians of the early twentieth century to its cause. Two of these individuals, the French mathematician Henri Poincaré and the German mathematician Hermann Minkowski, were particularly important in the development of the mathematics of the special theory of relativity—Poincaré, from the group theoretic point of view, and Minkowski, from the geometric aspect of relativity. We return to Poincaré's purely mathematical exploits later in this chapter, limiting ourselves here to his contributions to the mathematical facets of special relativity. Poincaré treated the special theory entirely from the point of view that all its physical features can be deduced from the mathematical properties of the transformation group that defines the invariants of the special theory. This group can be represented by matrices containing four rows and four columns. Poincaré studied the most general forms of such matrices in which the individual elements (terms) in each matrix are arbitrary, and showed that, for special choices of these elements, one obtains the Galilean group of transformations and for other choices one obtains the Einstein–Lorentz transformations. Owing to his extensive investigations into the group properties of these transformations, Poincaré was honored by his colleagues, who named the group of the Einstein–Lorentz transformations the Poincaré group. The importance of the Poincaré group for the mathematical development of the theory of relativity is that it does not distinguish in any way between space and time but instead describes the theory as a natural four-dimensional extension of the Galilean–Newtonian three-dimensional description of the universe, with time as a separate, unconnected parameter, which leads to Minkowski's four-dimensional geometry.

Newtonian dynamics is treated geometrically by introducing a three-dimensional coordinate system such as a rectangular Cartesian coordinate system defined by three mutually perpendicular axes X, Y, and Z; the position of a point-mass in such a coordinate system is then given by its three Cartesian coordinates x, y, and z, or by the three-dimensional vector $\mathbf{r}$ drawn from the origin of the coordinate system to the particle. As the particle moves, $\mathbf{r}$ changes from moment to moment so that we treat $\mathbf{r}$ as a function of the time t, which we read from a clock. Thus we write $\mathbf{r}(t)$ for $\mathbf{r}$ to indicate that t, the time, is a parameter to be treated quite differently from $\mathbf{r}$. Minkowski rebelled against this asymmetry between $\mathbf{r}$ and t, arguing that t must be treated mathematically exactly the same way as the three space coordinates x, y, and z are treated. Mathematically, space and time must be treated symmetrically, as an inescapable consequence of the constancy of the speed of light, as measured in all frames of reference (coordinate systems). But this approach immediately presented a physical difficulty; distance (space) and time are quite different physical entities. Distance is measured by laying a ruler along the distance to be measured whereas time is measured by the number of oscillations of a spring or pendulum in a given time interval. How can two such different kinds of measurements be combined into a single entity?

Another difficulty, which seemed insurmountable from the physical point of view, arose in trying to set up a coordinate system with a time axis on the same footing as the three space axes X, Y, and Z. One simply could not write x, y, z, and t as the four coordinates of an event, with t as the fourth coordinate of a point (the event) in a four-dimensional space-time manifold because it would be like adding apples and oranges. Moreover, it was unclear how to represent the time axis in such a four-dimensional coordinate system in which the three space axes x, y, and z cover all of space. No direction was available for the time axis because the three space axes preempt all directions of space.

Although this may have seemed to pose an insurmountable obstacle to the physicist, it presented no difficulty to Minkowski, one of the top mathematicians of that time, with an extensive background in every phase of pure mathematics. His work on the analytic theory of numbers is an excellent example of his great mathematical talents. He was probably attracted to the theory of relativity because he had been Einstein's teacher at the University of Zurich, which he subsequently left to become professor at the University of Göttingen. To merge space and time into a four-dimensional space-time manifold (with time on the same footing as space, as demanded by Einstein's theory), Minkowski first multiplied the time t by a speed to change it into a distance. But he could not simply multiply t by any speed; it had to be a speed that is the same for all observers (all frames of reference or coordinate systems). Minkowski was thus led to the product ct, where c is the speed of light, as the correct distance (speed times time is a distance) to assign to the temporal coordinate in the coordinate system of the four-dimensional manifold that correctly represents the space-time of special relativity. But Minkowski still had one other hurdle to overcome before he could complete his construction of a correct four-dimensional space-time; he needed to find a direction for the axis along which to measure the distance ct. Clearly, this axis cannot lie along any real direction because it is not a real distance. Moreover, the fact that the three real axes X, Y, and Z are mutually perpendicular to each other means that the fourth axis must lie along a direction perpendicular to three-dimensional real space (the X, Y, Z space) itself. Unfortunately, no such real direction exists. Therefore, the direction of the fourth axis must be imaginary.

After considering all the features that a fourth dimension must have,

Minkowski decided to multiply ct by $i = \sqrt{-1}$ (the imaginary unit) and define ict as the fourth dimension. This innovation represented an extension of Gauss's complex plane to a complex plane of four dimensions, consisting of the three real dimensions X, Y, and Z and the fourth imaginary dimension ict. Introducing i as a factor of the fourth coordinate had the additional advantage of rotating the fourth coordinate axis perpendicular to real space. This four-dimensional complex space is called the Minkowski space of special relativity. All the physical consequences of this theory can be deduced from the geometry of the Minkowski space. Thus a rotation of this space through an imaginary angle around the imaginary time axis is equivalent to an Einstein–Lorentz transformation. Here we see the great importance of pure mathematics for discovering physical phenomena which could not have been surmised otherwise; we must allow the mathematics to lead us to a greater understanding of the physics. But under no circumstances must the mathematics replace the physics.

We may summarize the difference between Newtonian dynamics and relativity dynamics by noting that in the Newtonian dynamics of a particle, three-dimensional vectors such as distance and momentum are the basic measurable entities. In relativity theory, however, four-dimensional vectors such as space-time intervals and momentum–energy concepts are basic. This leads to the equation Energy^2 = momentum^2 + $m_0^2 c^4$, which relates the energy of a moving particle to its momentum and its rest mass m_0 and the speed of light c. This shows that even if the particle is not moving (its momentum is zero) it still has energy, namely energy = $m_0 c^2$. If E is the energy of the particle and p is its momentum, the algebraic formula for the above equation is $E^2 = c^2 p^2 + m_0^2 c^4$.

We have gone to some lengths to present the relationship between the mathematics and the physics of the special theory of relativity to emphasize the power of the mathematics to reveal features of the physics that—in the absence of the mathematics—would have been extremely difficult to discover. This discussion also showed that these profound insights into nature can be obtained using the most elementary algebra and geometry. Indeed, the very elementary quality of the mathematics contributes to its beauty and to our own wonder that so much can be deduced from so little. Mathematically speaking, the entire theory is almost childish, but from the point of view of universal truth it is perhaps the greatest revelation of modern science.

To justify these bold words in praise of mathematics, we point to three other profound discoveries about matter and energy revealed by the elementary mathematics of the special theory of relativity. The invariant equation $p^2 - E^2/c^2 = -m_0^2 c^4$ can also be written as $c^2 p^2 - E^2 = -m_0^2 c^4$. Since the momentum p of the particle is mv, where v is its speed and m is its moving mass (to be distinguished from its rest mass m_0), we have $c^2 m^2 v^2 - E^2 = -m_0^2 c^4$.

By rewriting the basic equation in the form $E^2 = c^2 p^2 + m_0^2 c^4$ and taking the square root of both sides, we obtain

$$E = \pm\sqrt{c^2 p^2 + m_0^2 c^4}$$

The positive sign for the energy was the only one that had any meaning in classical physics. But in modern physics, the minus sign must also be considered. At first sight, this idea seems ridiculous because we cannot easily see how a particle can have negative kinetic energy, because kinetic energy depends on the square of the speed of the particle and a square is always positive; the kinetic energy must always be positive, at least from the classical point of view. But despite this absolute prohibition of negative kinetic energy in classical physics, it is not forbidden in quantum physics. In fact, the British theoretical physicist Paul Dirac showed in 1930 that negative kinetic energy must be included in a complete quantum description of particles. Dirac and other physicists of that period then went on to predict that particles with negative E are antiparticles. Shortly thereafter, the first antiparticle, the positron (the antielectron) was discovered in cosmic rays.

Without the mathematics, the nature of antimatter could not have been understood even if positrons had been discovered before special relativity and quantum mechanics had been developed. The mathematical prediction of the existence of antimatter illustrates the beauty of mathematical physics. Theoretical physicists such as Maxwell, Einstein, and Dirac have argued that a good gauge of the correctness of a mathematical theory of physics is its beauty. If one has a choice of two theories to explain certain physical phenomena, we should choose the more beautiful one as being closer to the truth. The mathematics of the special theory of relativity and its prediction of antimatter is beautiful because it is simple, free of ugly mathematical formulas, and brief. It is also very profound. One never knows what to expect from the mathematics nor what great revelations it holds for us. These revelations are valuable even though they often conflict violently with what our "common sense" tells us. But mathematicians and physicists do not concern themselves with such inconsistencies, viewing "common sense" in most cases as nothing more than the cumulative prejudices of society.

Just as the special theory of relativity generated a great deal of mathematical activity, so, too, did the general theory but the resulting mathematical activity was much more geometrical than analytical. Because the general theory is a geometrical theory of gravity, non-Euclidean geometry became the intellectual arena of the outstanding mathematicians for the first four decades of the twentieth century. The mathematical emphasis in this work was on developing the theory of coordinate transformations in its most general form. Einstein pointed out that the Einstein–Lorentz transformations constitute a very restricted group of transformations (hence the restricted theory of relativity) as they are limited to the transformations of inertial frames of reference—frames moving at constant speeds in a straight line. He insisted that the theory of relativity must be enlarged (generalized) to allow all frames of reference moving in any way whatsoever in free space and in gravitational fields. To Einstein this meant that the Minkowski four-dimensional space must be altered to four-dimensional Riemannian space. Thus Riemannian space became the mathematics of general relativity theory; it describes gravity not as a force but as the curvature produced in four-dimensional space-time by matter. If space is empty, its geometry is four-dimensional Minkowski geometry but it immediately becomes Riemannian when matter is present. The invariants (the laws) in this geometry are tensors; this importance of tensors led to a tremendous increase in tensor analysis, with mathematicians from all countries contributing to its growth. The geometry itself was expanded as differential geometry which, as we have already indicated, began with Gauss.

The theory of invariants had already been developed to a considerable extent by Cayley and Sylvester in England. In 1870, E. B. Christoffel and R. Lipschitz had constructed from tensors certain important mathematical quantities called Christoffel symbols that can be interpreted, in Riemannian space, as equivalent to the gravitational force in Newtonian gravity. Everything was thus in place for the mathematicians who followed to develop Einstein's general theory of relativity into a completely equivalent mathematical theory of Riemannian geometry. The geometers of that period accomplished this task by simply altering Minkowski's four-dimensional space-time geometry—which is essentially a four-dimensional Euclidean (flat space-time) geometry—by introducing curvature to obtain a curved four-dimensional space-time. These geometers were not primarily concerned with the geometry of space-time but rather with the properties of a general four-dimensional geometry which did not deal with space-time explicitly. But Einstein saw that this was precisely the kind of geometry that his general theory of relativity required; he adopted it and thus created the geometrical tensor form of the general theory of relativity—Einstein's theory of gravity.

There is an important difference between Einstein's special theory (Minkowski's flat four-dimensional space-time) and his general theory (non-Euclidean four-dimensional space-time). The difference lies in the two different four-dimensional theorems of Pythagoras for the two different geometries. If r is the vector position of a particle measured from the origin at the time t in Minkowski geometry, then its space-time interval squared is $s^2 = r^2 - c^2t^2$ (c = the speed of light). If r is given in terms of the three space coordinates x, y, and z, then $s^2 = x^2 + y^2 + z^2 - c^2t^2$; this is the theorem of Pythagoras in four-dimensional space-time. We may write this equation in a general four-dimensional form without referring to space or time explicitly by placing $x = x_1$, $y = x_2$, $z = x_3$, and $ict = x_4$ so that space and time are treated equally. If a particle shifts its position by an amount dr in a time dt (infinitesimal space displacement and time interval) then s changes by the infinitesimal amount ds and the theorem of Pythagoras in Minkowski space becomes $ds^2 = dx_1 + dx_2 + dx_3 + dx_4$. To introduce curvature in this space mathematicians multiply each term on the right-hand side by a different coefficient g, which is a function of space and time. In its most general four-dimensional mathematical form, the theorem of Pythagoras is written as

$$ds^2 = g_{11}dx_1{}^2 + g_{12}dx_1dx_2 + g_{13}dx_1dx_3 + g_{14}dx_1dx_4 + g_{21}dx_2dx_1 + g_{22}dx_2{}^2 + g_{23}dx_2dx_3 + g_{24}dx_2dx_4 + d_{31}dx_3dx_1 + g_{32}dx_3dx_2 + g_{33}dx_3{}^2 + \ldots + g_{44}dx_4{}^2$$

where the sixteen coefficients g_{ik} determine the curvature of space-time.

In this form coordinate systems of any kind, without restriction, may be used, which is an essential requirement of the general theory of relativity. All the physics—particularly the law of gravity—of general relativity, is contained in the sixteen coefficients g_{ik}, where the subscripts i, k are just identifying indices (like the first and second names of a person). Although ds^2 contains sixteen coefficients formally, only ten are different because $g_{ik} = g_{ki}$, and these ten are determined by the distribution of matter in space. These coefficients are all very formal but they acquired an incredible reality for cosmology and importance for understanding and probing the structure and dynamics of the universe when Einstein wrote out a set of the ten partial differential equations for the ten different coefficients g_{ik}—the famous Einstein field equations of the gravitational field—which are determined by the distribution of matter in space.

The physical deductions that flow from these equations are stupendous and their importance for every phase of physics is incalculable. The famous mathematical astronomer Karl Schwarzschild at Göttingen solved Einstein's field equations for a very special case shortly after Einstein announced them in 1915. Schwarzschild's solution led physicists and astronomers to the concept of the black hole. Einstein later applied his field equations to the entire universe, thus initiating modern cosmology, from which the expansion of the universe was deduced. This expansion was later verified observationally by Edwin Hubble.

Fortunately for Einstein, the mathematical basis for his field equations had already been laid by the mid-nineteenth-century French mathematician Poisson, whose work we have already discussed, and by the late-nineteenth-century Italian mathematician Gregorio Ricci-Curbastro. Poisson had discovered the famous "Poisson partial differential equation" of the Newtonian gravitational field; Einstein simply adapted this equation—which became his ten equations—to conform to the non-Euclidean geometry of four-dimensional space-time containing matter. Ricci, in 1884, had brought Riemannian geometry into the tensor form that is ideal for its application to four-dimensional space-time and to the description of the gravitational field. Ricci's mathematical work led Einstein to the discovery of the "Einstein–Ricci" tensor which he placed equal to a matter–energy tensor to obtain his field equations. The most complete exposition of the mathematics of the general theory of relativity was published in the early 1920s by the Italian mathematician Tullio Levi-Civita in a book called *Absolute Differential Calculus*; he presented the tensor calculus in its most elegant and useful form, not only for the theory of relativity but also for crystal structure, elasticity, and hydrodynamics.

With its merging of space and time into a single four-dimensional space-time manifold and its presentation of gravity, not as a force, but as a consequence of the curvature of space-time, the general theory of relativity attracted many pure mathematicians to its field. Outstanding among these was the German mathematician Herman Weyl, whose 1919 book *Raum, Zeit, und Materie* (*Space, Time, and Matter*) is a classic in the literature of relativity theory. Weyl became interested in general relativity because he believed that if he could properly enlarge the Einstein–Ricci tensor he would then be able to incorporate the electromagnetic field into the non-Euclidean geometry of four-dimensional space-time. After a thorough analysis of the general theory as Einstein had formulated it, Weyl saw that he could extend it mathematically by requiring that the basic invariants of the theory be invariant, not only to a transformation of coordinates but also to the change in the basic units of length and time we use in our measurements. This type of transformation, called a "gauge transformation," is used extensively now in modern field theory.

Weyl's introduction of gauge invariance into general relativity did, indeed, lead to an enlarged Einstein–Ricci tensor, and the additional terms he obtained do lead to Maxwell's electromagnetic field equations. But we obtain no insight into the relationship between the gravitational field and the electromagnetic field; gauge invariance is just a formal way (a purely mathematical procedure) for incorporating gravity and electromagnetism in the same mathematical formalism. Following Weyl's work, Einstein began his famous search for a single mathematical field theory that would unify the gravitational and electromagnetic fields. Unfortunately, this search for a unified field theory, which has occupied many outstanding mathematicians and physicists to the present day, has been fruitless. Today we are no closer to understanding the relationship between gravity and electromagnetism than we were before Einstein's discoveries but mathematicians are still busy analyzing Einstein's general theory and its consequences.

We noted in our introduction to this chapter that the first three decades of the twentieth century are not noted for great mathematical innovations or discoveries but rather for consolidating the known mathematics, extending its foundations, and thoroughly analyzing the logical structure and self-consistency of the axioms and definitions on which the fundamental mathematical theorems rest. In France, Henri Poincaré, standing like a guard at the gates of the sacred realm of pure mathematics, set the standard of research for all mathematicians. But his great devotion to pure mathematics did not restrict him in his enjoyment of mathematical physics, as we have already noted.

Poincaré found both pure mathematics and theoretical physics irresistible; in his complete dedication to these closely related intellectual activities he was closer to Gauss than any other mathematician. He was born on April 29, 1854, in Nancy, France, the son of Léon Poincaré, a distinguished physician. Henri's mother took it upon herself to educate Henri and his younger sister in the home. Although his mother's constant supervision might have cowered most children, Henri remained an eager student who absorbed his lessons with little apparent effort. Although he was bright, his coordination was poor. A bout with diphtheria robbed him of his voice for nearly a year and left him with a fragile constitution that hampered his physical activities for the rest of his life. But the boy was apparently not discouraged by his physical limitations; instead he turned to the world of books and knowledge. Blessed with an ability to read staggeringly fast and recall, almost word for word, the contents of any book, Henri swept through the family library like a German panzer division, absorbing books on subjects ranging from classical literature to natural history.

Henri was an outstanding student at the local grammar school but he did not initially demonstrate great promise in mathematics. Perhaps his emerging genius went unnoticed by his teachers because Henri was attempting to master all of his subjects instead of focusing on a single one. In any event, his academic talents made it possible for him to win many honors during his early school career and excel in all of his subjects except art. Henri would have been the first to admit that his drawing was atrocious; this flaw in his otherwise sterling academic record may have been rooted in his ambidexerity.

Only as a teenager did Henri realize that mathematics was his true calling. But his education was interrupted by the Franco–Prussian War in 1870 which saw the German army sweep across northern France in a fanlike movement of death and destruction that was stopped just outside the city of Paris. Although his poor health kept him out of combat, Henri did see much of the devastation firsthand as he often accompanied his father on his rounds. The ensuing four years of trench warfare laid waste to much of the countryside and instilled in Henri both an ardent patriotism for his beloved homeland and a disdain for the thuggery of Kaiser Wilhelm's Germany. But Henri never confused Germany's politics with its mathematics. In later life, he was among the first to recognize the important contribution made to physics by an obscure German scientist named Albert Einstein.

Poincaré entered the École Polytechnique in 1872 where he concentrated his considerable talents on the study of mathematics. Despite his high academic standing, he was admired and liked by his fellow students. But his utter incompetence in the school's military drills quickly ended any thought he might have had about following Descartes' example and pursuing a military career. Besides, Poincaré's strengths clearly lay in mathematics and not battlefield strategy. Accordingly, he entered the School of Mines to study engineering but gradually drifted back toward pure mathematics and managed, during his spare time, to write a thesis which he submitted for his doctorate in mathematics in 1878. Gaston Darboux, a professor of mathematics who was charged with the task of evaluating Poincaré's work, praised Poincaré's efforts, saying that the dissertation "contained results enough to supply material for several good theses." But Darboux also questioned Poincaré's instinctive, arguably unstructured, approach to his subject, declaring that "if an accurate idea of the way Poincaré worked is wanted, many points called for corrections or explanations" because "having once arrived at the summit he never retraced his steps."

After receiving his doctorate, Poincaré worked for two years as a professor of mathematical analysis at Caen before accepting a position at the University of Paris, where he remained for the rest of his career. During the early years of his career, he devoted much of his time to the theory of differential equations, continuing the attacks on the subject he had first mounted during the preparation of his dissertation. But Poincaré, like Gauss, was never content to limit himself to a single area of mathematics for a given period of time. Instead he dabbled in topics ranging from analysis and mathematical astronomy to the theory of numbers and higher algebra. His work in mathematics was recognized while he was still a young man of thirty-two with his election to the Academy of Sciences. The Academy recognized that the young Poincaré was already setting a blistering pace in his mathematical publications that totaled nearly 500 papers and book-length memoirs by the time of his death. Had Poincaré lived fifteen or twenty years longer, he might well have vied with Euler for the title of history's most prodigious mathematician.

A professor at Paris for thirty years, Poincaré charmed his students with the brilliance of his lectures in such diverse fields as potential theory, optics, electromagnetism, heat, celestial mechanics (particularly the gravitational n-body problem), thermodynamics, and hydrodynamics in physics, and probability theory, infinite series, differential equations, function theory, algebraic curves, the theory of orbits, and integral invariants in pure mathematics. Many of the problems he dealt with in pure mathematics grew out of his search for the solutions of problems he encountered in mathematical physics.

Poincaré spent much of his career grappling with some of the key

Henri Poincaré (1854–1912)(Courtesy AIP Emilio Segrè Visual Archives)

problems of mathematical astronomy. In particular, he tried, unsuccessfully, to solve the three-body gravitational problem that had bedeviled scientists for centuries. But his examination of the process by which such a problem could be attacked was so impressive that Poincaré was awarded a cash prize offered by the Swedish throne to anyone who could solve the problem outright. Poincaré also published several multivolume treatises dealing with celestial mechanics in the 1890s and 1900s which established him as the leading mathematical astronomer of his day. According to Darboux, Poincaré's influence on the study of celestial dynamics was revolutionary in that his methods could be applied both to dynamical problems such as mechanics and to problems involving periodic phenomena.

As Poincaré passed through middle age, he continued to be regarded as France's greatest and most versatile mathematician. He also began to gear more of his writings toward the general public, believing that the education of the citizenry was a necessary precondition to increased government support for scientific research. Poincaré immersed himself in the philosophy of mathematics and science; he wrote extensively, in a popular vein, about the foundations of mathematics and science and raised profound questions which are still unanswered to this day. In his most famous popular book, *The Value of Science and Science and Hypothesis*, he analyzed such basic concepts as truth in science and mathematics, measurement in science, space and time, and so on.

Poincaré's approach to both pure mathematics and mathematical physics and astronomy was a model of intellectual excellence for all mathematicians and scientists. One of his favorite subjects was the issue of mathematical creativity, a topic that required him to delve into such unfamiliar areas as psychology. He also wrote about the newer discoveries in science, especially physics, and, for a time, was one of the best-selling authors in France. His books were read by young and old alike and he was among the first to teach the general public about Einstein's mysterious space-time continuum. His death at the age of fifty-nine on July 17, 1912 after a long illness, ended any pretensions the French mathematical community might have still entertained about challenging Germany for primacy in the world of mathematics.

The German mathematician David Hilbert (1862–1943), a native of Königsberg, East Prussia, was considered almost universally as Poincaré's successor after Poincaré's death. Like Poincaré, Hilbert was a transitional figure but his was much more a twentieth-century orientation than that of Poincaré and Hilbert was much more comfortable with the uses to which mathematics was put by physicists after 1900. He shared with Felix Klein the honor of leading the study of mathematics at the University of Göttingen, becoming the senior professor after Klein's death in 1925.

Unlike Poincaré, Hilbert devoted most of the time in his mathematical research to pure mathematics, particularly to the basic problems presented by the foundations of mathematics. But his work in pure mathematics became very useful in physics, particularly his founding of what became known as "Hilbert space," which is the basic mathematics of quantum mechanics. His most famous work in pure mathematics is his book *The Foundations of Geometry*, which was published in 1900. This book deals primarily with the axioms of Euclidean geometry; Hilbert subjected these axioms to the most critical analysis possible and showed the minimum requirements that any set of geometrical axioms must fulfill. This work was in line with his basic philosophy of mathematics which he expounded at the

International Congress of Mathematics held in Paris in 1900. There he listed a series of twenty-three mathematical projects that he believed mathematicians should pursue.

Deeply concerned about maintaining the high quality of mathematical research that was flourishing at the time, Hilbert felt that future mathematicians should continue developing the new ideas of Cauchy, Cantor, Riemann, Lie, and Ricci. Although Hilbert was primarily a pure mathematician with a sense of responsibility for the development of the new mathematics, he did spend some time with mathematical physics and produced two very useful mathematical techniques for theoretical physics. The first of these is the application of the calculus of variations to the general theory of relativity; this enables the theoretical physicist to deduce Einstein's field equations from a generalized principle of least action, in line with Hamilton's principle of least action for Newtonian mechanics. Hilbert's second mathematical invention for theoretical physics was the "Hilbert space" which is of particular importance in the mathematical structure of quantum mechanics.

The name "Hilbert space" is a misnomer because it is not a geometrical space in any real sense, similar to Euclidean space, Riemannian space, or Einsteinian space; it is a fictitious space of functions in which the functions play the role of vectors. Because we can have an infinite number of independent functions, Hilbert space has an infinite number of dimensions. In quantum mechanics we introduce the concept of the state of a system, such as the state of an electron in an atom. This state, which is the wave function of the electron, is pictured as a "vector" in Hilbert space. If we know the "geometric" laws that govern Hilbert space, we can reduce the mathematics of the quantum mechanics to the geometry of Hilbert space. As the state of the electron changes, the vector that represents it in Hilbert space changes so that all the features of the changing state of the electron can be described by the changing Hilbert state vector. The concept of the quantum mechanical state of a system as a vector in Hilbert space became the basis of a complete, self-consistent mathematical structure of quantum mechanics. This concept unified the Heisenberg matrix mechanics and the Schrödinger wave mechanics; they are basically the same from a purely mathematical point of view but they present in different mathematical forms the way the state vector changes in Hilbert space. Hilbert's work in the mathematics of quantum mechanics stimulated some of the outstanding younger twentieth-century mathematicians to contribute to this very exciting field of mathematical physics. Most notable are Hermann Weyl, whose book on quantum mechanics and group theory, *The Theory of Groups and the Quantum Mechanics*, is a classic in this area, John von Neumann, who established a standard of excellence in the field with his book *The Mathematical Foundations of Quantum Mechanics*, and the algebraist B. L. Van der Waerden, whose book *Sources of Quantum Mechanics* presents a critical analysis of the basics of quantum mechanics.

In his later years, Hilbert became interested in developing a rigorous algebra of logic along the lines first proposed by Leibnitz and later carried on by George Boole, whose investigations into the "laws of thought" became the basis for all future research in this field. In his development of an algebra of theoretical logic, Hilbert was greatly influenced by the *Principia Mathematica* of Bertrand Russell and Alfred North Whitehead in which the authors set out to show that all mathematics can be reduced to pure logic. This is an extension of Gottlob Frege's work *The Foundations of Arithmetic* in which he deduced the fundamental rules of arithmetic from logic. In his 1928 book with W. Ackerman, *The Basis of Theoretical Logic*, Hilbert did just the reverse and showed that logic can be reduced to mathematics. All of these treatises on the relationship among mathematics, logic, and language led to the development of the language of the modern electronic computer.

Hilbert may be considered as the last of the great nineteenth-century and the first of the great twentieth-century mathematicians who bridged the gap between classical and modern mathematics. He saw clearly the new directions mathematics had to take to keep mathematics on its throne as the "Queen of the Sciences," and to make it increasingly more indispensable to the rapidly developing technology of the twentieth century. But he also knew that mathematics would remain valuable to physics only if mathematics remained true to its historical legacy.

CHAPTER 12

The Evolution of
Modern Mathematics

Men have become the tools of their tools.

—HENRY DAVID THOREAU

During the early decades of the twentieth century, mathematical research was driven by two different demands made on the mathematician: (1) the need of the mathematician himself to place mathematics on a flawless, logical base and to free it of all inner contradictions and philosophical inconsistencies and (2) the demand of the theoretical physicists for the advanced mathematics they required for the proper development of their physical theories. Having dealt primarily with the influence of theoretical physics on the development of modern mathematics in the preceding chapter, we now turn to the contribution that pure mathematics itself made to its own evolution. The difference between this aspect of the evolution of mathematics and that evolution driven by theoretical physics is that the latter is constrained by nature itself whereas the former is constrained only by the limits of the individual mathematician's imagination. This difference was already evident in Riemann's introduction of n-dimensional non-Euclidean space as a pure mathematical abstraction; that it met Einstein's needs in his development of a geometric theory of gravity was purely coincidental. The critical analysis of pure mathematics, which concerned such mathematicians as Klein, Poincaré, and Hilbert, began with Georg Cantor's concept of number, which led him to the theory of transfinite numbers. Because number had always been associated with quantity before Cantor began his analysis, he considered it necessary to introduce number in its most abstract sense, more as a symbol than as a quantitative entity; arithmetic, as a mathematics, would then be generated by the rules that are introduced to enable us to pass from one number to any other number. To rid number of any specific quantitative meaning, Cantor introduced the concept of the "aggregate" or "set," which led to modern "set theory." The set is defined as an aggregate of elements, each of which may refer to anything we please, which, of course, includes numbers. Set theory thus became the basis of modern mathematics and the analytical tool for structuring all other branches of mathematics. It is the introductory course one takes for the study of all higher branches of mathematics. By introducing general rules for the addition, subtraction, and multiplication of sets, one can obtain the rules that govern the specific mathematical operations in all branches of mathematics from arithmetic through calculus.

Bertrand Russell took the first concrete step, after Cantor, to place mathematics on a flawless foundation, free of abstract philosophical assumptions and logical inconsistencies. He was joined in this task by the British philosopher–mathematician Alfred North Whitehead; their collaboration produced one of the most famous works of the twentieth century, *Principia Mathematica*, a three-volume work that appeared in 1910–1913. Their project was monumental and their attempt to reduce mathematics to an "absolute logic" was heroic, but within a few years after their *Principia* was published, two British mathematicians, Ramsey and Chuistek discovered a number of contradictions in the logic of the *Principia*. These contradictions caused mathematicians to raise some fundamental questions concerning the usefulness and even the possibility of developing a complete mathematics that is free of inner logical contradictions.

Whether a completely self-consistent mathematics can be developed at all has remained an open question whose answer, one way or the other, does not bother mathematicians as long as it does not affect or interfere with the usefulness of mathematics as a tool. Logicians and philosophers, however, were greatly disturbed that so precise an intellectual discipline as mathematics might be flawed, even if only in a nonsubstantive way. In 1931, however, the Austrian mathematician and logician Kurt Gödel, then at the Institute of Advanced Studies at Princeton, proved that it is possible to construct an axiom that is neither provable nor disprovable in any system based on a complete set of self-consistent axioms. This is known as Gödel's theorem of the "excluded middle." It violates our sense of propriety because it states in essence that a mathematical theorem within a given discipline can be constructed which can both be true and false at the same time. But this should not diminish our faith in mathematics because we are confronted with a situation in physics similar to the violation of the excluded middle in mathematics. In our discussion of the motion of electrons, we saw that the mathematics of quantum mechanics states that the electron, though indivisible, can simultaneously be traveling along two different, partially sepa-rated, paths. We are thus confronted in physics by the excluded middle: a particle which is both divisible and indivisible at the same time.

Although the attempts to reduce all mathematics to a rigorous logic and, conversely, to reduce logic to a rigorous mathematics were only partially successful, they led to very important developments in the theory of language, information theory, servomechanisms, the theory of games, operations research, and computers. All of these disciplines have one important feature in common: they require a mechanism for translating commands expressed in some kind of symbolic (mathematical) language into meaningful actions or into an understandable language.

Because every language has a logical structure (a syntax), we should be able to establish a one-to-one correspondence between any language and logic and, therefore, between any language and mathematics. This corre-spondence became the basis of information theory and computer software, without which the modern electronic computer would have relatively little value. The ability to express any number on the binomial base (only the digits 0 and 1 are used) permits us to store vast amounts of information on silicon chips, which can be retrieved by the flow and registration of currents through microscopic circuits.

Once we have transformed a statement—an image, a sound, some words, or a mathematical formula—into stored information, such as a memory bank or data base on an electronic device (a silicon chip), or an optical device (a film strip or slide), or acoustically (a magnetic tape), we must have a way to retrieve this information. This need brings us into the realm of information theory because the question immediately arises as to whether we lose any information about our statements when we try to retrieve the information. We can best illustrate this loss of information by considering an image on a photographic plate. Owing to the aberrations which the lenses in our optical system impose on the rays of light passing through them, the image on the photographic plate is not sharp so that some information is lost. Note that all the information that the rays had about the source of the light when they left the source is still present in the rays, but the lenses interfere with the rays and thus reduce the sharpness or preci-sion of the message carried by the rays.

Mathematicians have contributed a great deal to information theory by first defining the basic elements that are essential to the theory and then constructing mathematical models that can be used to maximize the information that the retrieval system can produce. Two general elements are involved in the theory of information: the signal, which contains all of the information, and the ambient "noise," any phenomena that interfere with or distort the signal. The sole purpose of the mathematics of information theory is to maximize the ratio signal/noise or S/N, which means reducing the noise as much as possible because we have no control over the signal itself. In an optical system, for example, we reduce the "noise" produced by the lenses by combining a group of different lenses in the correct manner and carefully designing the contours of their surfaces to reduce the distortions (aberrations) they produce on the rays passing through them. Mathematicians have worked out the theory of lens aberrations so that the optical designer can follow the rays of light as they pass through the lenses, and then alter the contours of the surfaces of each lens to give the best image when the rays are focused on the photographic film or plate. The beauty and usefulness of the mathematics in this process (called ray tracing—a purely trigonometric procedure) is such that the mathematics can be done with pencil and paper, or a computer, without going near a lens. Moreover, the mathematics is a powerful analytical or diagnostic tool for pinpointing the cause of the optical noise in the lens system, and indicating the lens corrections that must be applied. The mathematical theory from which lens design stems goes back to Hamilton's optical theories. We recall that he developed the mathematical basis of Newtonian dynamics in its fullest and most useful form, as described in an earlier chapter. Hamilton's optics, in turn, rests on Fermat's principle of least time for the path of a ray of light moving between any two points. From the point of view of optical design, Fermat's principle tells us that our lens system is properly designed if all the rays of light starting from a given point of the source of light spend exactly the same time in moving from that point to the image of the source produced by the optical system.

What we have said about light and the optical retrieval systems applies, with some minor variations, to retrieval systems designed for any type of electromagnetic waves, such as radio and television sets. The noise may then be external noise, static, or internal noise arising from the electrical circuitry within the radio set or the faulty electronics that is used to amplify the signal. All these phenomena can be analyzed mathematically and the circuitry and the electronics can be designed to reduce the internal noise to a minimum and the amplification to its optimum value.

Our discussion of optical and electronic retrieval systems clearly reveals the importance of mathematics for the development of the technology of information retrieval systems. Without mathematics, the amazing growth of all such systems during the last half century would have been impossible even though all the basic physical principles and laws that govern these systems have been known for nearly a century. Astronomers are particularly grateful for these technological advances which have been guided by mathematics. For many years astronomers had only one window open to the universe at large—their earth-based optical telescopes which revealed to them an important but restricted part of the universe. Because the visible rays of light coming from certain parts of our own galaxy are blocked from reaching us by dust particles, optical telescopes cannot show us the core of our galaxy, at a distance of some 30,000 light years. But radio telescopes, developed with ever increasing sensitivity, permit astronomers to penetrate the dust and see the center of our galaxy; this galactic core is revealed by the radio signals that are emitted by the electrons in the hydrogen atoms when they flip over like spinning tops. Radio telescopes opened up a radio window to the universe, through which astronomers can look farther into space than ever before. The sensitivity of the modern radio telescope is so great that a radio telescope with a 100-foot paraboloid dish can detect a radio source that is about 1,000 times fainter than the faintest light source that an astronomer can detect with the largest optical telescopes. Radio telescopes have revealed such esoteric astronomical objects as quasars, pulsars (neutron stars), radio stars, and active galaxies. Because the frequencies of faint radio waves can be measured with much greater accuracy than the frequencies of faint optical waves, the radio telescope is an ideal instrument for analyzing radiation from the most distant objects that can be detected in the universe, and thus determining, with great accuracy, the rate of expansion of the universe.

The theory of information retrieval led to the discovery of the mathematical relationship of knowledge to the concept of entropy in thermodynamics. To establish this relationship, Claude Shannon, in 1949, introduced a measure for the "amount of information," H, contained in a message carried by a current along a wire. Because some information in a transmitted wire is always lost, the negative of H invariably increases. Because this is also true of the entropy of a system in thermodynamics, a mathematical relationship should exist between the negative of H (that is, $-H$) and entropy. Shannon discovered this relationship and thus placed information theory on a mathematical basis that can be understood in terms of a fundamental physical law—the second law of thermodynamics (the law of entropy) and the irreversibility in nature.

How is the loss of knowledge related to the increase of entropy in a system? We consider a gas confined by a partition to half the volume of a cylinder. Each molecule in that gas is in that designated partitioned half of the cylinder. If we now remove the partition, the gas fills the entire cylinder, and we immediately lose half our knowledge about the position of a molecule within the cylinder because the space in which it might be located has just doubled in size. The entropy of the gas, however, has not doubled; the entropy increases as the logarithm of 2. This was the starting point of Shannon's information theory. To illustrate this remarkable relationship between information and entropy we consider knowledge, order, disorder, and entropy in general. To this end we observe, initially, a gas in our cylinder at a certain temperature, consisting of two kinds of atoms, A and B. The atoms at that moment are in a state of maximum disorder and maximum entropy. We return to the cylinder some time later and note that instead of free individual atoms (A and B only) we now have the combinations (AB molecules) mixed in with some free atoms A and B. The temperature of the gas has also increased, which means that the entropy of the gas has increased. But the presence of molecules (AB) means that the matter in the cylinder is in a higher state of order than initially because molecules are more highly organized structures than collections of individual atoms. This does not contradict the relationship between disorder and entropy increase because in calculating entropy we must take into account the energy in the cylinder as well as the matter. The temperature increase of the gas in the cylinder means that more energy is present than initially. Because the increase in the energy is produced by the atoms when they form molecules, and such energy is highly disorganized, the net disorganization of the gas in the cylinder increases. We have thus discovered that order and disorder go together; you cannot have one without the other. Concerning the relationship between disorder and loss of knowledge, we need only call upon our own experience to verify that disorder and loss of information go hand in hand. Almost everyone has been frustrated by the failure to find a document containing important information among papers strewn about a highly disorganized desk.

In our discussion of information theory above, we have limited ourselves to the problem of the retrieval of information from a message sent from a distant source to a receiver, but some of the most important and interesting aspects of information retrieval occur within single organisms such as our bodies or even within single living cells. As such examples, though very important and interesting, are extremely complex and only partially understood, we shall describe the simplest of such systems—ordinary machines which respond to messages as they are sent to them—to illustrate the mathematical principles they depend on or incorporate. The response is called an output signal. If we picture the commander as part of the machine, then the commander and the machine constitute what is called a servomechanism. This is an area to which some of the most outstanding mathematicians of the twentieth century, such as Norbert Wiener, have contributed. Wiener labeled the mathematical theory of servomechanisms "cybernetics." Because this term stems from the Greek word for "helmsman" or "governor," Wiener implied that the power of the command, a person or another device, is part of the servomechanism. Cybernetics differs, in one very important feature, from the theory of information previously discussed because it boasts a "feedback mechanism." The standard theory of information concerns itself primarily with retrieving information from a message without any kind of feedback to the sender of the message. In the servomechanism, by contrast, the feedback is essential if the machine that receives the command is to perform accurately.

As an illustration of this important difference, we note that we may at times receive a message which requires our immediate action and a response to the sender. This illustrates the biofeedback process and the essence of a servomechanism. If the information contained in the message is merely informative and adds to our knowledge but requires no action, then no feedback is involved. The human body is a vast collection of servomechanisms, in which the feedback mechanisms may be easily perceived or so very subtle that we are hardly aware of them. The moving automobile is a very remarkable servomechanism in which driver and machine are intimately paired. The driver is constantly sending messages to the wheels and motor; he carefully monitors their responses. These responses are then immediately transformed into corrected commands to keep the automobile moving as desired. An airplane flying on instrumentation, a space satellite, a missile, and a smart bomb are all examples of complex servomechanisms.

A number of distinct operations are involved in any servomechanism, each of which requires a program written in a "language" to which the servomechanism can respond. The first of these operations is sending the message containing the appropriate information. The second is the response of a trigger, for example, an electric switch which activates the mechanism, such as a motor or a muscle. The third is releasing the energy that the motor or muscle requires to operate. The fourth and final is the triggering of a feedback mechanism to insure the correct operation of the motor. These messages are encoded mathematically in a special language that is built into a computer or some other device. With the rapid growth of automation in all kinds of manufacturing processes which is one of the most important practical applications of servomechanisms, cybernetics became one of the most popular courses in graduate mathematics.

Cybernetics is one of the three remarkable mathematical disciplines that emerged from the theory of information; the other two are "operations research" and the "theory of games." Operations research is a general mathematical technology which began in World War II when it was necessary to devise a strategy for damaging the enemy during an attack when very little was known about the disposition of the enemy's forces. It is a mathematical technology to increase the efficiency of any kind of campaign—whether it be military, sales, political, or advertising—when the target of the campaign is not clearly delineated. Much of this technology and its success clearly depends on probability theory and statistical methods. Not surprisingly, operations research has some resemblance to gambling but its nature is essentially pragmatic. Whether a particular operations research project succeeds or not is measured by the final result and not by the theory. The proof is in the pudding and not in any built-in truth such as geometry that rests on basic axioms.

Recognizing this deficiency in operations research, we must accept the danger of being far off the mark for any particular application even though our mathematics is correct. From its original purpose to aid military strategy during World War II, operations research has become a business tool, particularly in the area of markets; it has also spread into various areas of research in science. But there are very important differences between operations research and standard mathematics. In mathematics, definite procedures and mathematical techniques are applied to solve any particular problem or to prove a theorem. We may not find the solution to the problem or a proof of the theorem but the mathematical procedure is clear. In operations research, however, we have no well-defined, logical procedure to reach our goal but considerable guesswork, statistical methods, and probability may be used in operations research to achieve some surprising successes. The prediction by economists of the direction an economy will take is a very important example of the application of operations research; that no two economists agree in their predictions of the future of an economy or on the actions to be taken to lift an economy out of a recession is evidence of the uncertainties associated with operations research. As a result, there are certain difficulties that operations research economists encounter.

Because the modern economist has no basic economic theory to guide him in his predictions, he uses statistical inference, based on previous recessions, to construct an economic model from which he predicts the duration of a current recession. He uses certain parameters to construct his model, such as the average duration of all previous recessions in a given time, the "unemployment rate," the confidence level of the population, and the state of the stock market. If the current values of these parameters are close to the values of the very same parameters when the economy was coming out of an average recession in the past, the economist, in general, predicts that the economy will begin to recover from its recession. This is all well and good if all recessions are cyclical in the same sense and if no parameter has been overlooked that is not cyclical but is cumulative as, for example, personal debt, which may increase from recession to recession. Because recovery from a recession in a typical free market economy depends on the increasing purchasing power of the average consumer, no recovery from a recession can occur if personal debt increases to a point at which purchasing power is severely restricted or diminished.

One of the dangers inherent in using statistical inferences based on a collection of supposedly independent parameters in the application of operations research to complex systems such as a modern economy is that the parameters, in general, are not independent of each other but instead are functions of each other. If a model or a prediction program for any system is based on the assumption that the variables that determine the evolution of the system are independent of each other and depend only on the first power (no squares, cubes, or higher powers) of each parameter, the program is called a "linear program." Each parameter that enters into an algebraic equation used to describe the program appears to the first degree. Suppose we construct or invent a program that is to guide us in increasing our wealth from month to month. This program must, of course, involve our salary, our savings account, our investments, our expenses, and so on. If we call the program P, and use the letters $x, y, z, \ldots, u$ to represent these various ways we allocate our income each month, the algebraic expression for a linear program for increasing our wealth is written as $P = ax + by + cz + \ldots + qu$, where a, b, c $\ldots$, u are numbers. Clearly such parameters as savings accounts and monthly expenditures depend on each other so that a linear program in this instance is only an approximation. But it might be a good approximation for simple problems in which only two or three independent parameters are involved. Finally, we note that hidden variables may be present in any phenomena which may greatly affect the accuracy of the program designed for it. This is best illustrated by long-range weather prediction programs, which cannot possibly take into account such random, weather-affecting phenomena as solar flares, sunspots, and the solar wind.

Operations research, which depends on guesswork, statistical infer-ences, and probability theory, is part of a larger mathematical discipline, the theory of games, which we discussed briefly in Chapter 7. We return to it now from a more sophisticated mathematical base because it has grown enormously and attracted some of the most gifted twentieth-century mathe-maticians to its cause, including John von Neumann, ranked with Poincaré and Hilbert, as one of the three greatest mathematicians of this century. Because we all gamble, in a sense, in our decisions to pursue one course of action or another in our daily lives, we would all be happy if we had an exact mathematical procedure to guide us in our decisions. Unfortunately, we have no such guide but most of the choices we must make are of no great consequence to the orderly and fruitful pursuits of our lives. But very critical choices are forced on us from time to time. Having a precise mathematical theory of games might then be very helpful. But no such theory has ever been constructed even though, as we have stated, a mathematical theory of games—which is an extension of the theory of probability—has been developed. The mathematics of the theory of games is essentially an extension of linear programming, which was first proposed in 1921 by the French mathematician Emile Borel. His ideas were analyzed in detail by von Neumann who proved the basic theorem—the Minimum Theorem—of game theory and, then, in 1944, with Osker Morgenstern, published a book on game theory which has become the "bible" of this mathematical discipline.

Von Neumann's main objective in developing a theory of games was to construct a strategy that can be applied to economics for the benefit of a complete society. His goal was thus a nonzero-sum game theory, which is a radical departure from the zero-sum orientation of classical game theory. We can best describe the difference between these two kinds of games by considering two players who are matching coins with the winner taking both coins at each throw. In this situation, the winner's gain exactly equals the loser's loss so that the sum of gain and loss is exactly zero. In a gambling casino the game is not a zero-sum game because the odds are set slightly to favor the house. Otherwise, the house would reap no profit and therefore would fail. This is also true in any free-market or capitalist economy, which is driven by the profit motive. The game in this kind of economy is between the workers, as a group represented by a union, and the entrepreneurs represented by management. A third party, the public (consumers), which influences the game enormously, is also involved though not as a direct player. This game would be a zero-sum game if the gross income from all the items sold by the enterprise exactly equaled the total cost of production of all the items (raw materials, packaging, labor, management, advertising) but it becomes a nonzero-sum game as soon as profit is included, which favors the entrepreneur.

This game is greatly complicated by the dual role played by the workers, the management personnel, and even the entrepreneurs; they are members of the consuming public and therefore have an interest in the outcome of the game beyond that of a disinterested player. Although management's concern is primarily to maximize the profit of the entrepreneur, the managers, as part of the work force, cannot act independently. Thus the economic game is very complex.

Von Neumann's treatise on the theory of games appeared early enough to be applied to military strategy during World War II. Because war has always been treated by the contending parties as a game, it was natural that von Neumann's work was used extensively by the Allies. War, of course, is not a zero-sum game because there is no military action in which one can decide unequivocally how much each side gains or loses. Nor can one decide or define a loss or gain in any unambiguous way. The best the "players" can do is introduce the constraints that are imposed on each side by the losses that each side can sustain. The aim of each side is to force unsustainable losses on the opposing side before suffering such losses itself. Game theory can help in developing such an "unsustainable loss" strategy.

With the vast burgeoning of technology after World War II, game theory became very useful and, indeed, well nigh indispensable in solving flow problems in communications, transportation, and marketing. To illustrate this kind of problem we consider briefly a flow we all contend with in our daily lives—traffic. Here we consider traffic from the point of view of

John von Neumann (1903–1957) (Permission by Alan W. Richards; Courtesy AIP Emilio Segrè Visual Archives)

the driver rather than that of the pedestrian and we limit ourselves to a simple rectangular street grid similar to that found in most of Manhattan. The obvious solution is to arrange the traffic lights to favor the flow of traffic along those streets and avenues where the traffic is heaviest. But we are immediately confronted by a difficult—if not insoluble—subsidiary problem. Because the streets and avenues are perpendicular to each other and the traffic flows in opposite directions as we go from street to street and avenue to avenue, we must either favor the avenues or the streets in arranging the traffic lights to optimize the flow. We thus see that what appears initially to be a simple problem in principle can be very difficult in practice.

The traffic problem is not really a game in the sense that contending players are involved but its solution requires the application of the same kind of mathematical reasoning and technique as used in game theory. The point here is that mathematicians have taken under their wing general problems that cannot be solved by the standard kind of mathematical analysis applied in algebra and geometry. The problems—such as the traffic problem—that fall into the game theory category are highly nonlinear so that they cannot be defined by simple algebraic equations involving a single unknown. In fact, many unknowns, which depend on each other, are involved, even in simple straightforward problems like the traffic problem. But certain basic mathematical principles such as maximum and minimum principles can still be applied. In the traffic problem, for example, one might want to maximize the flow of bus traffic along certain avenues and do so by restricting one lane in each such avenue to bus traffic alone, forbidding any private traffic along that lane. Such a restriction affects the traffic along the other lanes and may therefore be counterproductive to some extent. The problem, in a sense, is therefore a maxi–min problem: we wish to maximize the flow of bus traffic and at the same time minimize the congestion this maximization may produce in the private car traffic. This is a sort of zero-sum game because the flow advantage gained by the bus traffic is almost, but not quite, equal to the flow disadvantage incurred by the private car traffic. As each bus carries many more people than any one car, the bus traffic must be favored in the ratio of the number of bus passengers to the number of private car passengers. Because both of these things depend on the time of day, that factor must also be taken into account by the traffic engineer.

An outgrowth of game theory, developed by von Neumann and S. M. Ulam, which is used extensively in all phases of scientific experimental research, is called Monte Carlo simulation. It was first used by von Neumann and Ulam in the early 1940s in the analysis of complex nuclear phenomena which play an important role in nuclear bomb construction and nuclear reactors. The Monte Carlo simulation technique uses high-speed electronic computers to simulate actual experiments involving, for example, two different nuclei that interact with each other under various conditions. One set of important conditions, called the initial conditions, is a set of numbers which specify all the physical parameters that define the experiment. The final observations or measurements of the experimenter depend on the initial conditions which he imposes on his experiment. From these observations, he then constructs a physical model of the underlying phenomena which must be independent of the initial conditions. To check the validity of his model, the experimenter must perform many experiments under an array of different initial conditions and with changes in his apparatus. These changes and the necessary complications they entail can be very expensive and time consuming. In the Monte Carlo procedure, the physical laws that are required to construct a mathematical model to account for the observations are plugged into the computer as mathematical equations or formulas and the computer then simulates, in a very short time, as many different variations of the experiment as may be desired, performing all the required calculations.

Two very simple examples of the Monte Carlo simulation procedure illustrate its basic features without introducing a complex mathematical formalism. We apply it to the determination of a target area lying anywhere in a larger circle whose area we know. We spray the entire area of the circle with bullets that may hit anywhere at random within the large circle, but with the constraint that no bullet is permitted to hit outside the large circle. Attached to the target is a device that produces a short, sharp ring when the target is hit. By just counting the number of rings we detect in any sufficiently long series of shots, we can calculate the target area as the total area of the large circle multiplied by the ratio of the number of audible rings produced to the total number of shots fired.

A more sophisticated example of the Monte Carlo procedure as discussed previously is the calculation of the number π (pi) by counting the number of times a needle of a given length L allowed to fall freely, intersects one of a series of parallel horizontal lines on a flat horizontal surface if d is the space (larger than the length L) between any two neighboring lines. It can be shown that the probability P for a single such intersection equals $2L/\pi d$. Therefore π equals $2L/d$ multiplied by the ratio of the number of intersections to the total number of times we drop the needle. Actual experiments of this sort have been performed and very accurate values of π have been obtained.

Most experiments in high-energy particle physics today are based on the Monte Carlo model because many similar experiments, with variations of the parameters, are performed; appropriate averages of the data are used to construct models of the underlying phenomena. One of the most famous and important such experiments was performed by Geiger and Marsden in 1911 at Ernest Rutherford's Cavendish Laboratories in Manchester. These experiments, under the aegis of Rutherford, were designed to determine unequivocally the structure of the atom. Two different models had been proposed, one, the "raisin-bun model" proposed by the British physicist Sir J. J. Thomson and the other, the nuclear atom proposed by Rutherford and others. The decision as to which model is correct depended on determining the distribution of the positive electric charge in the atom. If it is distributed through the atomic volume, like raisins in a bun, the Thomson model is the correct one. But if the positive charge is concentrated in a tiny nucleus, then the nuclear model is correct. Rutherford saw that this question can be answered by bombarding the atom with tiny electrically charged bullets and then studying the paths of these bullets after they have passed through the atom. This experiment was an example of a "size of the target" exercise described above which was applied to an atom and was ideal for Monte Carlo simulation.

Rutherford outlined such an "atom shooting" experiment and asked Geiger and Marsden, two experimental physicists in his lab, to do the shooting. Rutherford, whose specialty was the study of radioactivity of such heavy elements as uranium, saw that the positively charged alpha particles (helium nuclei) emitted at very high speeds from uranium nuclei are ideal, electrically charged bullets for shooting into atoms. Rutherford chose gold atoms as the targets for the alpha particle bullets; Geiger and Marsden took a piece of pure uranium and directed the alpha particles they emitted at a piece of very thin gold leaf. Most of the alpha particles passed through the gold leaf and came out with their paths only slightly deviated from the straight lines along which they were moving before encountering the gold leaf. But every now and then an alpha particle bounced back from the gold leaf as though it had struck an impenetrable wall. Rutherford correctly interpreted this result as evidence that that particular alpha particle had been repelled by a very strong positive charge, and reasoned, then, that the nucleus of an atom is positively charged. The experiment itself was not particularly difficult but it had to be performed with great care. In any case, the information it conveyed was revolutionary because it led to the Bohr model of the atom, to quantum mechanics, and, in time, to nuclear physics.

Experiments of this sort have to be repeated many times before their results are acceptable. They are, therefore, ideal for Monte Carlo simulation, which can easily be applied with modern computers. All one needs to do is program the computer to calculate all possible orbits and actually exhibit them on the computer screen. Before the Monte Carlo method or simulation can be used, a specific law of interaction between the alpha particle and the gold atom must be plugged into the computer, which must in turn be programmed to solve the equations of motion of the alpha particle for the given interaction. One can program the computer to apply a variety of interactions with a variety of initial conditions and geometric arrangements. The Monte Carlo simulation can reveal no new laws; it can only show the deficiencies, if they are present, of known laws.

A clear and careful reading of our description of the theory of information and its various outgrowths, such as servomechanisms, cybernetics, operations research, and game theory, reveals that they did not lead to any new mathematics. They did stimulate refinements and extensions of the known classical mathematics. These extensions, of course, demanded that the various classical mathematical concepts and theorems be reexamined and, if necessary, altered to meet the new demands. But these new demands on classical mathematics did not decrease the intense activity in classical mathematics itself, which was fueled by Klein's Erlangen Program, by Poincaré's dedication to the solution of some of the most difficult classical problems, and Hilbert's insistence that the basic axioms of all mathematics be reexamined. In Germany, the classical theory of functions of complex variables was enlarged and made more rigorous by such mathematicians as E. Landau. At the same time, G. Hardy in England and J. Ritt in the United States were pursuing the same classical mathematical goals so that classical mathematics was as alive as ever. But if any recent branch of mathematics deserves to be designated as "new" it is what is now called "topology."

In 1932, Hilbert and one of his students at Göttingen, Cohn-Vossen, published a book, *Aschauliche Geometrie*, which was later translated into English (in 1952) with the title *Geometry and Imagination*. This book, which grew out of Hilbert's penetrating analysis of the axioms of geometry, was the starting point of topology, which Hilbert considered as the avenue to a deeper understanding of geometrical and spatial concepts, which goes far beyond our intuitive perception of space and geometrical relationships. The essential feature of topology is what mathematicians call mapping; one then considers those geometrical relationships which are conserved in the mapping of one space on another. Topology, then, is the study of spatial features or configurations that remain unaltered when a geometrical figure is stretched, compressed or distorted in any way. These features are invariant to stretching, change in shape, or change in size. Thus a cylinder is equivalent to a plane from the topological point of view but a sphere is not. The difference is that a carpet rolled up as a cylinder can be unrolled onto a flat floor, but a sphere of carpet cannot. Another important feature of topology is the concept of connectivity. Thus the points on a plane define a simply connected space, but this is not true of the points on a plane on which a closed curve is drawn. The plane is then said to be doubly connected. The extension of these ideas to n-dimensional configurations has enabled topologists to give precise definitions of dimensionality.

We complete this chapter with the discussion of two mathematical developments that are closely related to each other: fractals and chaos. The fractal concept, as introduced by the mathematician B. Mandelbrot, is closely related to the group concept in that fractals are produced by the unlimited iterations of a particular geometric operation. These operations then constitute what is known as a Mandelbrot group. We recall that the group concept, as introduced in algebra by Galois and Abel, is represented by mathematical entities (e.g., matrices or permutations) that obey definite rules of multiplication, where, by multiplication, we mean successive operations by group elements. The most important of these entities for dynamics are coordinate transformations, but, in general, we may consider the iteration of a given function, which is applied to the function itself. Thus suppose that the function $f(x)$ is the square of x; applying this function successively we obtain a sequence $f(x) = x^2$, $f(f(x)) = (x^2)^2$, $f(f(f(x))) + ((x^2)^2)^2)$, or x^2, x^4, x^8, ... Clearly the numerical value that the successive terms approach, as we continue with this iteration, depends on the initial value of x. For $x = 1$, all the terms in the sequence equal 1. For values of x larger than 1, the terms approach infinity; for values of x less than 1, the terms approach 0; and for values of x less than 0 (negative x), the terms oscillate between positive and negative values. Here we have used a very simple iterative process. But the iteration can be as complex as we please. Mandelbrot's contribution to this theory of iterations was his introduction of geometric iterations. His purpose in doing this was to see whether all types of contours (e.g., shorelines, foliage, patterns in a forest, and so on) can be reproduced by fractals, the name Mandelbrot gave to the end product of his geometric iterations.

A few examples of Mandelbrot sets show how they generate fractals of any desired design or complexity. We consider an equilateral triangle and construct at the center of each side of this equilateral triangle a smaller equilateral triangle and then on the sides of these triangles we construct still smaller triangles, with sides parallel to those of the original triangle. We continue this iteration with similar but smaller triangles at the centers of the sides of each of the first generation of smaller triangles indefinitely. By choosing various scales for the lengths of the sides of each succeeding generation of triangles and the orientation of these triangles we can produce fractals of any desired design.

Professor E. Spiegal of Columbia University has given examples of fractals that correspond in appearance to the large-scale distribution of galaxies in the universe. In these examples, the dimensions of the fractals are set by an interval of unit length, two of which are then reproduced as a pair on a smaller scale. The center of each member of the pair is then attached to the ends of the original interval at a certain angle, and this entire pattern, on progressively smaller scales, is continued very many times. The final result (set of points) produced or generated by this mapping has the kind of voids and dense filamentary distribution one finds for the distribution of galaxies. But we hasten to add that the correlations or similarities one obtains in the appearance of Mandelbrot fractals and the actual physical distribution of matter, whether molecules, stars, or galaxies, is no explanation for these distributions. It merely tells us that an orderly basis or prescription for constructing a hierarchy of forms can appear to be disordered or chaotic. Does this mean that chaos is a natural consequence of order or that order is always present but obscured by the apparent disorder or chaos that accompanies it?

The study of chaos arose from observations that dynamical systems, such as binary stars or planetary systems, which are governed by precise laws, can, under certain conditions, behave in an unpredictable way; in fact, become chaotic. That predictability of the behavior of a dynamical system is not always possible does not necessarily mean that the system is chaotic or that the precise mathematics used to describe the dynamical laws that govern the system is incorrect. Predictability is not only a matter of having the correct mathematical equations to describe the dynamics of the system but also a matter of knowing about the initial conditions of the system. An elementary example illustrating this point is the problem of determining the path of an object that is thrown in some direction from some point on the earth. We imagine this event as occurring in a vacuum and we neglect the rotation of the earth. Two laws govern the motion of the body: Newton's law of motion and his law of gravity. These two laws are combined into a single set of what are called the "differential equations of motion" of the body. To obtain the orbit of the thrown body we must solve these differential equations (integrate them) which we can do without too much difficulty. But the solutions do not give us the path of the body. Indeed, the solutions give us an array of different possible paths from which we must choose the path along which the body is really moving.

To explain this failure of the solutions of the correct equations of motion to give us the orbit of any given thrown object, we note that the equations of motion tell us only how the position and motion of the body change from moment to moment so that the solution contains two quantities called the initial conditions that must be known at the moment we throw the body: the position of the body at the moment it is thrown and its velocity when it is launched. Once these numbers are substituted into the mathematical expressions for the orbits, the orbits can be precisely expressed. But we see that this substitution introduces an element of unpredictability because our knowledge of the initial position and initial velocity cannot, by the very nature of measurement, be precise. This means that our predicted orbit will depart more and more from the actual orbit as time goes on. This departure becomes so pronounced in time that the predicted and actual orbits have no relationship to each other.

Pursuing this discussion a bit further, we consider an artificial satellite launched into orbit around the earth. It will move in a closed orbit around the earth only if its speed, on launching, is smaller than a critical speed called the speed of escape. If its speed exactly equals this critical speed of escape, the satellite moves off in a parabolic orbit, never to return; if its speed is smaller than the speed of escape, the orbit is closed and if the speed of launching exceeds the speed of escape, the orbit is open. With the launching speed below the speed of escape, the orbit is an ellipse, but the exact ellipse cannot be specified because the exact position and exact launching velocity cannot be determined. All we can say is that the orbit may be any one of an aggregate of ellipses whose sizes and shapes lie within certain restricted ranges. If the speed of the satellite at launching exceeds the speed of escape, the orbit is a hyperbola. It may be any one of an array of hyperbolas within a certain limited range of shapes. Our lack of knowledge about the exact orbit implies a kind of chaos which, as noted, stems from our uncertainty about the initial conditions of the launched satellite.

In Newtonian theory, the two-body problem, in principle, is exactly soluble, but this is not true of the n-body problem, where n is larger than 2. The simplest example in this dynamical category is the three-body problem which can be solved exactly only under very restricted initial conditions. To illustrate this point we consider the simple case when the three bodies are the sun, Jupiter, and a pebble. The pebble does not affect the orbits of the sun and Jupiter in any appreciable way but its initial conditions greatly affect its own orbit, which, depending on its initial conditions, can range from an almost closed orbit around the sun or Jupiter to a very complex orbit that weaves around both the sun and Jupiter. Thus the initial conditions determine whether we have order or chaos. If we start with three equally massive bodies, such as three stars, very dramatic changes in the orbits of all three bodies can occur if the initial conditions of any one of the bodies are changed.

Spiegal has investigated the role of chaos in the structure of astronomical systems such as individual stars and collections of stars, such as galaxies. In general, the astrophysicist constructs stable stellar models. In fact, the condition for stability is introduced as an axiom so that the final stellar model constructed by the astrophysicist cannot tell us anything about the onset of instability when the basic parameters of his model change. This is a shortcoming of the model, but Spiegal points out that instabilities may set in, which then lead to chaos. This is particularly true of stars in which certain cyclical phenomena occur, such as pulsations, rotations, and magnetic cycles like sunspot activity. Spiegal has argued that we can understand such phenomena only if we apply chaos theory to them. We note, however, that chaos is more a matter of definition than of the kind of reality that we assign to a law of nature. We cannot say that chaos is a phenomenon having a reality that goes beyond the definition of the word itself. Moreover, this definition is open to many interpretations. Chaos, as a mathematical discipline, does not exist; the mathematical description of chaos does not involve or require a new kind of mathematics. The equations that describe chaotic motion are the standard differential equations of motion that describe dynamical systems in equilibrium.

If we do accept chaos as a guide to a deeper understanding of nature, we must accept the testimony of nature itself about the role of chaos in its own evolution and present structure. Accepting the Hubble law of expansion as a correct description of the present state of the universe, we must also accept the "big bang" as the initial state of the expansion during which all the initial conditions of the particles in the universe were imposed. Whatever these conditions were, they led to a highly ordered universe as we see it today. Nature's evidence is that order—not chaos—is the rule in the universe. Though the equilibrium in the universe appears to be fragile, it is actually extremely stable as measured by the lifetime of such structures as planetary systems, stars, and galaxies. In the eighteenth century, the French mathematician Lagrange demonstrated mathematically that gravitational structures such as our solar system are incredibly stable, maintaining a state of equilibrium for billions of years. Even one of the restricted solutions of the three-body problem—the equilateral triangle solution—as manifested by Jupiter, the sun, and the Jovian asteroids, is exceedingly stable. The seething chaos on the solar surface and in solar sunspots, as described by Spiegal and his collaborators in their chaos theory, is but evidence of a higher overall state of order.

Of all the ordered systems in the universe, the most remarkable, persistent, most continuous, and, therefore, most stable is life. In a remarkable essay *What is Life?*, the Austrian physicist Erwin Schrödinger showed that the existence of life depends on two basic principles, the quantum theory, which is based on the existence of a quantum of action, and the "order-from-disorder" principle. The quantum theory establishes order on the atomic and molecular scale and the "order-from-disorder" principle establishes order on the macroscopic scale. This is related to the second law of thermodynamics (entropy) which establishes a direction (irreversibility) in all processes in nature. Two physical natural features must exist for order to arise from disorder: force fields (gravity, electromagnetism) and the release of energy (increase of entropy). Forces produce order out of disorder, but only if the production of order is accompanied by the release of energy. Schrödinger then went on to argue that a hierarchy of order arises with higher states of order arising from lower states. As this is unidirectional, in accordance with the second law of thermodynamics, chaos can never arise in living systems.

To emphasize the persistence of life and the continuity of inherited characteristics, Schrödinger argued, some twenty years before the discovery of DNA by Watson and Crick, that genes must be molecules which can change only discontinuously (quantum jumps). This discontinuity ensures that the order in genetic material will persist over hundreds of generations. Schrödinger's analysis of life and the permanence of inherited characteristics compels us to conclude that chaos has no relevance to the evolution of the most highly ordered structure in the universe—the living organism.

CHAPTER 13

Whither Mathematics?

The moving power of mathematical invention is not reasoning but imagination.

—AUGUSTUS DE MORGAN

If mathematicians did nothing more today than rest on their laurels, the world would applaud and say "Well done; you have given us all we need to know to pursue our science and technology with no need for any new mathematical discoveries." But we do not want mathematicians to sit idly by, relishing their past achievements, because we now know more about the nature of future mathematical discoveries than the pre-Gaussian public knew about the non-Euclidean geometries that came later and revolutionized not only geometry and other branches of mathematics but also laid the mathematical foundations for Einstein's great work. We should not expect the current and future generations of mathematicians to do such heroic things as Gauss, Riemann, Bolyai, and Lobachevski did in geometry or as Cantor did in his theory of transfinite numbers. What, then, is the future of mathematics? It is as glorious, though not as heroic, as the mathematics of the past. We speak here of past heroic mathematics because it took as much intellectual courage to pursue non-Euclidean geometry and transfinite numbers in the nineteenth century as it took to support Copernican cosmology in the sixteenth and seventeenth centuries. Giordano Bruno was burnt at the stake in Rome and Galileo Galilei was put under house arrest for supporting Copernicus. Immanual Kant was among the leading philosophers who argued that Euclidean geometry is the only certainty in the universe; taking a contrary position required great courage.

It is unclear whether discoveries as great as non-Euclidean geometry and transfinite numbers will be made in the future. But we can be sure of three mathematical activities that will be avidly pursued:

1. solving great problems that have challenged and defied mathematicians for generations in every branch of mathematics;

2. improving mathematical techniques and rigor in current mathematics;
3. improving current mathematics as a tool for scientists and engineers in almost every phase of science and technology.

As far as new, revolutionary discoveries go, the situation in mathematics is quite different from that in physics. Unlike mathematics, physics, including astronomy, has developed linearly in the sense that each new discovery in physics was a signpost pointing to further discoveries. Of course the path was not always uniformly straight and there were many diversions in physics from the mainstream of discoveries. But the line of discoveries from Newton's laws of motion and law of gravity to our present picture of the laws of nature, of the structure of matter, and of cosmology is fairly easy to trace. Moreover, the direction along which we must proceed to enlarge our understanding of space, time, and matter is also clearly indicated in what we know at present. This does not mean that discoveries as important as the theory of relativity and quantum theory will not be made again. Our certainty that the universe is rational, and, therefore, worth investigating scientifically, does not mean that it is infinitely fathomable. As two unfathomable problems in physics, we cite the nature of mass and the nature of electric charge. As of yet, nobody has the faintest notion as to the nature of these basic entities nor why mass—elementary particles traveling at less than the speed of light—always comes with electric charge.

As far as mathematics is concerned, we cannot discern, in current mathematical research, the new mathematics of tomorrow. The contemporary mathematician, concerned with pursuing significant original research, must answer the question of whether he or she should try to revolutionize mathematics, as Planck, Einstein, Bohr, and Schrödinger revolutionized physics, or take the mathematics as it is and simply try to fill in the gaps left by earlier mathematicians. The first choice is the more challenging and rewarding one, but having made that choice, the mathematician must then confront the problem of how to revolutionize any branch of mathematics if that can be done at all. This problem is very forbidding because, as far as we can see, the basic features of all branches of mathematics have been set. No new avenues seem open to the mathematician along which he can lead mathematics to new revolutionary heights. Gauss, Cantor, Riemann, Poincaré, Klein, Hilbert, and other mathematical giants have left no revolutionary mathematical feats for the young mathematicians of today to perform. But this does not mean that mathematical research is moribund. Quite the contrary, because, like all intellectual disciplines, mathematics presents constant challenges to its practitioners. These challenges, as noted before, are of a threefold nature:

1. The current mathematicians in each branch of mathematics have inherited numerous problems that the founding fathers did not or could not solve.
2. Physicists, astronomers, chemists, biologists, economists, and even sociologists look to the mathematician for ever-increasing precision in formulating their theories or models.
3. The study of the relationship of mathematics to computer technology and to the possible development of artificial intelligence.

The young mathematician, entering his or her professional career, after his or her prescribed course of study, is naturally attracted to the great unsolved problems in mathematics. After all, what can be more exciting or satisfying than solving such a problem and having his or her name associated with those of the great mathematicians of the past? Among the most tantalizing and challenging of these problems are those in the theory of numbers. Gauss ranked these problems not only as the most challenging but also as the most beautiful in mathematics. Referring to the theory of numbers as "higher arithmetic," Gauss wrote as previously stated: "The most beautiful theorems of higher arithmetic have this peculiarity, that they are easily discovered by induction while . . . their proofs lie in exceeding obscurity and can be revealed only by very penetrating investigations. This gives higher arithmetic the magic charms which makes it the favorite goal of leading mathematicians, not to mention its inexhaustible richness in which it far excels all other parts of mathematics."

Fermat's "great theorem," probably the most famous theorem in the history of mathematics, also ranks high in this realm of unsolved problems (as noted earlier, recently a proof has been offered by Dr. Andrew Wiles of Princeton University, but as of this writing has yet to gain general acceptance in the mathematical community). Fermat had devoted a great deal of time to studying Diophantus' analysis of whole number relationships and as stated previously had written, in the margin of a page of his Latin translation of Diophantus, dealing with the "division of a square number into two other square numbers" the following words: "To divide a cube into two other cubes, a fourth power, or, in general, any power whatever into two powers of the same denomination above the second is impossible, and I have assuredly found an admirable proof of this, but the margin is too narrow to contain it." One of the most remarkable and provocative statements in the history of thought, it has challenged every mathematician for more than 350 years to reproduce Fermat's proof. But no one has been able to do so. Expressed algebraically, this theorem states that the equation $x^n + y^n = z^n$ cannot be solved for integer values of x, y, and z, if n is larger than 2. This theorem has been proved for various specific values of n larger than 2, but no general proof has ever been found for any value of n larger than 2.

The theory of numbers has more unproved theorems and challenging problems than any other branch of mathematics including:

1. finding a general formula for the distribution of prime numbers (a formula that gives the number of primes below any given integer);
2. proving that, however far out one may go in the number system, two prime numbers exist that differ by 2 (e.g., the primes 29 and 31);
3. finding a formula for prime numbers.

In addition to the ageless unproved theorems about primes that still beckon mathematicians, there are many conjectures that are still unproved such as Goldbach's conjecture that every even integer is the sum of two primes (e.g., $12 = 5 + 7; 18 = 7 + 11$) and that every even integer is the difference of two primes in an infinitude of ways. These and other conjectures about prime numbers are very attractive to mathematicians because they are easily understood and appear to be deceptively easy to prove. Theorems in number theory are more difficult to prove because they cannot be attacked analytically as can problems in all other branches of mathematics.

Although no revolutionary breakthroughs have occurred in mathematics since Klein's Erlangen Program stimulated Poincaré and Hilbert to make their brilliant contributions to early twentieth-century mathematics, mathematicians have greatly diversified the domains of applicability of each branch of mathematics. Thus, the theory of groups, which originated with Galois in his analysis of the roots of algebraic equations, has been applied to all branches of mathematics and to theoretical physics. Indeed, it has become so important that mathematicians have devoted considerable time to the study of the representations of groups. By the representation of a group, mathematicians mean a mathematical scheme or aggregate of elements that obey the same law of multiplication as do the elements of the group.

In the same vein, the theory of sets, which Cantor introduced to generalize the number concept, has been extended to all branches of mathematics to separate these branches from their arithmetic base and make them as general as possible. An excellent example of this segregation is the extension of the integral concept from its original definition developed by Reimann to what is now called the Lebesque integral. The Riemann integral grew out of the need to find the sum of an infinite set of products of a function $f(x)$ and an infinitesimal dx: $f(x)\,dx$, which is obtained by changing x in very tiny steps dx from some initial value x_1 to some final value x_2. Lebesque generalized this idea (the Riemann integral) by replacing the x values by a set so that the integral is a sum over a set.

Independently of Lebesque, the Dutch mathematician Stieltjes introduced another definition of the integral—called the Stieltjes integral—which is important in physical problems. The Stieltjes integral stemmed from the averaging concept, which, from the arithmetic point of view, is a fairly simple concept if we seek the average of a finite number of discrete elements to which numbers have been assigned. We find the average height of a group of people, for example, by summing all the heights and dividing the total by the number of people in the group. But suppose now that we seek the average of a number of numerical entities to which we assign different weights. We then obtain what we call a weighted average which is given by the integral $\int f(x)w(x)\,dx$, where $w(x)$ is the weight assigned to the function $f(x)$ for the value x. The expression of $\int f(x)dw(x)$ is called the Stieltjes integral. Other mathematicians such as Dirichlet, Fresnel, Poisson, and Darboux, have their names attached to integrals that play important roles in mathematics and physics, but these are not new definitions of the integral.

An important new kind of equation, called the integral equation, evolved rapidly in the twentieth century from the application of the integral calculus to problems in physics and mathematics. Just as we have the differential equation in which a certain function of x [e.g., $f(x)$] and its derivative $df(x)/dx$ appear, so we can also have an equation in which $f(x)$ and the integral $\int f(z)K(x, z)\,dz$, where $K(x, z)$ is called the kernel of the integral equation. Such equations appear in almost every branch of physics. In general, they are more difficult to solve than differential equations.

Because all kinds of integrals appear in problems of physics, mathematicians have developed many techniques to integrate various functions. Indeed, the integral is so important in mathematics, physics, astrophysics, astronomy, and chemistry, that many volumes have been written on integration techniques and on special types of integrals. Of particular interest are the integrals of certain functions called elliptic functions, which are the solutions of important types of equations that describe complex vibrational motions such as the detailed oscillations of an ordinary pendulum. If the pendulum swings through a small angle, its oscillations can be fairly accurately described by ordinary trigonometric functions such as sines and cosines, but if the swings are through large angles, the solutions must be expressed in terms of elliptic functions. The integrals of such functions are called elliptic integrals.

Even though the young professional mathematician today may not find his dedication to pure mathematics as rewarding as he may have hoped it would be, he will still pursue it as his primary goal because of its purity and its freedom from the kind of restrictions that science and technology impose on applied mathematics. But science, particularly physics, can stimulate pure mathematical research of the most esoteric kind, so that pure mathematics cannot be totally independent of applied mathematics. Indeed, the history of physics and of mathematics shows that hardly a branch of pure mathematics exists some aspect of which has not been used by physicists and technologists. That scientists and mathematicians had a close professional relationship in the past is not surprising. Since the universe possesses mathematical features, it can be understood in all its measurable details only with mathematics. Thus, mathematics quickly became the language of science because we can check a mathematical prediction in science by measurement. Mathematics takes science from the realm of speculation and uncertainty to the realm of reality and certainty in so far as reality and certainty can have any meaning at all.

From the time of Leibnitz and Newton, when modern science was born, to the beginning of the twentieth century, science and mathematics were so intimately related that it was difficult to distinguish between the mathematical physicist and the pure mathematician. One need only consider the great contributions to mathematics and physics by Newton, Laplace, Hamilton, Fourier, Cauchy, Gauss, and Riemann to see the close relationship between science and mathematics. By the beginning of the twentieth century the distinction between pure mathematics and theoretical physics became quite pronounced and sharp. One could then say that Einstein was a physicist and Poincaré was a mathematician. At the same time, scientists and mathematicians saw that even though pure mathematics can be pursued without science, science itself cannot be pursued without mathematics. But in spite of the intellectual dominance of the mathematician, the physicist contributes the intellectual excitement. As a result, mathematicians, particularly since the advent of relativity and quantum theory, have flocked to mathematical physics and mathematical technology. The future of mathematics lies in both pure and applied mathematics. Mathematicians have recognized this bifurcation as being inevitable and have contributed extensively to the solution of difficult problems in theoretical physics, astronomy, and chemistry. Not all mathematicians are happy about the descent of mathematics from its Olympian status as the purest of all intellectual disciplines to that of a handmaiden of physics. G. Hardy, the dean of British mathematicians during the early decades of the twentieth century, vehemently opposed the watering down of pure mathematics by the straying of mathematicians from their role as purists to that of applied mathematicians. This point of view has had little effect on the collaboration between physicists and mathematicians, which is bound to increase as the discoveries in science increase.

As a guide to our forecasting of the future of applied mathematics, we briefly review its past. Going back to the early Greeks we note that Pythagoras, though primarily a pure mathematician, was the first applied mathematician in demonstrating that musical harmony depends on the ratio of the pitches to the sounds that we hear. For such combined sounds to be pleasant, the ratio of their pitches must be simple fractions such as $\frac{2}{3}$. Sounds may be defined as atonal if the pitch ratios depart even slightly from these simple fractions. Euclid, who, more than any other Greek, epitomized Greek mathematics, produced very little applied mathematics. But since geometry is a mathematical model of space, Euclid, in a sense, was the ultimate applied mathematician—even though he did not consider himself as such. Archimedes, perhaps the greatest mathematician of the Hellenistic period, was according to Plutarch's *Marcellus*, ambivalent about applied mathematics and the investigations on which his fame largely rests: "Although these inventions had obtained for him the reputation of more than human cleverness, he decided to leave no written record of his work on such inventions; regarding as ignoble and sordid the business of mechanics and every sort of art which is directed to use and profit, he placed his whole ambition in those speculations, the beauty and subtlety of which are untainted by any admixture of the common needs of life." In spite of his lofty ideals, however, Archimedes applied his mathematical skills to the solution of practical problems whenever they were required.

Apollonius, who, after Archimedes, was the outstanding early Greek mathematician, remained a pure mathematician, but his remarkable treatise on conics certainly helped Kepler nearly two millennia later to promulgate his three laws of planetary motion, the essence of applied mathematics. Up to this point in our story, pure mathematics and applied mathematics had no individualities of their own, but with the advent of Hipparchus, the greatest of the Greek astronomers, pure mathematics and applied mathematics began to grow apart and take on lives of their own. Hipparchus used geometry and trigonometry extensively to change astronomy from a set of haphazard, amorphous unrelated observations of the stars, sun, planets, and moon, to a precise observational science. But Hipparchus was not a mathematician.

Hipparchus and his astronomy are known to us primarily through the work of Ptolemy, who publicized Hipparchus's work and used his observations of the apparent motions of the planets to construct his own Ptolemaic geocentric model of the solar system. Ptolemy was a sufficiently good mathematician to construct his epicycle-laden model but his work was primarily applied mathematics. At the beginning of the Christian era, applied mathematics had thus become a discipline distinct from pure mathematics.

Kepler took the next important step in the development of applied mathematics as a distinct mathematical discipline. Having earned a degree in theology from the University of Tübengen in Germany, he turned away from theology and accepted a chair in astronomy and mathematics at the University of Graz. Copernicus's book on the revolution of celestial bodies convinced Kepler that his principal role in life was to prove that the heliocentric theory of the solar system as proposed by Copernicus was correct. To this end, he had to use all of his mathematical knowledge and skills. He labored at this task from 1600 to 1630, using Tycho Brahe's accurate observations of the motion of Mars. Discarding the circular orbits for the planets that Copernicus had proposed, Kepler used his knowledge of conic sections to propose elliptical orbits, one of the great strokes of genius in the history of mathematics which established mathematics as the language of science. Just the statement that the orbits of the planets are ellipses is sufficient to establish Kepler's other two laws of planetary motion. Kepler was not a pure mathematician but he was a master of applied mathematics. He had no physical laws to guide him which makes his discoveries all the more remarkable. Kepler died some thirteen years before Newton was born—which was the same year that Galileo died—so he did not know the law of gravity or the laws of motion. But he certainly influenced Leibnitz and Newton.

Galileo, too, began his scientific career as a mathematician, occupying the chair of mathematics at the University of Pisa at the age of twenty-five and then, three years later, accepting the chair of mathematics at the University of Padua, but his drive to perform experiments and to observe the heavens left him little time to pursue mathematics seriously. In any event, his study of motion clearly indicated to him that mathematics is indispensable for drawing any general conclusions or deducing any principles or laws from such studies. Galileo expressed this philosophy in his essays on astronomy and physics. Because physical measurements are numerical entities, relationships among such entities must be governed by the laws of arithmetic, which means that mathematics becomes the language of measurement and, hence, of science. In his study of motion, however, Galileo never allowed the mathematics he used to dominate the science he pursued avidly. One of the charges leveled against Galileo at the papal court was that he used mathematics to challenge Aristotelian philosophy, thus demeaning basic theological doctrine. Galileo had, indeed, used mathematics to prove that the orbit of a projectile is a parabola, in agreement with observations, but that truth was used as evidence against him.

Galileo was the last of the post-Copernican scientists who had to defend mathematicians against the Aristotelians, the postscholastics, and the theologians. With the advent of Leibnitz and Newtonian mathematics, mathematicians themselves became respectable and fully accepted as important, indeed, indispensable, in the pursuit of knowledge. Before Newton, scientists and pure mathematicians were lumped together as "mathematicians" because universities had no chairs in science; chairs in astronomy (a science) were listed as chairs in astronomy and mathematics. The Newtonian era marked the beginning of the formal separation between pure mathematics and science. Thus Leibnitz was a pure mathematician and considered as such by his contemporaries. Newton, on the other hand, was primarily a scientist, but accepted by his colleagues as both a scientist and a mathematician. Newton's impact on science was much greater than his impact on mathematics because higher mathematics could have developed or evolved from Leibnitz's calculus. But Newton's laws of motion and his law of gravity were critical for the growth of physics. These laws would have been discovered after Newton but that time lag would have certainly delayed the rapid development of both physics and mathematics that occurred in the post-Newtonian era.

One need only contemplate the names of the great mathematicians who flourished then and whom we have already discussed in detail in earlier chapters to see how rapid and rich the growth of both mathematics and physics was during that period. That such a burgeoning of mathematics and science could have occurred if mathematics and physics had not greatly influenced each other is doubtful. That mathematics can reveal scientific truths that cannot be discerned directly from experimental observations was already demonstrated in Newton's mathematical derivation of Kepler's three laws of planetary motion. That pure mathematics can represent or reproduce in formulas all the features of the universe in detail was a revelation of tremendous impact. The Newtonian derivation of Kepler's third law was the most striking example, at that time, of the power of mathematics to go beyond the direct observations of a phenomenon and add to knowledge. Kepler's statement of the third law, which he called the law of harmony, is that the squares of the periods (times of revolution around the sun) of the planets are proportional to their mean distances from the sun. If P is the period of a planet and r is its average distance from the sun, then Kepler's third law is expressed algebraically as P^2/r^3 = constant; the ratio of the square of the period of a planet to the cube of its mean distance from the sun is the same number whether the planet is earth, Mars, Jupiter, or any other planet. In other words, Kepler's statement of the law implies that the ratio above does not depend on the mass of the planet but only on the mass of the sun. Newton's derivation of the third law, however, shows that the ratio is not a constant but varies from planet to planet; it is slightly smaller for the Jovian (more massive) planets than it is for the four terrestrial planets (Mercury, Venus, Earth, and Mars).

Kepler failed to find this variation of P^2/r^3 from planet to planet because Tycho Brahe's data, which Kepler used, was not accurate enough to reveal this variation. Here we have an example of the power of mathematics, combined with great universal truths (Newton's laws of motion and his law of gravity), to lead us to new laws. Newton's mathematical derivation of Kepler's three laws and, in particular, his correction of the third law, in agreement with accurate observations, was a tremendous stimulant to mathematicians to apply their mathematics to the solution of physical problems. The impact of physics on mathematics was thus immeasurable because it is doubtful that mathematics would have evolved as rapidly as it did and in the direction it did without the pressure that physics exerted on it. But these new mathematical developments, in turn, accelerated the growth of physics. This symbiosis has grown increasingly stronger since Newton's time so that they are now absolutely essential to each other.

To the mathematicians of the post-Newtonian era, the mathematical problems presented by physics were particularly attractive because they could be solved, if at all, only by the application of pure mathematics and because they carried with them a kind of revolutionary excitement and a mark of distinction and immortality that pure mathematics did not. Thus Lagrange is more noted for his work in theoretical physics (Lagrangian mechanics and celestial mechanics) than for his work in algebra. This is also true of Laplace, d'Alembert, Poisson, Hamilton, and even of Gauss, the essential mathematician, who is, perhaps, better known for his work in electricity and magnetism than for his work in the theory of numbers, which Gauss considered his greatest mathematical achievement. The importance of mathematics to physics is that the mathematics permits one to replace the real phenomena with which physicists deal by ideal phenomena that are described in terms of such imaginary concepts as points, lines, and planes. The remarkable feature of this mathematical idealism is that it describes the real phenomena with amazing accuracy. This has prompted certain physicists like Eddington and Jeans to describe the power that guides the universe as a mathematician.

Newtonian dynamics and Newtonian gravitation presented mathematics with what is perhaps the greatest of all unsolved mathematical problems which challenge mathematicians today: the n-body (n mass points or perfect spheres) gravitational problem. Stated in its simplest form, the problem is to find the orbits of n bodies with arbitrary masses, moving under their mutual gravitational attractions. Once the law of gravity and the laws of motion are specified, the problem is a purely mathematical one. If n is 2 so that two bodies (mass points or perfect spheres) are moving under their mutual gravitational actions, the problem can be solved exactly; the solution, as we have seen, is one of the conic sections. But for n larger than 2, no general solution has ever been found. This is one of the great challenges that physicists have presented to mathematicians. From a practical point of view, the physicist is not really interested in a general solution because it would probably be far too complicated for the physicist to use. With modern computers, the physicist can obtain an approximate solution to any desired accuracy for any given value of n. That real solutions exist, is, of course, indicated, if not completely proved, by the many hundreds of billions of galaxies all around us in space, each of which contains hundreds of billions of stars, interacting gravitationally and moving in their respective gravitational orbits within their galaxy in such a way that the galaxy retains its structure for billions of years. Taking our own galaxy, the Milky Way, as a specific example, we note that our solar system revolves around the core of the galaxy, whose center is 30,000 light years from the sun, once every 250 million years. Using Newton's laws we can describe this phenomenon with fairly simple mathematics and thus deduce the total mass of the core of the galaxy.

Another example of nature's solution of the gravitational n-body problem is our solar system, which consists of one star, the sun, nine planets, a few dozen satellites, thousands of asteroids, and an outer shell of a few hundred billion incipient comets. Carrying out a detailed, exact mathematical analysis of the gravitational dynamics of such a system is as yet impossible; the best one can do is introduce approximations based on the gravitational dominance of the sun, owing to its enormous mass—it contains more than ninety-nine percent of the mass in our solar system. We neglect the gravitational effects of the planets on each other to obtain Keplerian orbits around the sun and then correct these results by successive numerical approximations, to take into account the gravitational influences of the planets. This numerical procedure can be made as accurate as we please. That a gravitational structure such as our solar system has remained the same for billions of years in spite of all the gravitational interactions of the bodies within it is quite remarkable. To understand why the solar system is so stable requires a careful mathematical analysis of its dynamical properties; this analysis was performed by Lagrange in 1788 as part of his great work on analytical mechanics. This is an excellent example of how mathematics, if properly applied, reveals physical truths we would not discover in any other way.

A more recent and very important example of this utility of mathematics was the discovery of the hidden or dark matter in the universe, which constitutes more than ninety percent, and possibly as much as ninety-eight percent, of all the matter in the universe. Its very name, "dark matter," indicates that its existence must be inferred mathematically from its gravitational effects or properties. To this end we study the dynamics of a large ensemble of physical bodies such as galaxies; the famous Virgo cluster of galaxies, at a distance of 62 million light-years from us, is an excellent model. We can count 2,500 galaxies in this cluster and measure their average, random velocities. From these numbers we discover that the gravitational "glue" supplied by the visible galaxies—each containing about 200 billion stars—is not sufficient to keep these galaxies together. A mathematical analysis of the gravitational dynamics of the Virgo cluster shows that about 100 times as much invisible as visible matter must be present in the cluster to keep it from dispersing owing to the motions of its individual galaxies.

Following the great impetus that Newtonian mechanics and gravitation gave to the growth of all branches of mathematics, electromagnetic theory and thermodynamics came on the stage and provided additional stimuli to this growth. Interestingly enough, in spite of the ever increasing intimacy between mathematics and physics, including astronomy and chemistry, the distinction between mathematicians and physicists as separate groups increased. The close dependence of physics on mathematics also contributed greatly to the division of physicists into theoreticians and experimentalists. The need for the theoretical physicist to master mathematics in all its branches, including number theory, and the need for the experimental physicist to master the very demanding techniques of complex instrumentation are so time-consuming that no physicist can be both a theoretician and experimentalist. James Clerk Maxwell, the outstanding theoretical physicist of the second half of the nineteenth century, had a complete mastery of all branches of mathematics of his day. But he was not a mathematician even though his mathematical skill extended beyond the mere knowledge of mathematics; he had to know how to apply particular mathematical techniques to the selection of particular physical problems. This ability is a rare gift which distinguishes the great theoretical physicist from his more run-of-the-mill colleagues.

With Maxwell's discovery of the partial differential equations of the electromagnetic field and the wave equation, a remarkable relationship between mathematics and physics was revealed: different branches of physics can be described by the same mathematical equation. Thus, the same general partial differential equation that describes the propagation of a light wave also describes the propagation of a sound wave and the propagation of heat in a solid. Of course, the wave function that obeys, or is the solution of, the wave equation represents different physical entities in each of the three examples cited above. But, regardless of these differences, the mathematics, as represented by a single wave equation for diverse physical phenomena, establishes the remarkable mathematical unity of nature. This unity also illustrates the frugality of nature, which uses the same technology to produce a great variety of phenomena.

The mathematics of thermodynamics also revealed a property of nature which could not have been discerned completely without mathematics. This property, as expressed in the second law of thermodynamics, is that natural phenomena in the universe are irreversible. To complete this concept of irreversibility, a new, purely mathematical function—the entropy—as previously defined and discussed, had to be introduced into physics. This was done by physicists—not by mathematicians—even though the entropy itself cannot be observed or measured directly in the same way as temperature or distance. Nevertheless, the entropy, despite its purely mathematical nature, plays a profound role in nature because it establishes a direction for the occurrence of events in nature or, more generally, the direction of the evolution of the universe. Indeed, the mathematics shows us that events in the universe must progress in such a direction that the total entropy increases. This is also true of isolated chemical reactions, nuclear reactions, thermodynamical reactions, life processes, the evolution of living organisms, and even the evolution of society. Here we see that mathematics gives us a deeper understanding of nature than we can obtain from laboratories alone. Because most of what we know about the structure of matter (molecules, atoms, nuclei) and about stellar structure has not been acquired by direct observation or measurement of the temperatures at the centers of stars, mathematics is our intellectual telescope and microscope.

The beginning and the early years of the twentieth century saw the marriage of mathematics and physics develop such a strong bond that thinking of one without the other was and is practically impossible. This does not mean that mathematicians are physicists or vice versa; they both retained their identity even though mathematicians were attracted more strongly than ever before to theoretical physics owing to the emergence of the quantum theory and the theory of relativity, which forced us to alter our concepts of space, time, matter, energy, measurement, and even causality. So subtle and profound are the consequences of these theories for our understanding of space, time, and matter that mathematics became and has remained, our indispensable guide in probing nature along the paths dictated by these theories. Some of the new kinds of mathematics that had been developed by the beginning of the twentieth century were just what was needed to express these theories in their simplest, most comprehensive and comprehensible form. Thus the theory of relativity, in merging space and time into a single space-time manifold, required a four-dimensional complex geometry as developed by Minkowski. With Einstein's merging of gravity with geometry, Minkowski's four-dimensional flat geometry had to be replaced by Riemannian four-dimensional non-Euclidean, curved, space-time.

The great mathematicians Poincaré and Hilbert contributed significantly to the theory of relativity—Poincaré to the special theory (the Minkowski geometry) and Hilbert to the general theory (the Riemannian geometry)—not as physicists but as mathematicians. The general theory of relativity greatly spurred the growth of the tensor calculus, an extension of vector analysis, without which the most important physical deductions from the general theory would have been very difficult, if not impossible. Without the general theory of relativity, tensor calculus may have remained an esoteric mathematical curiosity, with few practitioners, instead of becoming, for physicists, the powerful mathematical tool that it is now.

One of the most promising and gifted German mathematicians of that period, Herman Weyl, was truly a mathematical physicist as much as a pure mathematician; he made very important contributions to the physics itself instead of to the mathematical aspects of physics. In 1918, some three years after Einstein had promulgated his general theory of relativity, Weyl indicated that the next great step in merging geometry and physics would be the development of a unified theory of electromagnetism and gravity. The idea that spurred Einstein was to extend the four-dimensional Riemannian geometry to encompass electromagnetism as well as gravity. Weyl was the first to show how this might be done by enlarging the concept of invariance. Einstein's special and general theories of relativity were based on the important principle that the laws of physics must be invariant (independent of) the type of coordinate system one uses to describe nature. The mathematical form of a law must not change if one goes from one type of coordinate system to another. That principle of invariance alone carries with it all the physical consequences of the general theory of relativity—in particular, Einstein's law of gravity. The profound impact that a very simple mathematical concept—that of coordinate invariance—had on the nature of the laws of physics prompted Weyl to introduce the concept of gauge invariance of the laws of physics in addition to that of coordinate invariance.

To explain the concept of gauge invariance we consider transporting a given length, for example, a vector, from one point in space-time to another point. We then have no way of determining whether or not the length of the vector in its new position is different from its length in its original position. Gauging (measuring) the length of the vector in its new position does not help us in answering this question because any ruler we use to measure the length must change in the same way as the length of the vector does. Weyl therefore argued that the laws of nature must be so formulated as to be independent of the gauge we use to measure distances or lengths. This concept, known as the principle of gauge invariance, has played an extremely important role in the modern theories of particles and of fields. Weyl showed that if this principle is applied to Einstein's general theory of relativity it does, indeed, lead to a new field in addition to Einstein's gravitational field. This field, as Weyl further demonstrated, has all the properties of the electromagnetic field. However, Weyl's principle gives no insight into the relationship between gravity and electromagnetism so that it has lain fallow all these years and been treated as no more than a mathematical formalism. Weyl developed his concept of gauge invariance in his classic book *Space, Time, and Matter*.

Concerning the influence of the quantum theory on mathematics and vice versa, we have already noted that Hilbert constructed the basic mathematical framework—the famous Hilbert space—in which quantum mechanics must be pursued. Quantum mechanics is the complete mathematical formalism that started with Planck's discovery of the quantum of action to explain the emission of "black body" radiation and ended with the work of de Broglie, Heisenberg, Schrödinger, and Dirac. Quantum mechanics appeared in two different mathematical forms: the matrix mechanics (its algebraic form) and the wave mechanics (its differential equation or mathematical operator form) which greatly stimulated the development of the algebra of matrices and of the mathematics of differential mathematical operators. No new kind of mathematics had to be invented to pursue quantum mechanics; the known mathematics only had to be adapted to the special demands of quantum mechanics.

Because the theory of groups is very useful in simplifying the mathematics of quantum mechanics, that theory enjoyed considerable growth with the evolution of quantum mechanics. Herman Weyl contributed to that growth with the publication of his book *Quantum Mechanics and Group Theory*. In his book, Weyl discusses all phases of quantum mechanics with special emphasis on symmetry, which plays an important role in all branches of physics, particularly the standard theory of high energy particle physics. The study of symmetry and its relationship to the laws of nature reveals most fully the impact of mathematics on physics. Symmetry can be defined in many ways, but in physics it is defined in terms of the way the laws of nature are related to time and to the geometry of space. Thus, Newton's law of gravity is called spherically symmetric because the force it describes is the same at all points on the surface of a sphere with the source of the gravitational field, a mass particle, at the center of the sphere. One can apply symmetry arguments to solve complex physical problems in a simple way without using formal mathematics. As an elementary example, in gravity, we consider the gravitational field on the surface of a sphere, or outside the sphere, within which the matter is distributed uniformly or in uniform concentric shells, like the layers of an onion. One can find the gravitational field of such a spherical distribution of matter in the standardized mathematical way by using the integral calculus. But one can dispense with this formal mathematical procedure by using symmetry arguments which show immediately that the gravitational field of this sphere of matter is exactly the same as it would be if the mass of the sphere were concentrated in a point at the sphere's center.

The most fruitful application of symmetry principles to physics, however, has been in the discovery of the relationship of the basic conservation principles in physics to the invariance of the laws of the physics to the transformation from one coordinate to another. Suppose that we translate (move) our coordinate system from one point in space to another and find that the laws of nature are unaltered—the laws are symmetrical with respect to a translation of our coordinate system. We can then deduce mathematically the principle of the conservation of momentum. If the laws are symmetrical in time, meaning that the laws will remain the same in the future as they were in the past, we deduce the principle of the conservation of energy. Finally, if the laws are invariant to a rotation, we deduce the principle of the conservation of angular momentum. Establishing these relationships between symmetries, which are purely mathematical properties, and conservation principles, which are purely physical properties, shows the simplicity and beauty of the laws of nature when they are expressed mathematically.

The theory of relativity is the expression of the most profound of all symmetries in nature—the symmetry between space and time. Einstein's insistence that the mathematical formulation of the laws of nature must incorporate space and time into a single four-dimensional space-time geometry led to all the consequences of the theory of relativity. In particular, it replaced the three-dimensional vectors of Newtonian classical physics, changing with time, as entities independent of space, by four-dimensional space-time vectors which do not differentiate mathematically between space and time. Here we see the profound difference between our perception of space and time and the four-dimensional mathematical portrayal of it. We see a real difference between space and time in our daily activities because whereas we can walk in any direction we please in space—space motion is reversible—we cannot go backward in time since time motion is irreversible.

The question of time reversal arises quite naturally in cosmology because all phenomena in the universe are governed by the quantum theory and the theory of relativity, which do not forbid time reversal, considered from the mathematical point of view above. But the second law of thermodynamics, as we have already noted, tells us that natural events are irreversible, as indicated by the increase in entropy associated with all natural events. Physicists and cosmologists have therefore suggested that the increase of entropy establishes the direction of time. Cosmologists have related this arrow of time to the expansion of the universe, which has prompted some cosmologists and physicists to suggest that if the universe were to stop expanding and begin to collapse, time itself would reverse. This conclusion is clearly incorrect; even if the universe were to collapse on itself the entropy in the universe would continue to increase because all of the natural processes in the universe would still proceed irreversibly. This does not mean that the universe could not, upon collapsing, return to its initial state just before the "big bang." It could, indeed, through a series of irreversible states, along an irreversible path. Its final state of collapse would, of course, be identical to its initial state just before the "big bang" if the radiation now in the universe were transformed completely into mass. Another "big bang" would then follow. This means that the dynamics of the universe are reversible as demanded by time symmetry.

With this discussion of symmetry, a purely mathematical concept, we end our story of mathematics. We consider this an appropriate concluding topic of our story because symmetry, more than any other mathematical concept, unifies all of mathematics and is the intellectual bridge that connects mathematics to all aspects of the universe, including life itself. The evidence for the symmetry of natural laws is present at all levels of natural structures, from atomic nuclei to galaxies. The protons and neutrons in atomic nuclei arrange themselves in symmetrical dynamic patterns; their symmetry is governed by simple relationships among a set of integers called magic numbers. The electrons, swirling around the nucleus of any atom, arrange themselves in groups which are defined by sets of four integers (positive and negative) called quantum numbers; these sets, each with four integers, determine the chemistry of atoms. The great physicist, Wolfgang Pauli, discovered a special symmetry that governs the various sets of four quantum numbers that nature assigns to the electrons in any given atom. This symmetry states that no two sets of quantum integers assigned to the electrons in any atom can be identical. This is called the Pauli "exclusion principle"; without it atoms, as we know them, could not exist and chemistry could not occur. In the carbon atom, for example, six electrons circle the nucleus, and the two electrons closest to the nucleus have the two sets of quantum numbers $(1, 0, 0, 1)$ and $(1, 0, 0, -1)$. The four electrons circling the nucleus at a greater distance than the two inner electrons have such sets (one for each electron) of quantum integers as $(2, 0, 0, 1)$, $(2, 0, 0, -1)$, $(2, 1, 0, 1)$, and $(2, 1, 1, 1)$. The choice of integers in a set for these four electrons must be the integer 2 and triple combinations of the integers -1, 0, and 1, but no two sets can be identical. This scheme is amazing in its simplicity and beauty; we wonder at the ingenuity of nature that can perform its magic with elementary arithmetic.

Proceeding from atoms to molecules we discover that symmetry again plays the dominant role in molecular structure. The electrons in the atoms that form the molecules, from the simplest to the most complex, are governed by a remarkable symmetry that arises from the complete identity of all electrons. No two electrons can be distinguished from each other. The laws of nature then dictate that we must construct models of molecules that take this electron symmetry into account. Indeed, this symmetry describes the glue that keeps the atoms in a molecule stuck together.

The next stage in this hierarchy of symmetries is the structure of crystals, which are large identical geometrical arrangements of atoms of the same kind like diamonds or of different kinds like ice. Here we see the beginning of what we call beauty; beauty is the manifestation to our minds of the highest form of symmetry. The most sublime example of this beauty is the human mind itself, which consists of many millions of subunits, all interrelated to each other by a complex symmetry that we perceive and understand very vaguely at the present time. Unless we can construct computers with the same kind of complex symmetry that govern the molecules in our brains, creating artificial intelligence may be an impossible dream.

Returning to symmetry and beauty, we note that the living organisms in nature that are most appealing to our sense of beauty are those that are most symmetrical in their structures and colors. Thus, beauty and symmetry go together so that, in a sense, we can trace beauty back to mathematics. May we go beyond this point and argue that all aesthetics and even ethics stem from symmetry and, therefore, from mathematics? This extrapolation may be too bold and extravagant because, as we have seen, mathematics is the study of the interrelationships of ideal elements: points, lines, planes, numbers, and so on, whereas beauty, ethics, and aesthetics, in general, are amorphous concepts that have as many different definitions and interpretations as there are people willing to hazard such definitions. In spite of these difficulties associated with establishing any kind of precise definition of such concepts as aesthetics, they may stem from a rigorous base. If that is so, we may hope that mathematicians will ultimately establish or develop a mathematics that can be applied to a precise explication, not only of concepts that arise from our emotions, but also the various societal relationships that govern our lives. This is a noble goal for mathematicians.

Bibliography

Ball, W. W. Rouse, *A Short Account of the History of Mathematics*. New York: Dover, 1960.

Bell, Eric Temple, *Men of Mathematics*. New York: Simon & Schuster, 1937.

Boorse, Henry A., Lloyd Motz, and Jefferson Hane Weaver, *The Atomic Scientists*. New York: John Wiley, 1989.

Boyer, Carl B., *A History of Mathematics*. Princeton: Princeton University Press, 1985.

Clark, Ronald W., *Einstein: The Life and Times*. New York: Avon, 1971.

Einstein, Albert, and Leopold Infeld, *The Evolution of Physics*. New York: Simon & Schuster, 1938.

Kasner, Edward, and James R. Newman, *Mathematics and the Imagination*. New York: Simon & Schuster, 1940.

Kline, Morris, *Mathematics and the Physical World*. New York: Dover, 1959.

Motz, Loyd, and Jefferson Hane Weaver, *Conquering Mathematics*. New York: Plenum, 1991.

Motz, Lloyd, and Jefferson Hane Weaver, *The Story of Physics*. New York: Avon, 1989.

Newman, James R., *The World of Mathematics*. New York: Simon & Schuster, 1956.

Randall, John Herman, Jr., *The Making of the Modern Mind*. New York: Columbia University Press, 1976.

Singh, Jagjit, *Great Ideas of Modern Mathematics: Their Nature and Use*.
    New York: Dover, 1959.

Smith, D. E., *History of Mathematics*. New York: Dover, 1951.

Strunk, Dirk, *A Concise History of Mathematics*. New York: Dover, 1987.

Index

Abel, Niels Henrick, 69, 187, 191–194, 208

Abelian groups, 190–191, 208

Aberration
  chromatic, 119
  of light, 89
  spherical, 119

Abscissa, 81, 107

Absolute Differential Calculus (Levi-Civita), 286

Abstract mathematics, 57–58

Acceleration, 67
  definition, 134
  of gravity, 104–105, 276
  mass/force relationship, 29, 72–73, 134–138, 273–274

Ackerman, W., 293

Acoustics, 225

Addition
  algebraic, 59–60
  arithmetic, 34, 35, 38
  associative law of, 38
  commutative law of, 38
  of congruences, 199
  of fractions, 40, 41–42, 45
  of vectors, 249

Aesthetics, 335–336

Aggregate, 295; *see also* Set theory

Agriculture, 4–5

Ahmes, 82–83

Alchemy, 130, 132–133

Alembert, Jean d', 99, 159, 162, 167–168, 175, 178, 204, 223, 327

Alexandrian school, of mathematics, 24–28

Algebra, 57–77
  abstract nature of, 57
  ancient Greeks' use of, 30
  astronomy and, 125
  Boolean, 32, 76–77, 271–272
  "father of," 58
  fundamental theorem, 69–70
  of groups, 74–75
  of matrices, 76, 251
  multiplication rule, 32
  noncommutative, 243, 274
  notation, 59–60
  of propositions, 76–77
  relationship with geometry, 113–117, 209
  rules of, 32, 59–61
  of sets, 73–74, 77
  transformation theory of, 249
  of vectors, 72–73

Algebraic equations, 60–62, 64–73, 114
  coefficients of, 59, 63, 68–71
  cubic, 68–69, 161
    roots of, 187
  definite, 71–72
  of first law of thermodynamics, 257
  hypotenuse of right triangle, 15–16
  linear (first-degree), 64, 68
  nth degree, roots of, 186–189
  quadratic, 68, 161
    for conic sections, 117
    roots of, 187
  quartic, 69
  quintic, 187

Algebraic equations (*Cont.*)
  simultaneous, 75–76
Algebraic numbers, 48
Aliquot, 11
Al-Khwarizmi, 58
Almagest (Ptolemy), 26, 27–28, 83
Ampere, 226
Anaximander, 13
Angle, 18, 49–50
  cosine of, 81–84, 102
  definition, 18, 81
  distance and, 79–82
  of incidence, 90–91
  magnitude, 80
  navigational, 5
  sine of, 81, 82, 83–85, 86, 88, 102
  tangent of, 81, 82, 85, 102
  unit, 23
Anne, Queen of England, 132
Annuity tables, 203, 204
Antimatter, 283
Antiparticles, 283
Apollonius of Perge, 19–20, 24, 323
Arabs
  algebra development by, 58
  arithmetic development by, 39
  integer use by, 30
  numeral system of, 30, 33
  trigonometry development by, 83
Arago, François, 97
Archimedes, 4, 19, 20–24, 30, 83, 110
  as applied mathematician, 323
  On the Equilibrium of Planes, 21
  On Floating Bodies, 21
  Sand Reckoner, 20, 21–22
Archimedes principle, 20
Archimedian screw, 20, 21
Area, 50
  as geometric concept, 7
Aristarchus of Samos, 18–19, 22, 25
Aristotelians, 126
  Galileo's challenge of, 325
Aristotle, 6, 126, 135
Arithmetic, 31–55
  as abstract mathematics, 57–58

Arithmetic (*Cont.*)
  development of
    by ancient Greeks, 29–30
    by Arabs, 58
    by Hindus, 29–30
  as decimal system, 34–35
  of planes, 48–53
  as "queen of mathematics," 33–34
  relationship to geometry, 45–55
  rules of, 37–38
Arithmetica (Diophantos), 58, 168
Arithmetization, of mathematics, 263–264
Ars Conjectandi (Jakob Bernoulli), 203
Artificial intelligence, 335
Aschauliche Geometrie (Hilbert and Cohn-Vossen), 310
Associative law, 38
Asteroids, orbits of, 212, 229
Astronomy
  Alexandrian school of, 24–28
  geometry and, 7
  trigonometry and, 83, 86–90
  See also Motion, planetary
Astrophysics, 2
Atomic structure, 308–309
  symmetry of, 334–335
Attalus, 19
Attempt to Deal with the Intersection of a Cone with a Plane, An (Desargues), 269–270
Averaging concept, 321
Axioms
  of Euclidean geometry, 8–11, 232, 233, 291
  fifth, 10, 232, 233
  relationship with theorems, 9
  as "self-evident truth," 8, 9–10
Ayscough, Hannah, 127
Ayscough, William, 127–128

Barrow, Isaac, 128, 130, 191
Bartels, Johann, 210–211
Basis of Theoretical Logic, The (Hilbert and Ackerman), 293
Berlin Academy of Sciences, 98, 144, 157, 159

Berlin school of mathematics, 263–264

Bernoulli, Daniel, 95, 97, 176–177, 223, 260

Bernoulli, Jakob, 95, 175–176, 202–203

Bernoulli, Johann, 95, 133, 175, 176, 182

Bernoulli, Nikolaus, 95, 97, 176, 182

Bernoulli family, 157

Bernoulli lemniscate, 176

Bernoulli principle, 176–177

Bessel, F. W., 89

"Big bang" theory, 314, 334

Binomial theorem, 129, 213

Black body radiation, 332

Black hole, 166, 286

Blue prints, 270–271

Bode's law, 162

Bohr, Niels, 274

Boltzmann, Ludwig, 200, 262

Bolyai, Janos, 182, 227, 230, 232–233

Boole, George, 32, 76–77, 271–272, 293

Boolean algebra, 32, 76–77, 271–272

Borel, Emile, 304

Boyer, Carl B., 14–15, 28

Boyle, Robert, 107, 118, 201, 255

Boyle's law, 107, 118, 201, 255

Brachystochrone, 176

Bradley, James, 89–90

Brahe, Tycho, 89, 103, 104, 105, 126, 140, 324, 326

Briggs, Henry, 54

Broglie, Louis Victor de, 332

Bruno, Giordano, 317

Brunswick, Duke of, 211, 212–213

Brunswick family, 143, 144

Budget of Paradoxes (de Morgan), 99

Buffon, Georges de, 206

Building construction, geometry and, 4

Buoyancy, 21

Calculating machine, 174

Calculus, 125–150

  differential, 22, 139–149; *see also* Differential equations

  thermodynamics and, 249

  integral, 22, 23, 222

Differential equations (*Cont.*)

  Newton's development of, 125–126, 129, 139–141

  Leibnitz's development of, 73, 130, 132, 141, 143–144, 145–146

  precursor of, 23, 75

  relationship with physics, 125–126

  tensor, 286

  of variations, 152–153, 157, 176, 292

  vector, 250

  zero use in, 34

Calendar

  Babylonian, 88

  Julian, 25

Camera, 5

Cantor, Georg, 73, 209, 239, 263, 264, 295–296, 320

Cardano, Geronimo, 68–69, 187, 194–195, 201

Cardano's formula, 69, 187

Cardinal numbers, 266–268

Carnot, Sadi, 226

Cartographic projection, 28

Catapult, 20–21

Cat's cradle proof, 173

Catenary, 176

Catherine the Great, 95, 99, 100

Catholic Church, 109–110, 144, 173

Cauchy, Augustin-Louis, 185, 193–194, 214, 215–223

  function theory development by, 154, 214, 220–222

  life of, 215–220

Cauchy integral theorem, 222

Cavendish, Henry, 139

Cayley, Arthur, 76, 209, 246, 249, 251, 284

Chaos theory, 312–315

Charles, Jacques Alexandre Cesar, 255, 258

Charles X (of France), 217, 219

Chords, 26–27, 83

Christian era, of mathematics, 27–29

Christine of Sweden, 112–113, 174

Christoffel, E. B., 284

Christoffel symbols, 284

Circle, geometry of, 20, 22–23
  circumference, 22–23, 24, 232, 233, 234
  radius, 22–23
Circular functions, 82
Circumference
  of circle, 22–23, 24, 232, 233, 234
  of earth, 25–26
Clairaut, Alexis Claude, 167
Clausius, Rudolf, 257
Clifford, William Kingdon, 209, 276
Coefficients, algebraic, 59, 63, 68–71
Colburn, Zerah, 240
Columbus, Christopher, 28
Combinations, algebra of, 202, 203
Combinatorial analysis, 144
Comet
  Halley's, 131
  orbits, 205
Commutative law, 38
Complex numbers, 51–53, 154–155
  on a plane, 52–53
  trigonometry and, 93–94
  vectors and, 252
Complex plane, 51, 52–53, 93, 220
  of four dimensions, 282
Complex variables, functions of, 94, 213,
      220–223, 236, 238, 240
Compound interest, 94, 154
Comptes Rendus, 217
Computers, 319
  Boolean logic and, 77
  software, 297
Concise History of Mathematics, A (Struik),
      231
Congruences, 199–200
  quadratic binomial, 213
Conics/conic sections, 19–20, 116–117,
      118–119, 323
Conjugate dynamical variables, 245–246
Conon, 22
Conservation laws
  of energy, 62, 177, 248; *see also*
        Thermodynamics, first law of
  of momentum, 248, 333
  See also Relativity theory

Continuity, geometric, 7–8
Convergence, 195–196, 208
  of infinite series, 220, 221–222
Coolidge, J., 203
Coordinate(s), celestial, 87
Coordinate axes, 81
Coordinate invariance: *see* Invariants
Coordinate systems, 46, 48
  abscissa of, 81, 107
  Cartesian, 104, 106–107, 119–124, 137,
        249–250
  differential, 147–148
  of Fermat, 104
  four-dimensional, 280–282
  functional relationships of, 113–116
  Gaussian, 227–229
  ordinate of, 81, 107
  polar, 120–121
  rotations, 251–252
  three-dimensional, 119
  transformation of, 120–124, 249, 251
Copernicus, Nicolaus, 19, 27, 89, 324
Cosine
  of the angle, 81, 82–84, 102
  de Moivre's theorem of, 101
  of partial differential equations, 223,
        224
Cosine law, 86
Cosmology, 2
  "big bang" theory, 314, 334
  Cartesian, 109–110
  Copernican, 317
  Einstein and, 286
  Euclidean geometry and, 9
  time reversal and, 334
Counting, 267–268
Creativity, mathematical, 291
Crelle, August, 193, 194
Crystals, 335
Cube roots, 54
Cubes, of distances, 50
Cubic equations, 68–69, 194
  roots of, 187
Cybernetics, 300–302
Cycloids, 174

Darboux, Gaston, 289, 290
Darboux integral, 321
Dark matter, 328–329
Decimal number system, 33–34
Decimal point, 42–43
Decimal representation, of fractions, 42–45
Dedekind, Richard, 209, 239, 263, 265
Deductive reasoning, 15
Definition, geometric, 8
Degree, 18, 80
Del Ferro, S., 187
"De Mensura Sortis" (de Moivre), 203
De Moivre, Abraham, 100–102, 167, 203
  Doctrine of Chances, 203
de Moivre's equation, 252
de Moivre's formula, 260
de Moivre's theorem, 203
de Morgan, Augustus, 271–272
  Budget of Paradoxes, 99
Denominator, 36, 37, 40
  common, 41–42
Desargues, Gerard, 175, 269
  Attempt to Deal with the Intersection of a Cone with a Place, An, 269–270
Desargues theorem, 269–270
Descartes, René, 48, 103, 169–170, 173
  analytic geometry development by, 75, 77, 169–170
  coordinate system of, 104, 106–107, 119–124, 137, 249–250
  death of, 174, 202
  life of, 107–113
  vortices theory of, 170
Descriptio (Napier), 53–54
De Stella Nova (Kepler), 201
Dickson, L. E., 198
Diderot, Denis, 99, 167–168
Differential equations, 141, 148, 149–150, 167
  of motion, 312–313
  partial, 223, 329
    Hamiltonian, 245–246
    Poisson, 225, 254, 286
Diffraction, optical, 177
Dimensionality, 49, 50–51

Diophantus, 30, 58, 319
Diophantine analysis, 30
Dirac, Paul Adrian Maurice, 225, 274–275, 283, 332
Direction, as geometric concept, 7
Dirichlet, Peter Lejeune, 209, 239
Disquisitiones Arithmeticae (Gauss), 212, 265
Distance
  angles and, 79–82
  cubes of, 50
  as geometric concept, 7, 9
  measurement, 280–281
  stellar, 88–90
  between two points, 8–9
Divergence, 195
  of infinite series, 221
Divine Comedy (Dante), 201
Division
  arithmetic, 34, 37, 38
  factors in, 38–39
  of fractions, 40–41, 45
  of polynomials, 66
Division (/) sign, 38
Doctrine of Chances (De Moivre), 203
Dürer, Albrecht, 270
Dynamics, 104

e (transcendental number), 55, 94, 101, 154, 252
Earth
  circumference, 25–26
  orbit, 88–89
  rotation, 86–87
Eclipse, solar, 13
Ecliptic, celestial, 87
Economics, operations research and, 303
Eddington, Arthur S., 2, 327
Egyptians, ancient, 4–5
  fractions development by, 35–36
  geometry development by, 5
  numeral system of, 33
  trigonometry development by, 82–83
Einstein, Albert, 288
  $E = mc^2$ equation of, 61

Einstein, Albert (*Cont.*)
 gravitational field equations of, 286, 292
 light corpuscularity theory of, 244
 Minkowski and, 281
 probability theory and, 200
 thought experiments of, 117
 See also Relativity theory
Einstein–Lorentz transformations, 279,
  280, 284
Einstein–Ricci tensor, 286, 287
Eisenstein, Ferdinand Gottfried Max, 235
Electric charge, 46–47, 318
Electricity, 225
Electromagnetics, 329
 equations, 250, 273–274, 275, 287
 fields, 225, 252–254
 quantum electrodynamics of, 275
 unified field theory of, 287, 331
Electron, Bohr's model, 274
Electrostatics, 229–230
Elements (Euclid), 5, 6–7, 15, 21
Elizabeth of Bohemia, 110, 112
Elliptic functions, 230
Energy
 conservation of, 62, 177, 248; *see also*
  Thermodynamics, first law of
 momentum relationship of, 282
Entropy, 249, 254, 257, 258–259, 330
 irreversibiilty and, 334
 knowledge and, 299–300
Epicycle model, of planetary motion, 19,
  27, 28
Equality (=) sign, 35
Equator, celestial, 87
Equinoxes, 26, 87–88
Equivalence, principle of, 275–276
Erastosthenes, 22, 24, 25–26
Erlangen Program, 208, 268, 269, 271,
  310, 320
Errors
 exponential law of, 200, 206, 261–262
 theory of, 260–262
Essay on Conics (Pascal), 173
Euclid of Alexandria, 4, 5–13, 23, 24, 29, 37
 as applied mathematician, 323

Euclid of Alexandria (*Cont.*)
 Elements, 5, 6–7, 15, 21
 fifth axiom, 232, 233
 geometry development by, 5–10, 12–13
 number theory, 10–12
Euclid of Megara, 6
Eudemus, 19
Eudoxus of Cnidus, 5, 13, 17–18, 23–24
Euler, Leonhard, 93–101, 151, 157, 159,
  178, 181, 223
 complex number theory of, 154–155
 errors theory of, 260
 life of, 95–100
 probability theory of, 204
Euler–Lagrange equation, 152–153
Euler's equation, 93–94, 154, 178
Euler's number: *see e*
Excluded middle, theorem of, 296–297
Exclusion principle, 335
Exhaustions method, 23–24
Expansion, law of, 314
Experimental results, 204
Exponential law of errors, 206
Exponents
 algebraic, 60

Factors
 in division, 38–39
 of perfect numbers, 11
 of polynomials, 66–67
 of prime numbers, 12
Faraday, Michael, 252–253
Feedback mechanism, 301, 302
Fermat, Pierre de, 31, 75, 91, 103
 death of, 202
 life of, 168–171
 probability theory, 201–202
 unproved equation of, 159
Fermat numbers, 31
Fermat's last theorem, 168, 319–320
Fermat's principle of least time, 146–147,
  243, 244
Ferrari, Kudovico, 68–69
Field equations
 of the gravitational field, 286, 292

Field equations (*Cont.*)
  relativistic, 225
Field theory, 248, 275
Fifth degree, general equation of, 192,
    194–195
Force, mass/acceleration relationship, 29,
    134–138, 273–274
Force fields, 225, 315
Foundations of Arithmetic (Frege), 293
"Foundations of a General Theory of
    Domains" (Cantor), 266
Foundations of Geometry, The (Hilbert),
    293
Fourier, Jean Baptiste Joseph, 28, 83, 215,
    223–224
Fourier integral, 224
Fourier series, 223–224
Fractals, 311–312
Fractions, 29
  addition of, 40, 41–42, 45
  decimal representation of, 42–45
  denominators, 36, 37, 40
    common, 41–42
  development of, 35–36
  distribution on a line, 47–49
  division of, 40–41, 45
  logarithms of, 55
  multiplication of, 40, 42, 45
  numerators, 36, 40
  subtraction of, 40, 45
  symbols for, 35–36
Frederick the Great, 98–99, 157, 159
Frege, Gottlob, 271–272, 293
French Academy of Sciences, 95, 157, 158,
    160, 163, 164
Fresnel, Augustin Jean, 226
Frobenius, Georg, 209, 263
Functional relationships, of coordinate
    systems, 113–116
Functions, 63–64, 67
  of complex variables, 51, 52, 94, 213,
    220–223, 236, 238, 240, 310
  derivatives, 145
  elliptic, integrals of, 321–322
  general theory of, 154

Functions (*Cont.*)
  graphs of, 113–116
  maxima, 115
  minima, 115
  of real variables, 114
  theory of, 208
  trigonometric, 72, 83–85

G, 139
Galaxies
  core mass, 328
  gravitational dynamics, 327–329
Galileo Galilei, 103, 104–105, 133, 168,
    317, 324–325
  probability theory of, 201
  telescope development by, 119
  trial of, 109, 325
Galois, Evariste, 208
  algebraic equation development by, 60,
    64, 65, 69, 269
  group theory of, 75, 161, 194, 269, 320
  life of, 182–186, 189–191
Game theory, 2–3, 167, 171–172, 203–204,
    205, 304–310
  Monte Carlo simulation, 307–310
  nonzero-sum versus zero-sum, 304–305
  use in World War II, 305
Gas(es)
  Boyle's law, 107, 108
  equation of state of, 255
  general law of, 255
  kinetic theory of, 201, 225, 249, 254–
    255
  probability theory of, 200
  state functions, 258
  thermodynamics of, 254–260
Gas constant, 201
Gas laws, 61, 255
Gas pressure, 67, 107, 118
Gas volume, 107, 118
Gauge invariance, 287, 331–332
Gauss, Karl Friedrich, 33–34, 52, 114,
    121, 154, 181, 182, 192–193, 226–
    233, 236, 271, 327
  algebraic equations and, 69–70

Gauss, Karl Friedrich (*Cont.*)
  complex plane theory of, 220
  Disquisitiones Arithmeticae, 212, 265
  exponential law of errors of, 200, 206,
      261–262
  geometry and, 226–229, 270, 284
  life of, 209–215
  on number theory, 319
  probability theory and, 300
  theory of numbers and, 197–200
Gauss (G), 229
Gauss's theorem, 229–230
Gay-Lussac, Joseph Louis, 255, 258
Geiger, Hans, 308–309
Generalization, 41
Genes, 315
Geodesy, 214, 229
Geography (Ptolemy), 28
Geometric definition, 8
Geometric series, 195
Geometry, 4–30
  algebra's relationship with, 58, 77, 79,
      209
  analytic, 75, 77, 103–124, 270
    complex, 220
    of conic sections, 116–117, 118–119
    Descartes and, 108–109, 110, 119–120,
        170–171
    Fermat and, 170–171
    Galileo and, 103, 104–105
    Kepler and, 103–104, 105, 117
    Newton and, 105–106
  ancient Greeks' development of, 31–32
  arithmetic's relationship with, 45–55
  of complex numbers, 52–53
  differential, 147–148, 228, 270, 284
  Euclidean, 5–10, 12–13
    axioms, 8–11, 32, 344, 234, 291
    comparison *n*-dimensional geometry, 234
    vector analysis and, 248
  four-dimensional, 280–282, 284, 330
  games and, 2–3
  non-Euclidean, 230–235, 317
    elliptical, 226–227
    hyperbolic, 226–227

Geometry (*Cont.*)
  non-Euclidean (*Cont.*)
    n-dimensional, 50–51, 227, 229, 230,
        233–235, 234, 239, 275–279, 280–
        282, 284, 330
    relativity theory and, 50–51, 275–279
  plane: *see* Geometry, Euclidean
  projective, 269–271, 271
  Pythagoras's influence on, 15–16
  vector, of functions, 275
Geometry and Imagination Geometrie
      (Hilbert and Cohn-Vossen), 310
Gibbs, Josiah Willard, 248–249, 262
God
  algebraic proof of existence, 99
  as hypothesis, 165–166
  as mathematician, 2; 264–268
  perfect number preference of, 11
Gödel, Kurt, 296
Golden Age, of mathematics, 208
Graphs, 49, 105
  of coordinate systems, 113–116
  pressure-volume, 107, 118
Grassmann, Hermann, 248–249
Gravitational field, 151–152
  equations of, 286, 292
  symmetry of, 332–333
  unified field theory of, 287, 331
Gravity/gravitation
  acceleration of, 104–105, 276
  measuring device, 214
  Newton's law of, 53, 138–139, 325
  potential theory of, 167
  universal law of, 129, 130, 131
  See also Relativity theory
Greeks, ancient
  arithmetic development by, 29–30
  geometry development by, 5–30
Green, George, 209, 246
Grimm, Peter, 100
Grossman, Hermann, 239
Group theory, 122–123, 187, 188–191, 194,
      269, 320
  Abelian groups, 190–191, 208
  algebra of, 74–75

Group theory (*Cont.*)
 finite groups, 194
 fractals and, 311–312
 infinite groups, 194
 of permutations, 188–191
 of quantum mechanics, 332–333
 relativity theory and, 278–279
 of transformation, 269–271, 278–279
Gsell, Catherine, 97, 100
Gustav, Carl, 209

Halley, Edmund, 131, 156
Hamilton, William Rowan, 76, 114, 209,
  214, 239–246, 276, 298, 327
Hamiltonian, 245
Hamilton–Jacobi equation, 239, 246
Hardy, G. H., 198–199, 310, 323
Harmonic motion, simple, 85
Harmonic series, 195
Harmony, 323
 Pythagorean concept of, 13
Harmony law, 326
Heaviside, Oliver, 167
Heisenberg, Werner, 245–246, 274, 332
Hermite, Charles, 193, 222
Herodotus, 13
Hero of Alexandria, 83
Hertz, Heinrich, 253–254
Hiero, King, 21
Hilbert, David, 272, 275, 291–292, 295,
  310, 320
 Aschauliche Geometrie, 310
 The Basis of Theoretical Logic, 293
 The Foundations of Geometry, 291
Hilbert space, 275, 291, 292
Hindus
 fractional symbols of, 36
 numeral system of, 33
 trigonometry use by, 83
Hipparchus of Nicaea, 19, 24, 26, 27, 28,
  83, 323–324
Hippasus, 14
History of Mathematics (Smith), 17
History of the Theory of Numbers
  (Dickson), 198

Hitler, Adolf, 271
Holmboe, Bernt, 191
Hooke, Robert, 130, 133
Hospital, Marquis de l', 175
Hubble, Edwin, 286
Hubble law of expansion, 314
Humboldt, Alexander, 213
Huygens, Christian, 127, 130, 142, 175
 death of, 202
 probability theory of, 202
 wave theory of, 177, 178, 226
Hydraulics, 20, 30
Hydrodynamics, 176–177
Hydrostatic principle, 21
Hypotenuse, of right triangle, 15–16, 30,
  58

Imaginary axis, 53
Imaginary numbers, 52–53
Inertia, 126–127, 134, 135
Infinite power series, 84–85, 148
Infinite series, 148, 153–155, 192, 195,
  208, 265
 convergence of, 220, 221–222
 divergence of, 221
Infinitesimal time interval, 140–141
Information retrieval systems, 299–300
Information theory, 297–299
 cybernetics and, 300–302
 thermodynamics and, 299–300
Integers
 Arabs' development of, 30
 numerological use of, 16–17
 as sum of two primes, 320
Integral equations, 166, 321
Integrals, 320–322
Integrating factor, 259
Interest rates, 44
*Introduction to the Analysis of Infinites,
  The* (Euler), 153
Invariants, 122, 123, 124, 269, 284–285, 331
Inversion rule, 41
Irrational numbers, 22, 43, 48, 55
 cosine as, 85
 sine as, 85

Irrational numbers (*Cont.*)
  tangent as, 85
Irreversibility concept, 330
Isochrone, 176
Isothermal, 118
Iterations, geometric, 311–312

Jacobi, Karl Gustav Jakob, 193, 194, 209,
    235, 239
James II, 132
Jansen, Cornelius, 173
Jeans, James Hopwood, 2, 327
Jesuits, 108, 173, 174
Jupiter, asteroids of, 155–156

Kant, Immanuel, 161–162, 163, 317
Kant–Laplace hypothesis, 161–162
Kepler, Johann, 54, 133, 168
  death of, 324
  law of harmony of, 326
  laws of planetery motion of, 53, 55, 92,
      103–104, 117, 125, 126, 142, 323,
      324, 326
  probability theory of, 201
  theoretical physics and, 125–126
Kinetic energy, 62
  negative, 283
  of the particle, 151
Kinetic theory, of gases, 201, 225, 249,
    254–255
Klein, Felix, 208, 268–269, 271, 276, 291,
    295
Knowledge, entropy and, 299–300
Kronecker, Leopold, 209, 239, 263–266
Kummer, Ernst, 209, 263

Lagrange, Joseph Louis, 99, 151, 165, 175,
    178, 181, 216, 223, 315, 327
  algebraic equation development by, 60,
      64, 65, 69, 183, 187–188
  errors theory and, 206, 260–261
  life of, 156–161
  Mécanique Analytique, 157, 161
  probability theory of, 204
  three-body problem solution by, 155–156

Lagrange interpolation formula, 65
Lagrangian of the particle, 152
Lagrangian points, 156
Lambert, Johann Heinrich, 231
La Methode (Descartes), 110
Landau, E., 310
Language, mathematics and, 297
Laplace, Marquis Pierre Simon de, 144,
      175, 178, 181, 207–208, 213, 215–
      216, 327
  errors theory and, 261
  life of, 162–166
  nebular theory of, 161–162
  probability theory and, 165, 166–167,
      204–205
  Theorie Analytique des Probabilities,
      165, 166–167, 205
  Traité de Mécanique Céleste, 163, 165–
      167, 207
Laplace transform, 166, 167
Latitude, 26, 88
Law of universal gravitation, 129, 130, 131
Laws of nature, unification of, 244; *see
      also* Unified field theory
Least action, principle of, 147, 152, 176,
      244–245, 292
Least squares, theory of, 167
Least time, principle of, 243, 244
Legendre, Adrien Marie, 167, 193, 222,
      231
  Theory of Numbers, 235
Leibnitz, Gottfried Wilhelm, 65, 117, 127,
      130, 133, 151, 178, 325
  differential calculus development by, 73,
      130, 132, 141, 143–144, 145–146
  on game theory, 203–204
  life of, 141–145
  notation system of, 182, 209
Le Monde (Descartes), 109–110
Lemonnier, Pierre Charles, 160
Lenses
  aberrations of, 298
  light passage through, 90–91, 243, 297,
      298
  See also Optics

Lever, 20
Levi-Civita, Tullio, 246
   Absolute Differential Calculus, 286
Lie, Sophus, 269, 276
Life, 315
Life insurance tables, 203, 204
Light
   aberration of, 89
   corpuscularity of, 129–130, 177, 244
   electromagnetic theory of, 249, 250,
      252–253
   Huygens's theory of, 130
   Newton's theory of, 129–130, 132,
      177
   passage through lenses, 90–91, 243,
      297, 298
   path of, 146–147
   polarization of, 226
   principle of least time of, 91, 146–147,
      176, 243, 244, 298
   Snell's law of, 90–91
   speed of, 279
   wave theory of, 177, 225, 226
   See also Optics
Limit, 222
Line
   definition, 45–46
   as geometric concept, 7
Linear programming, 303–304
Lippershey, Hans, 119
Lipschitz, R., 284
Lobachevski, Nikolai, 10, 182, 227, 230,
   232, 233, 234
Logarithmic spiral, 176
Logarithms, 39, 53–55
   natural, 55
Logic
   axioms and, 9
   symbolic, 76–77
Longitude, 88
Lord Kelvin: *see* Thomson, William
Louis Philippe, 186, 219
Louis XIV, 142–143
Louis XVI, 160
Louis XVIII, 165

Magnetic field, 229
Magnetism, 225, 252–253
Magnetometer, bifilar, 214
Mandelbrot, B., 311
Mandelbrot groups/sets, 311–312
Manifold, 49, 50–51
Maps, 271
   conformal, 214
Marcellus, 21
Marcellus (Plutarch), 323
Marie Antoinette, 160
Mars, orbit, 103–104, 105
Mass
   force/acceleration relationship, 29, 134–
     138, 273–274
   inertia of, 126–127
   nature of, 318
*Mathematical Foundations of Quantum
     Mechanics, The* (von Neumann),
     275, 293
Mathematical operator, 32
Mathematical Theory of Probability, The
     (Coolidge), 203
Mathematicians
   greatest, 22
   See also names of individual
     mathematicians
Mathematics
   nonsequential development of, 31
   pure versus applied, 322–323
Matrix(es), 32
   algebra of, 76, 251
Matrix mechanics, 274
Matter, molecular theory of, 262
Maupertuis, Pierre-Louis, 147, 152, 175,
     178, 243, 244–245
Maxima, 115, 146
Maxwell, James Clerk
   distribution theory of, 262
   electromagnetic theory of, 249, 250,
     252–253, 273–274, 275, 287, 329
   gas studies of, 200, 201, 262
Mayer, Julius, 255
Mean, error in, 261
Measurement, 46, 325

Mécanique Analytique (Lagrange), 157, 161
Mechanics, 20, 22, 76, 104
  Lagrangian, 327
  matrix, 274
  statistical, 147, 172, 200, 201, 206, 225,
    249, 260
Meditationes (Descartes), 110
Mémoir sur le calcul integral (d'Alembert), 167
Mensura Sortis, De (Descartes), 203
Mersenne, Marin, 108, 109, 173
Mesopotamians, 33
Méthode, La (Descartes), 110
Michelson, Albert, 279
Microscope, 5
Milo, 14
Mind
  mathematical, 1–2
  symmetry of, 335
Minima, 115, 146
Minimum theorem, 304
Minkowski, Hermann, 279, 280–282, 284,
    330
Minkowski space of special relativity, 282
Minus (−) sign, 35
  algebraic, 60
Miscellanea Taurinensia, 157
Möbius, August, 209, 239
Möbius strip, 239
Modeling, mathematical, 271
Moivre, A. de: *see* de Moivre, Abraham
Momentum
  conservation of, 248, 333
  relationship to energy, 282
Monde, Le (Descartes), 109–110
Monge, Gaspard, 217, 270
Morgenstern, Oskar, 304
Morley, Edward, 279
Morse, Samuel, 230
Motion
  force and, 126
  Fourier's theory of, 28
  Galileo's experiments with, 104–105,
    126–127
  Maupertuis's principle of least action,
    147, 152, 176, 244–245

Motion (*Cont.*)
  planetary
    epicyle models, 19, 27, 28
    Eudoxus's theory of, 17, 18
    Kepler's theories of, 53, 55, 92, 103–
      104, 125, 126, 142, 323, 326
    Lagrange's theories of, 155–156
    Newton's theories of, 131; *see also*
      Motion laws, Newtonian
    Ptolemy's theory of, 28
    trigonometric measurement of, 86–88
    See also orbit
  simple harmonic, 85
Motion laws, Newtonian, 29, 72–73, 127,
    117–118, 134–138, 325
  algebraic expression of, 61–62
  differential equations of, 312–313
  Hamilton's reformulation of, 245–246
  Lagrangian of the particle and, 152
  motion invariants, 278
  Poisson brackets, 225
  second law, 126, 134–138, 141
Multiplication
  arithmetic, 34, 37, 38
  of congruences, 199
  of fractions, 40, 42, 45
  of polynomials, 66
  of vectors, 73, 249–250

Napier, John, 53–55
Napoleon Bonaparte, 160, 161, 164–166,
    213, 216, 217
Nasir-al-din, 231
Natural law, geometry and, 7
Natural logarithms, 55
Navigation, celestial, 5
n-body gravitational problem, 155–156,
    290, 313–314, 315, 327–328
Nebular hypothesis, 161, 162, 163
Negative numbers, 34, 46–47
  square roots of, 52
Newton, Isaac, 22, 71, 89, 103, 105–106,
    114, 162–163, 177–178, 191, 213
  binomial theorem of, 106
  death and burial of, 145, 151

Newton, Isaac (*Cont.*)
　derivation of Kepler's laws by, 326
　differential calculus development by,
　　125–126, 129, 139–141
　life of, 127–134
　Optics, 132
　theoretical physics and, 125–135, 136
　two-body problem of, 155
　See also Motions laws, Newtonian
Newtonian mathematics, importance of,
　325
Nicomachus, 11
Noncommutative algebra, 243, 274
Noncommutative law, 38
North celestial pole, 5, 87
Nuclear bomb, 307
Numbers
　as abstract symbols, 57–58
　cardinal, 266–268
　large, 21–22
　ordinal, 45, 46, 267–268
Numbers, theory of, 17, 29, 168, 197–200
　Archimedes's, 21–22
　Diophantine analysis of, 30
　Euclidean, 10–12
　Fermat's contribution to, 31
　Gauss on, 319
　Pythagorean, 13
　theorems of, 171
　unproved theorems of, 319–320
Numeral systems
　Arabic, 30
　decimal, 33–34
　Greek, 29, 30
　Hindu, 29–30
　Roman, 29
Numerator, 36
Numerology, 14, 16–17

Oersted, Hans Christian, 252, 253
Ohm's law, 61
One, cube of, 71
On Floating Bodies (Archimedes), 21
On the Equilibrium of Planes
　　(Archimedes), 21

Operations research, 302–304
Optical devices, 5
Optical diffraction, 177
Optical interference, 177
Optical retrieval system, 298–299
Optics, 30, 76, 225, 239, 240–241
　geometrical, 90–91, 243–246
　physical, 243
Optics (Newton), 132
Orbit, 92–93
　algebraic expression of, 104
　of asteroids, 212, 229
　of comets, 205
　differential equations of motion and,
　　312–313
　elliptical, 103–104, 105, 117, 125, 140,
　　313
　hyperbolic, 313
　of Mars, 103–104, 105
　n-body problem of, 155–156, 290, 313–
　　314, 315, 327–328
　Newton's theory of, 139–140
　parabolic, 313, 325
　perturbations of, 229
　trigonometric formula for, 92
Order, of the universe, 314–315
Order-from-disorder principle, 315
Ordinal numbers, 45, 46, 267–268
Ordinate, 81, 107

Paciola, Luca, 58–59
Parabola, 119
Parabolic orbit, 313, 325
Parallax, of stars, 89
Parallel axiom, 227, 230–231, 232, 233–
　234
Paris Academy of Sciences, 95, 157, 158,
　160, 163, 164
Partial derivative, 150
Pascal, Blaise, 171–175, 201–202
Pascal, Jacqueline, 172, 173, 174–175
Pauli, Wolfgang, 334–335
Pauli exclusion principle, 335
Pensées (Pascal), 173–174
Percentages, 44–45

Perfect numbers, 11–12
Periodic phenomenon, 83
Permutations, 188
    Bernoulli's formula for, 202–203
    group theory of, 188–191
Perspective, 269–270
Perturbations, 229
Phase space, 262–263
Pherecydes, 13
Physical laws, mathematical interpretation
        of, 179
Physicist-mathematicians, 114, 177–178
Physics
    Aristotelian, 104, 126
    comparison with mathematics, 226
    probability use in, 205–206
    relationship to mathematics, 5, 51–52,
        104
    theoretical, 2, 114, 117, 151
    theory/experimental dichotomy, 106
    trigonometry use in, 83
pi (π), 4, 20, 22–23, 24, 31, 206, 308
Piazzi, G., 155
*Plaine Discouery of the Whole Reuelation of
        Saint John* (Napier), 53
Planck, Max, 244, 332
Plane
    arithmetic of, 48–53
    complex, 93, 220
        of four-dimensions, 282
    complex numbers on, 52–53
    definition of, 50
    motion of particle on, 72–73
Plane Loci (Apollonius), 169
Plato, 6, 17, 22, 264
    "second," 25
Pliny the Elder, 22, 27
Plus (+) sign, 35
Plutarch, 21
Poincaré, Henri, 272, 295, 310, 320
    life of, 287–291
    relativity theory and, 279–280, 331
Poincaré group, 280
Point, 7–9
    coordinates of, 81

Point (*Cont.*)
    as geometric concept, 7
    on a line, 46–49
    on a plane, 48–49
    definition, 45–46
Poisson, Simeon-Denis, 222
Poisson brackets, 225
Poisson partial differential equation, 225,
        254, 286
Polar coordinates, 176
Polygon, regular, 24, 31
Polynomials, 63, 65–72, 75, 77
    algebraic, 83
    finite, 84
    graph of, 116
    infinite, 84
Poncelet, Jean Victor, 270
Positive numbers, 34, 46
Positron, 283
Potential energy of the particle in the field,
        152
Potential theory, 167
Powers, of numbers, 39
    algebraic, 60
    of congruences, 199
Power series, 84–85
    convergence, 196–197
Prime numbers, 12, 29
    sieve of Erastosthenes of, 25
    unproved theorems of, 320
Principia Mathematica (Newton), 131
Principia Mathematica (Russell and
        Whitehead), 272, 293, 296
Principia Philolsophiae (Descartes), 110
Principle of equivalence, 275–276
Principle of least action, 147, 152, 176,
        244–245, 292
Principle of least time, 91, 146–147, 176,
        243, 244, 298
Principle of virtual work, 167
Probability, mathematical definition of,
        205–206
Probability theory, 165, 167, 171–172, 176,
        197, 200–206, 225
    applications to physics, 260–263

Probability theory (*Cont.*)
  binomial coefficients of, 176
Problem of Points, 202
Product, of numbers, 190
Proof
  "cat's-cradle," 173
  as Pythagorean concept, 15
Propositions, algebra of, 76–77
Provincial Letters (Pascal), 174
Ptolemy, 19, 24, 231
  Almagest, 26, 27–28, 83
  Geography, 28
Pulsar, 299
Puzzles, 2, 3
Pythagoras, 5, 6, 13–17, 323
Pythagorean formula, 152–153
Pythagoreans, 13–14, 16–17
Pythagorean theorem, 15–16, 58, 82, 86, 147
  of non-Euclidean geometries, 271, 285

Quadratic equations, 68, 161
  for conic sections, 117
  roots of, 187
Quantum electrodynamics, of electromagnetism, 275
Quantum mechanics, 51, 206, 239, 274–275, 292–293
  group theory and, 332–333
  Hilbert space and, 292
  matrices, 251, 332
  state of the system concept of, 292
  wave mechanics and, 217, 220, 274, 332
Quantum Mechanics and Group Theory (Weyl), 332
Quantum numbers, 334–335
Quantum theory, 2, 247–248, 315
  Hilbert space and, 332
Quartic equations, 69
Quasars, 299
Quaternion, 243, 276
Quintic equations, 187–188

Rabdologia (Napier), 54
Radian, 18, 23, 81

Radiation, 253
  black body, 332
Radius, 22–23
Rates, 44
Raum, Zeit, und Materie (Weyl), 286–287, 332
Ray, characteristic function of, 243
Ray tracing, 298
Reasoning
  deductive, 15
  mathematical, 1–2, 31
Reductio ad absurdum, 12
Refraction, 90
Relativistic field equations, 225
Relativity theory, 2, 50, 51, 134, 153, 214, 247–248, 271, 275–287
  calculus of variations and, 292
  general theory, 275–279, 284–287, 292, 331
  mathematics of, 275–284
  non-Euclidean geometry and, 50–51, 275–279, 280–282, 284, 330
  principle of equivalence and, 275–276
  special theory, 278–284, 285, 331
  tensor analysis and, 276–277
  transformation theory and, 278–279, 278–282
Religion
  mathematics and, 11
  See also Catholic Church; Jesuits
Residues, theory of, 222
Ricci-Curbastro, Gregorio, 246, 286
Richard, Louis-Paul-Emile, 184–185
Riemann, George Friedrich Bernhard, 10, 209, 214, 233–239, 248
  life of, 235–238
  n-dimensional geometry of, 227, 229, 230, 233–235, 276, 330
Ritt, J., 310
Roman numerals, 29
Romans, ancient, 27
Roots, 39–40
  of algebraic equations, 70–72
  of linear functions, 115
  logarithms and, 54, 55
  of one, 71

Roots (*Cont.*)
   of polynomials, 70–72
   of quadratic equations, 68–69
Rotation
   of earth, 86–87
   trigonometric description of, 92
Ruffini, Paolo, 69, 187
Russell, Bertrand, 272, 293, 296
Rutherford, Ernest, 308–309

St. Augustine, 11
St. Petersburg Academy, 95, 97, 98
Salmon, George, 209
Sand-Reckoner (Archimedes), 20, 21–22
Schrödinger, Erwin, 76, 274, 315, 332
Schwarzschild, Karl, 286
Science
   Greek versus modern, 126
   relationship with mathematics, 322–323
Scientific theory, 28
Seneca, 27
Servomechanisms, 301–302
Set, definition, 295–296
Set theory, 31, 265–268, 295–296, 320–321
   algebra and, 73–74, 77
Shannon, Claude, 299–300
Short History of Mathematics, A (Ball), 15
Sieve of Erastosthenes, 25
Signal/noise ratio, 298
Simple harmonic motion, 85
Simultaneous equations, 75–76
Sine
   of the angle, 81, 82–85, 86, 88, 102
   de Moivre's theorem of, 101
   law of, 86, 88
   of partial differential equations, 223, 224
Six, as perfect number, 11
Slide rule, 55
Smith, Barnabas, 127
Smith, David, 6, 17
Snell's law, 90–91
Solar cycle, 17
Solar day, 88
Solar system
   geocentric model of, 27, 28, 324

Solar system (*Cont.*)
   gravitational structure of, 328
   heliocentric model of, 18–19, 25, 89
   nebular hypothesis of, 161–162, 163
Soul, transmigration of, 14
Sources of Quantum Mechanics, The (Van der Waerden), 293
Space
   geometric continuity of, 7–8
   mathematical concept of, 79
   three-dimensionality of, 80
Space, Time, and Matter (Weyl), 286–287, 332
Space-time, four-dimensionality of, 276, 277, 284–285, 295, 333–334; *see also* Relativity theory
Spectroscope, 5
Sphere, homocentric, 17
Spiegal, E., 312, 314, 315
Sports, mathematics and, 3
Square, as regular polygon, 24
Square roots, 39, 54
   of fractions, 41
   of negative numbers, 52
   of quadratic equations, 68–69
Stars
   binary systems, 92
   classification, 28
   distance measurement, 88–89
   neutron, 299
   parallax, 89
Stella Nova, De (Kepler), 201
Stieltjes integral, 321
Stirling, James, 260
Straight line
   axioms of, 10
   definition, 8
   as number systems, 33
   on a plane, 233
Struik, Dirk, 231
Substitution, 269
Subtraction
   arithmetic, 34, 35
   of congruences, 199
   of fractions, 40, 45

Sun, diameter, 89–90
Sundial, 18
Supernova, 201
Surface, as geometric concept, 7
Surface radiation, 166
Surveying, 85–86, 88, 229
Suspension bridges, 176
Sylvester, James Joseph, 76, 209, 249, 251, 284
Symmetry, 332–333, 334–336

Tangent, 81, 82, 85, 102
Tartaglia, Niccolo, 68–69, 187
Technology, mathematics and, 4
Telegraphy, 214, 230, 253–254
Telescope, astronomical, 5, 91, 119, 130, 299
Tensor, Einstein–Ricci, 286, 287
Tensor analysis, 239, 276–277, 284
Thales, 5, 13, 29
Theano, 13–14
Theorems
  Cauchy integral, 222
  of excluded middle, 296–297
  Gauss's, 229–230
  integral, 222
  Minimum, 304
  in number theory, 171, 197–198
  proof of, 10
  Pythagorean, 15–16, 58, 82, 147, 271, 285
  relationship with axioms, 9
Theorie Analytique des Probabilities (Laplace), 165, 166–167, 205
Theory of Extensions (Grassmann), 248–249
Theory of Groups and the Quantum Mechanics, The (Weyl), 293
Theory of Numbers (Legendre), 235
Theory of Systems of Rays, A (Hamilton), 241
Thermodynamics, 249, 254–260, 329–330
  first law, 255–257, 258–259
  second law, 257–258, 329–330
    information theory and, 299–300
Thomson, J. J., 309

Thomson, William, 257
Thought, 1
Thought experiments, 117
Three-body gravitational problem, 290, 313–314, 315
  restricted, 155–156
Time
  measurement, 280–281
  reversal, 334
Topology, 238, 310–311
Traité de Dynamique (d'Alembert), 167
Traité de Mécanique Céleste (Laplace), 163, 165–167, 207
Transcendental numbers, 20, 48
Transfinite numbers, 32, 48, 239, 264, 266–268, 295, 317
Transformation theory, 250–251, 269–271
  algebra of, 249
  of coordinates, 120–124, 249, 251
  relativity theory and, 278–282
Treatment of the Projection of Figures, A (Poncelet), 270
Triangle
  equilateral, 24
  "solving" the, 86
  sum of angles of, 10
  sum of sides 0f, 232, 233, 234
  trigonometry of, 85–86
Trigonometric identities, 102
Trigonometry, 4–5, 18, 79–102, 125
  spherical, 88
  use in astronomy, 26–27, 83, 86–90
Turin Academy of Sciences, 157
Two-body gravitational problem, 155, 313; *see also* n-body gravitational problem

Ulam, S. M., 307
Uncertainty principle, 245–246, 274
Unified field theory, 287, 331
Unit angle, 23
Unit circle, 80–82, 83–84, 85
Universal gravitation, law of, 129, 130, 131
Universe
  chaos of, 314–315

Universe (*Cont.*)
  collapse of, 334
  entropy of, 330
  expansion rate of, 299
  stability of, 314–315

Value of Science and Science and
     Hypothesis, The (Poincare), 291
Van der Waerden, B. L., 293
Variations, calculus of, 152–153, 157, 176,
    292
Vector(s), 50
  algebra of, 72–73
  electromagnetic, 253
Vector analysis, 119, 137–138, 239, 248–
    252, 276
Vibration, 167
  elliptic integrals of, 321–322
  mathematical analysis of, 223–224
Virtual work, principle of, 167
Voltaire, François Marie Arouet de, 22, 99
Volume
  as geometric concept, 7

Volume (*Cont.*)
  of spatial figures, 50
von Neumann, John, 114, 304–305, 306, 307
  The Mathematical Foundations of
    Quantum Mechanics, 275, 293
von Walterhausen, Sartorius, 213
Vortices theory, 131, 170

Wallis, John, 175, 202
Wave equations, 76, 225, 274, 329
Wave function, 63
Wave mechanics, 217, 220, 244, 274, 332
Wave motion, 85
Wave theory, of light, 177, 178, 225, 226
Weber, Wilhelm, 214, 230, 236
Weierstrass, Karl, 209, 239, 263, 264–265
Weyl, Herman, 114, 331–332
  Space, Time, and Matter, 286–287, 332
Whitehead, Alfred North, 296
Wiener, Norbert, 301
Work-on-the-body concept, 62

Zero, 33, 34